METHODS IN MOLECULAR BIOLOGY™

Series Editor
John M. Walker
School of Life Sciences
University of Hertfordshire
Hatfield, Hertfordshire, AL10 9AB, UK

For other titles published in this series, go to
www.springer.com/series/7651

METHODS IN MOLECULAR BIOLOGY™

Mitosis

Methods and Protocols

Edited by

Andrew D. McAinsh PhD

Marie Curie Research Institute, Surrey, UK

Humana Press

Editor
Andrew D. McAinsh
Chromosome Segregation Laboratory
Marie Curie Research Institute
The Chart
Oxted
Surrey RH8 OTL
United Kingdom
a.mcainsh@mcri.ac.uk

ISSN 1064-3745 ISSN 1940-6029 (eBook)
ISBN 978-1-62703-850-8 ISBN 978-1-60327-993-2 (eBook)
DOI 10.1007/978-1-60327-993-2
Springer Dordrecht Heidelberg London New York

Cover illustration: Figure 2 in Chapter 9.

Printed on acid-free paper

Springer is a part of Springer Science+Business Media (www.springer.com)

Preface

Mitosis is a dynamic and stochastic process that has evolved to accurately segregate the genetic material into two daughter cells, thereby preventing genomic instability and the development of disease. Mitosis is a phenomenally complicated process that involves hundreds, if not thousands, of protein components and regulatory steps that operate in both space and time to drive cell division. Modern methods make it possible to ask for mechanistic principles underlying this bewildering complexity. The main purpose of this volume is to provide an up-to-date collection of methods and approaches that are used to investigate the mechanism of mitosis at the molecular level. While many of these methods are focused on mammalian cells, we have, where appropriate, included chapters using model organisms. We hope to capture both current approaches and the future direction of method development, with contributions from both established researchers and emerging young scientists.

This book is designed with two general groups of readers in mind: First, graduate students and postdoctoral researchers who are beginning work for the first time in a mitosis laboratory. Second, researchers who are already working in the mitosis field who require a resource for both established and newly developed methodologies. To achieve this, the organization of this book developed into three general areas: First, we cover methods that can be used to inactivate your gene of interest, or deplete proteins of interest (chapters 1–3). Second, we learn about specific biochemical and microscope-based methods (chapters 4–9). Third, we discover approaches to monitor and measure key mitotic processes (chapters 9–20). Given the complexities of mitosis, it seems highly probable that such a combination of imaging, biochemical and genetic methodologies will be crucial to our future understanding of mitotic regulation.

I would like to thank all of the authors for their enthusiasm and effort in putting together this set of methods, as well as all the members of my laboratory for proof reading and correcting the chapters and especially Sarah McClelland who read all the chapters at least twice at various stages. Finally, I wish to thank Lyndy Rasmusen for her help in administering the project and keeping me organized.

Andrew D. McAinsh

Contents

Contributors

VINCENT ARCHAMBAULT • *Cancer Research UK Cell Cycle Genetics Research Group, Department of Genetics, University of Cambridge, Cambridge, UK*
ALEXIS BARR • *Department of Oncology, Cancer Research UK Cambridge Research Institute, University of Cambridge, Cambridge, UK*
ELI BERDOUGO • *Molecular Biology Program, Memorial Sloan-Kettering Cancer Center, New York, USA*
MÓNICA BETTENCOURT-DIAS • *Cell Cycle Regulation Lab, Instituto Gulbenkian de Ciência, Oeiras, Portugal*
SATYAREBALA BORUSU • *Institute of Biochemistry, ETH Zurich, Zurich, Switzerland*
PIER PAOLO D'AVINO • *Cancer Research UK Cell Cycle Genetics Research Group, Department of Genetics, University of Cambridge, Cambridge, UK*
BARBARA DI FIORE • *Gurdon Institute, University of Cambridge, Cambridge, UK*
MICHAEL J. EMANUELE • *Department of Biochemistry and Molecular Genetics, University of Virginia, Charlottesville, VA, USA*
ANDREW M. FRY • *Department of Biochemistry, University of Leicester, Leicester, UK*
JOSEFIN FERNIUS • *Wellcome Trust Centre for Cell Biology, University of Edinburgh, Edinburgh, UK*
FANNI GERGELY • *Department of Oncology, Cancer Research UK Cambridge Research Institute, University of Cambridge, Cambridge, UK*
DANIEL GERLICH • *Institute of Biochemistry, ETH Zurich, Zurich Switzerland*
MARCO GEYMONAT • *Stem Cell Biology and Developmental Genetics, National Institute for Medical Research, London, UK*
DAVID M. GLOVER • *Cancer Research UK Cell Cycle Genetics Research Group, Department of Genetics, University of Cambridge, Cambridge, UK*
GOHTA GOSHIMA • *Institute for Advanced Research, Nagoya University, Nagoya, Japan*
MARGARET GRIMALDI • *Cell Cycle Control Laboratory, Marie Curie Research Institute, Surrey, UK*
KEVIN G. HARDWICK • *Wellcome Trust Centre for Cell Biology, University of Edinburgh, Edinburgh, UK*
PETER K. JACKSON • *Departments of Cellular Regulation, Genentech Inc., South San Francisco, CA, USA; Department of Pathology, Stanford University School of Medicine, Palo Alto, CA, USA*
PRASAD V. JALLEPALLI • *Molecular Biology Program, Memorial Sloan-Kettering Cancer Center, New York, USA*
LINGYAN JIN • *Department of Molecular and Cell Biology, University of California at Berkeley, Berkeley, CA, USA*
YUU KIMATA • *Cell Cycle Control Laboratory, Marie Curie Research Institute, Surrey, UK*
CHRISTIANE KLEBIG • *Institute of Biochemistry, ETH Zurich, Zurich, Switzerland*
ERNEST D. LAUE • *Department of Biochemistry, University of Cambridge, Cambridge, UK*

MARIANA LINCE-FARIA • *Instituto de Biologia Molecular e Celular, Universidade do Porto, Porto, Portugal*
CATHERINE LINDON • *Department of Genetics, University of Cambridge, Cambridge, UK*
HELDER MAIATO • *Instituto de Biologia Molecular e Celular, Universidade do Porto, Porto, Portugal*
IRINA MATOS • *Instituto de Biologia Molecular e Celular, Universidade do Porto, Porto, Portugal*
PATRICK MERALDI • *Institute of Biochemistry, ETH Zurich, Zurich, Switzerland*
ANDREW D. MCAINSH • *Chromosome Segregation Laboratory, Marie Curie Research Institute, Surrey, UK*
SARAH MCCLELLAND • *Chromosome Segregation Laboratory, Marie Curie Research Institute, Surrey, UK*
THOMAS MÜLLER-REICHERT • *Max Planck Institute of Molecular Cell Biology and Genetics, Dresden, Germany*
ANDREW W. MURRAY • *Department of Molecular and Cellular Biology, Harvard University, Cambridge, MA, USA*
EILEEN O'TOOLE • *The Boulder Laboratory for 3-D Electron Microscopy of Cells, University of Colorado, Boulder, CO, USA*
ANTONIO J. PEREIRA • *Instituto de Biologia Molecular e Celular, Universidade do Porto, Porto, Portugal*
SUZANNA L. PROSSER • *Department of Biochemistry, University of Leicester, Leicester, UK*
MARCIN R. PRZEWLOKA • *Cancer Research UK Cell Cycle Genetics Research Group, Department of Genetics, University of Cambridge, Cambridge, UK*
MICHAEL RAPE • *Department of Molecular and Cell Biology, University of California at Berkeley, Berkeley, CA, USA*
MASAMITSU SATO • *Department of Biophysics and Biochemistry, Graduate School of Science, University of Tokyo, Tokyo, Japan*
MICHAEL SCHMITZ • *Institute of Biochemistry, ETH Zurich, Zurich, Switzerland*
SCOTT C. SCHUYLER • *Department of Molecular and Cellular Biology, Harvard University, Cambridge, MA, USA*
STEVE SEDGWICK • *Stem Cell Biology and Developmental Genetics, National Institute for Medical Research, London, UK*
ADONIS SPANOS • *Stem Cell Biology and Developmental Genetics, National Institute for Medical Research, London, UK*
P. TODD STUKENBERG • *Department of Biochemistry and Molecular Genetics, University of Virginia, Charlottesville, VA, USA*
MATTHEW K. SUMMERS • *Departments of Cellular Regulation, Genentech Inc., South San Francisco, CA, USA*
KOZO TANAKA • *Wellcome Trust Centre for Gene Regulation and Expression, College of Life Sciences, University of Dundee, Dundee, UK; Institute of Development, Aging and Cancer, Center for Research Strategy and Support, Tohoku University, Miyagi, Japan*
TOMOYUKI U. TANAKA • *Wellcome Trust Centre for Gene Regulation and Expression, College of Life Sciences, University of Dundee, Dundee, UK*
MARIE-EMILIE TERRET • *Molecular Biology Program, Memorial Sloan-Kettering Cancer Center, New York, USA*
TAKASHI TODA • *Laboratory of Cell Regulation, Cancer Research UK, London Research Institute, London, UK*

ALBERTO TOSO • *Institute of Biochemistry, ETH Zurich, Zurich, Switzerland*
MIKA TOYA • *Laboratory for Developmental Genetics, RIKEN Center for Developmental Biology, Kobe, Japan*
MICHELLE TRICKEY • *Cell Cycle Control Laboratory, Marie Curie Research Institute, Surrey, UK*
ADAM WILLIAMSON • *Department of Molecular and Cell Biology, University of California at Berkeley, Berkeley, CA, USA*
HIRO YAMANO • *Cell Cycle Control Laboratory, Marie Curie Research Institute, Surrey, UK*
WEI ZHANG • *Department of Biochemistry, University of Cambridge, Cambridge, UK*
DEBORAH ZYSS • *Department of Oncology, Cancer Research UK Cambridge Research Institute, University of Cambridge, Cambridge, UK*

Chapter 1

Knock-in and Knock-out: The Use of Reverse Genetics in Somatic Cells to Dissect Mitotic Pathways

Alexis R. Barr, Deborah Zyss and Fanni Gergely

Abstract

Reverse genetic methods, such as homologous gene targeting, have greatly contributed to our understanding of molecular pathways in mitosis, especially in yeast. The chicken B-lymphocyte line, DT40, represents a unique example among vertebrate somatic cells where homologous gene targeting occurs at very high frequency. DT40 cells therefore provide a useful and accessible somatic genetic system for wide-ranging biochemical and cell biological assays. In this chapter, we describe the main principles of homologous gene targeting, the concept of targeting construct design and the detailed experimental protocol of how to achieve successful knockouts. We also mention methods for conditional disruption of essential genes and conclude with specific procedures for the study of mitosis in DT40 cells.

Key words: DT40, homologous recombination, knockout, gene targeting, mitosis, reverse genetics, cre-loxP, Tet-off inducible.

1. Introduction

1.1. Background

The main aim of reverse genetic methods is to identify phenotypes associated with the function of a specific gene product. Currently, RNA interference is the fastest and most popular reverse genetic method to diminish the expression of a gene. Certain questions, however, such as the effects of well-defined mutations or the roles of specific domains within a protein can be addressed in a much more precise and sophisticated way by gene targeting. Until a decade or so ago, the use of gene targeting was restricted to yeast and mouse genetics, but in recent years, somatic cell genetics has become a reality. Below, we describe

Andrew D. McAinsh (ed.), *Mitosis: Methods and Protocols*, Methods in Molecular Biology *545*,
DOI 10.1007/978-1-60327-993-2_1, © Humana Press, a part of Springer Science+Business Media, LLC 2009

an experimental protocol to create targeted gene knockouts in vertebrate somatic cells, an approach successfully utilized by a number of laboratories studying mitosis. A partial list of examples includes the identification and subsequent characterisation of kinetochore components *(1–4)*, studies on chromosome condensation and architecture during mitosis *(5)* and dissecting the role of mitotic kinases in centrosome duplication and cell cycle checkpoints *(6–8)*.

Targeted gene disruption is based on homologous recombination (HR), a mechanism with an important physiological role in meiosis and DNA repair *(9)*. During HR, homologous DNA molecules align and exchange fragments in a double crossover event. Exogenous DNA molecules are also capable of crossing over with highly similar sequences, thus allowing the introduction of engineered pieces of DNA into specific loci within the genome of a cell. Homologous gene targeting has turned out to be an extremely useful method in yeast and mouse genetics. In most vertebrate somatic cells, however, the ratio of homologous to nonhomologous recombination is very low. Although promoter trap targeting vectors and recombinant adeno-associated viral delivery have improved gene targeting efficiency in some human cell lines, such as HCT116, DLD1, and fibroblasts, targeted events are still relatively rare, making the approach rather labour-intensive *(10–13)*. Therefore, with the notable exceptions of mouse embryonic stem (ES) cells, mammalian cells in culture remain a feasible but not overly popular choice for gene targeting. They, however, rapidly lose their appeal when the combined disruption of two or more genes is required within a cell.

The chicken B-lymphocyte cell line, DT40, provides a much more convenient way to deliver targeted gene mutations *(9)*. The highly recombinogenic DT40 cells exhibit a ratio of up to 1:2 of homologous versus nonhomologous targeting events, as a result of elevated levels of HR and reduced frequency of random integrations *(14)*. This remarkable targeting efficiency is orders of magnitudes higher than those observed in mammalian cells, but its molecular basis remains poorly understood *(15)*.

DT40 cells are derived from an avian leucosis virus-induced lymphoma in White Leghorn Chicken (*Gallus gallus domesticus*). DT40 cells display a stable karyotype of 80 chromosomes, including 11 autosomal macrochromosomes, the ZW sex chromosomes, and 67 microchromosomes. They are near diploid except for the trisomy of two chromosomes: chromosome 2 and one minichromosome. The size of the chicken genome is about one-third of that of mammalian genomes. Consistently, chicken introns are also much smaller, a clear advantage for gene targeting approaches. Other benefits of DT40 cells include a fast generation time of 8–10 h and the relative ease of propagating them under laboratory culture conditions. Last, but not least, the design of targeting constructs has become considerably easier now that the sequence of the chicken genome is available. Although the genome sequenced is that of the red jungle fowl (*Gallus gallus*), and DT40 originates from the domestic chicken, only a small degree of sequence variation and polymorphism is present between these two chicken strains. Overall, the combination of accessible genome sequence, efficient gene targeting, and fast generation time provides a powerful reverse genetic system in vertebrate cells, where homozygous knockout cell lines can be generated within 10–12 weeks.

1.2. Basic Concepts of Homologous Gene Targeting in DT40 Cells

The experimental strategy and the main stages of gene targeting in DT40 are illustrated in **Fig. 1.1**. The first and foremost part of the process is the design of a targeting construct. The targeting construct should recombine into a specific genetic locus and hence disrupt or modify a gene of interest when introduced into a cell. Naturally, the design of the construct depends on the aim of the experiment: achieving a full knockout, deleting a specific

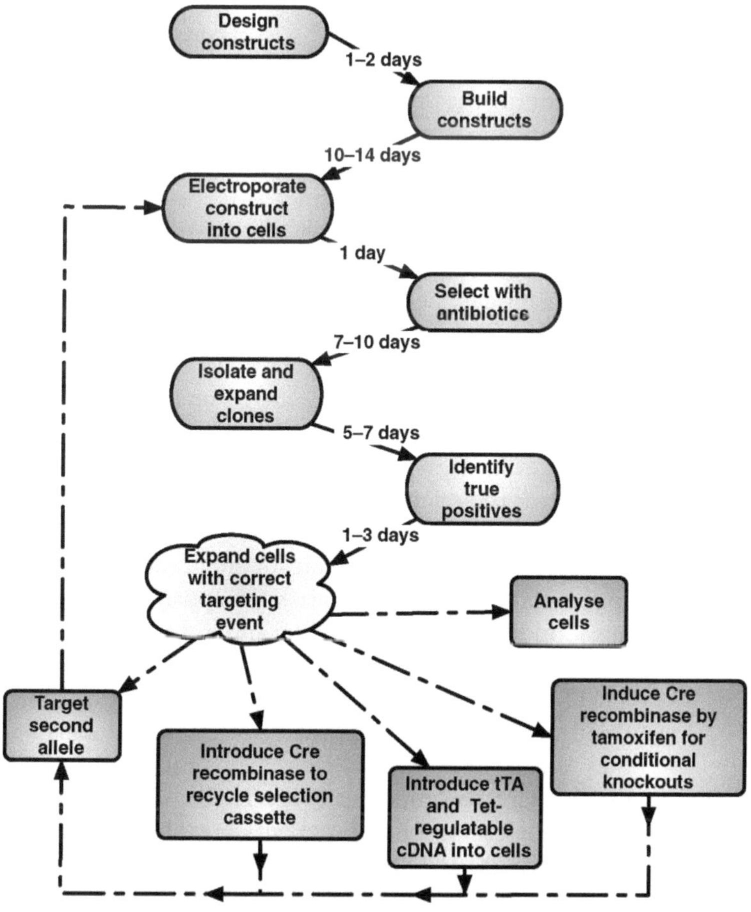

Fig. 1.1. Work flow diagram of gene targeting in DT40 cells.

domain, introducing point mutations, or inserting a tag. Here, we mention a few general guidelines of construct design, however, for more complex knockout constructs we recommend that the readers consult the literature.

First, the number and size of exons and introns that constitute the target area should be examined. An important consideration when deciding on targeting strategies is the size of the targeted region, because regions exceeding 4 kilobases (kb) recombine less frequently. This limitation can be overcome by the use of the tamoxifen-inducible Cre-loxP system. Here, loxP sites are introduced to either side of the target area, hence allowing Cre recombinase-mediated deletion of larger regions *(16)*. DT40 cells that express inducible Cre recombinase greatly facilitate this approach *(17)*.

In the case of shorter deletions (<4 kb), once the domain to be removed, altered, or tagged is identified, and the targeting strategy is established, the design of the targeting construct is relatively straightforward as illustrated in **Fig. 1.2A**. Briefly, the final targeting construct should consist of an antibiotic selection cassette, which is to replace an essential region of a gene, flanked by two homology arms (1–3 kb). The construct must have a unique restriction enzyme site outside the homology arms to allow linearization of the vector (*see* **Note 2** for more information). Furthermore, the design of the construct should also allow screening for positive targeting events. Screening with PCR is straightforward, but if Southern blotting is to be used, unique restriction sites must be identified in the genomic region outside the homology arms (*see* **Fig. 1.2A**). When possible, probes should also be made against sequences outside the construct. Once the construct design satisfies the criteria above, the homology arms are PCR amplified and cloned from genomic DNA from wild-type DT40 cells. If both alleles of a gene are to be targeted, a construct with identical arms but different selection cassette is used for the second targeting event. It is also possible to reuse the same construct, but only if the selection cassette is flanked by loxP sites that allow the recycling of the cassette prior to targeting the second allele.

Because many of the mitotic genes are essential for cell survival, a straight genetic knockout approach is not always feasible. Conditional gene targeting and inducible gene expression are two ways to overcome this limitation. Conditional gene targeting in DT40 closely mimics the approach taken in mouse ES cells. Briefly, this involves the placing of essential exons between loxP sites, ensuring that the allele remains functional until the expression of Cre recombinase (**Fig. 1.2B**). Two important requirements for conditional knockouts are (1) loxP flanked selection cassettes and (2) DT40 cells expressing tamoxifen-inducible Cre recombinase *(17)*. After targeting the first allele, tamoxifen is used to induce partial recombinants that have lost the selection cassette

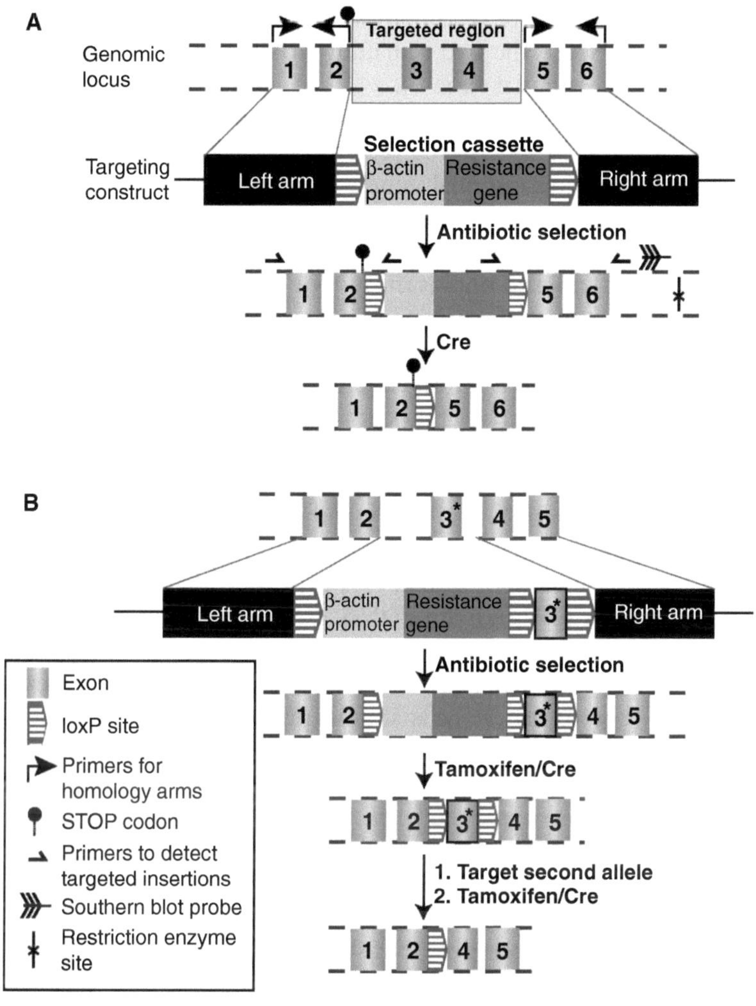

Fig. 1.2. (**A**) A schematic diagram of the gene targeting process. The aim of this construct design is to create a truncated gene product containing exons 1 and 2. The *left* and *right* homology arms consist of exons 1 and 2, and exons 5 and 6, respectively. A set of two PCR primers is designed to amplify each arm. These contain restriction enzyme sites and one of the primers also contains an in-frame STOP codon. Upon integrating into the locus, the antibiotic selection cassette replaces exons 3 and 4. Correct targeting events are identified using Southern blotting (probe and restriction enzyme site) or PCR (primers for detection). The antibiotic selection cassette can be excised by transiently expressing Cre recombinase. (**B**) A typical conditional gene targeting strategy. The aim of this design is to introduce loxP sites around an essential exon 3* that do not interfere with normal gene expression but render exon 3* excisable by Cre. The construct contains homology arms, selection cassette, and the essential exon 3*. Cells with correct insertion are selected and treated with tamoxifen to induce Cre. Expression of Cre leads to three distinct recombination events, one of which is a partial event that results in cells that have lost their resistance cassette but retained exon 3*. Subsequently, the second allele of the gene is targeted by standard methods (as in **A**). Timely disruption of the gene is initiated by tamoxifen, which induces Cre and leads to the excision of exon 3*.

but retained the essential exons (*see* **Fig. 1.2B** for details). These events should be distinguishable by PCR or Southern blotting. Subsequently, the second allele of the gene can be disrupted by a direct knockout approach. Once both alleles are targeted, gene disruption is achieved in these cells by tamoxifen treatment that induces Cre recombinase and hence catalyses the recombination between the two remaining loxP sites, leading to the excision of the essential exon(s).

As an alternative to conditional gene targeting, inducible gene expression systems (e.g., Tet-on or Tet-off) are also used for studying essential genes. Prior to the disruption of the genomic locus of an essential gene, its cDNA is placed under a regulatable promoter and introduced into cells. Subsequently, both alleles can be knocked out, as the cDNA compensates for their absence. The knockout phenotype can be studied by turning off the expression of the cDNA. The Tet-off inducible system is based on the activity of a tetracycline-controlled transactivator protein (tTA). In the absence of the tetracycline analogue doxycycline, tTA binds to and activates expression from TRE (tetracycline response element) promoters *(18)*. The gene of interest can be placed under the control of a TRE promoter and introduced into the DT40 genome via random integration. tTA is either co-transfected into these cells, or stable, tTA expressing DT40 clones are used *(19)*. The tTA and the cDNA are introduced either as the first step of the process or following the successful disruption of the first allele. Then, the remaining allele(s) of the gene are targeted as before. These cells should propagate in the absence of tetracycline, but when tetracycline is added, the expression of the cDNA is switched off leaving the cells without the essential gene.

As soon as cells are generated with the desired genetic mutations, the full spectrum of cell biology and biochemical methods used in mammalian tissue culture cells is available in DT40 cells. These, combined with powerful gene targeting, make DT40 a unique, accessible, but still surprisingly underused, tool for studying mitosis.

2. Materials

2.1. Making the Targeting Construct

1. DT40 cell line
2. Puregene cell and tissue DNA isolation kit (Gentra Systems)
3. Takara LA Taq
4. Primers for PCR amplification of left and right arms
5. 0.2 ml PCR tubes
6. pLoxNeo, pLoxPuro, pLoxBlasti vectors *(20)*
7. Reagents for molecular cloning (T4 ligase, restriction enzymes, buffers, chemically competent *E. coli*, LB plates, antibiotics)
8. Reagents for DNA electrophoresis (agarose, buffer, ethidium bromide, DNA ladder)
9. Reagents for purifying and extracting plasmid DNA from bacteria and agarose gels

2.2. Transfecting DT40 Cells

1. DT40 growth medium:
   i. 500 ml RPMI Medium with Glutamax (Invitrogen)
   ii. 50 ml Foetal Bovine Serum (Invitrogen)
   iii. 5 ml Chicken Serum (Sigma)
   iv. 10 ml Penicillin/Streptomycin (Invitrogen)
   v. 4 μl β-mercaptoethanol
2. Phosphate-Buffered Saline (PBS), sterile
3. Restriction enzymes and buffers
4. Electroporation cuvette 4 mm
5. Biorad Genepulser electroporator
6. Flat-bottomed tissue-culture treated 96-well plates
7. Haemocytometer
8. Ethanol
9. 3 M Sodium Acetate pH 5.2
10. G418 (Invitrogen) 50 mg/ml, store in dark, 4°C
11. Blasticidin (Invitrogen) 0.5 mg/ml, –20°C
12. Puromycin (Clontech) 10 mg/ml, store at –20°C

2.3. Selecting Targeted Transformants

1. 24-well tissue culture-treated plates
2. Puregene cell and tissue DNA isolation kit
3. Reagents for DNA electrophoresis as in **Section 2.1**

2.3.1. Long-Range PCR to Screen for Targeted Insertions

1. Takara LA Taq
2. Primers for checking integration site

2.3.2. Southern Blotting to Check for Targeted Insertions

1. Depurination solution: 250 mM HCl (in distilled water)
2. Denaturation buffer:1.5 M NaCl; 0.5 M NaOH (in distilled water)
3. Church and Gilbert buffer: 7% (w/v) SDS, 0.5 M phosphate buffer pH7.2, 10 mM EDTA
4. Stringency wash solutions:
   Low: 2 × SSC, 0.1% (w/v) SDS
   Medium: 1 × SSC, 0.1% (w/v) SDS
   High: 0.1 × SSC, 0.1% (w/v) SDS
5. Whatman paper and tissues for Southern blot sandwich
6. HyBond N^+ membrane 17 × 20 cm (positively charged nylon transfer membrane from Amersham Biosciences)
7. NEBlot Kit (New England Biolabs), labelled dNTPs
8. DNA crosslinker
9. Medical X-ray film

2.4. Targeting the Second (or Subsequent) Alleles

Refer to materials in **Sections 2.2** and **2.3**.

2.5. Inducible Tet-off DT40 Cell Lines and Conditional DT40 Knockouts

2.5.1. Inducible Tet-Off DT40 Cell Lines

1. Tetracycline-repressible vector pUHD10.3. *(21)*
2. Vector containing modified transactivator, tTA3. *(22)*
3. Hypotonic buffer: 10 mM Tris-HCl pH 7.4, 10 mM KCl, 1.5 mM $MgCl_2$,
   10 mM β-mercaptoethanol.
4. High-salt buffer:15 mM Tris-HCl, pH 7.4, 1 mM EDTA, 500 mM NaCl, 1 mM $MgCl_2$, 10% glycerol, 10 mM β-mercaptoethanol, protease inhibitor cocktail
5. 1 ml syringe
6. 19G needle
7. 2 × Laemmli SDS sample buffer
8. Nitrocellulose membrane
9. Doxycycline, 50 mg/ml in H_2O; store in dark at 4°C
10. Trypan blue (0.4% solution in H_2O)
11. G418 (Invitrogen) 50 mg/ml, store in dark, 4°C

2.5.2. Conditional DT40 Knockouts

1. 4-hydroxy-tamoxifen, 0.05 M in ethanol; store at –20°C

2.6. Recycling the Resistance Cassettes

1. Amaxa nucleofector II
2. Amaxa nucleofection kit T
3. Plasmid with Cre-recombinase
4. Haemocytometer
5. 6-well tissue culture-treated plates
6. 96-well flat-bottomed tissue culture-treated plates
7. Selective antibiotics (*see* **Section 2.2** Transfecting DT40 cells)
8. Puregene cell and tissue DNA isolation kit
9. Takara LA Taq
10. Primers to amplify across deleted region
11. Reagents for DNA electrophoresis as in **Section 2.1.**

2.7. Analysing Targeted Clones by Immunofluorescence

1. Poly-L-Lysine (Sigma) 2 mg/ml in H_2O; store at –20°C
2. Glass slides
3. Cytospin
4. Glass coverslips
5. Prolong antifade reagent, with DAPI (Invitrogen)
6. 100% Methanol (store at –20°C)
7. Ethanol
8. 4% Formaldehyde (make fresh before use in PBS)
9. PBS/0.1% Tween20
10. 1×PBS/0.2% Triton X-100
11. Blocking buffer (store at 4°C): 5% BSA, 1 × PBS, 0.02% sodium azide
12. Primary antibodies
13. Fluorescently labelled secondary antibodies

2.8. Live Cell Imaging of Mitosis

1. Chambered coverglass (Lab-Tek, Nuncbrand)
2. Poly-L-lysine (Sigma) 2 mg/ml in H_2O. Store at –20°C
3. Hoeschst 33342 (Sigma), 1 mg/ml stock in H_2O. Store in the dark at –20°C
4. Fluorescent subcellular markers such as EGFP-Tubulin (Clontech) or GFP-Histone-H2B.

3. Methods

3.1. Making the Targeting Construct

Once a targeting strategy has been established (*see* **Fig. 1.2** for specific examples), the next step is to design a gene targeting construct. Both these steps require the use of the *Gallus gallus* genome browser (http://www.ensembl.org/Gallus_gallus/index.html). The DT40 website is also a good source of information with topics ranging from cell culture conditions to advice on construct design (http://pheasant.gsf.de/DEPARTMENT/dt40.html). When construct design is completed, order or synthesise the appropriate pair of left and right arm oligonucleotide primers and proceed to next step.

Extract genomic DNA from DT40 cells using the Puregene Tissue and Cell DNA isolation kit according to the manufacturer's instructions. (For alternative methods to extract genomic DNA *see* **Note 1.**) PCR amplify the left and right arms of your targeting construct from the genomic DNA (*see* **Fig. 1.2A** and **Note 2**). For PCR amplification use LA Taq from Takara. Make up the PCR mix in 0.2 ml PCR tubes as follows: 1 μg genomic DNA, 1 μl of each of the primers at 10 μM, 5 μl 25 mM $MgCl_2$, 5 μl 10 × buffer, 8 μl dNTPs, 0.5 μl LA Taq, 28.5 μl H_2O. Cycling conditions are 98°C 1 min, followed by 30 cycles of 94°C 30 s, 68°C × min (where × depends on length of amplified region; allow 1 min/kb), 72°C 10 min. *See* **Note 3** for alternative conditions.

1. Run the PCR product on an agarose gel by DNA electrophoresis.
2. Excise bands corresponding to the left and right arms. Build targeting construct by sequentially introducing the arms into pLoxNeo, pLoxPuro, or pLoxBlasti *(20)* using standard cloning methods.

3.2. Transfecting DT40 Cells

This protocol is adapted from Dr. Jean-Marie Buerstedde *(23)*.

1. Maintain DT40 cell line in appropriate DT40 growth medium (see Materials) in a tissue-culture incubator at 39°C, 5% CO_2.
2. Linearise 60 μg of the targeting vector, using the appropriate restriction enzyme (*see* **Note 2**). Set up the restriction digest in 500 μl total volume with 5 μl of restriction

enzyme. Digest for at least 3 h, or overnight (if the restriction enzyme does not have star activity) to ensure all DNA is linearised. Run 10 μl of the digest on an agarose gel by electrophoresis to check for linearisation.

3. Add 3 volumes of absolute ethanol and 50 μl of 3 M sodium acetate, pH 5.2 to the restriction digest mixture. Put at –20°C for at least 1 h, or overnight.
4. Centrifuge the sample at 13,000 × g, 15 min.
5. Remove and discard the supernatant.
6. Wash the DNA pellet in 0.5 ml 70% ethanol. Flick the tube gently to dislodge the pellet from the side of the tube.
7. Centrifuge at 13,000 × g, 15 min.
8. From this point on, treat the DNA pellet as sterile. Remove all the supernatant and allow the pellet to air-dry in a laminar flow hood.
9. Resuspend the pellet in 300 μl 1 × PBS. If the pellet is slow to dissolve, place at 55°C for 1 h. (*see* **Note 4**).
10. Count DT40 cells using a haemocytometer.
11. Centrifuge 2 × 10^7 cells at 800 × g, 5 min, 4°C.
12. Remove supernatant and gently resuspend the cells in ice-cold 1 × PBS.
13. Centrifuge cells again.
14. Repeat steps 12 and 13 to ensure all traces of serum are removed.
15. Resuspend the cell pellet in 300 μl ice-cold 1 × PBS and transfer to a prechilled 4 mm sterile electroporation cuvette on ice.
16. Add the linearised DNA to the cells in the cuvette.
17. Vortex on medium speed to mix. Keep on ice for 15 min. Vortex gently every 2–3 min.
18. Electroporate the cells using a Biorad genepulser at 25 μF, 550 V.
19. Keep on ice for a further 10 min.
20. Add 60 ml of prewarmed DT40 medium (39°C) to a 100 ml sterile reservoir.
21. Transfer the contents of the cuvette into the warm media (39°C) and mix well.
22. Aliquot 100 μl of the cell-containing medium into each well of six flat-bottomed 96-well plates (*see* **Note 5**).
23. Place the 96-well plates in a tissue-culture incubator at 39°C, 5% CO_2 for 24 h.
24. After 24 h, add the appropriate selection drug at the following concentrations: 1.5 μg/ml G418, 30 μg/ml Blasticidin, 0.5 μg/ml Puromycin. Make up double the concentration of drug required in 60 ml DT40 medium and aliquot 100 μl into each well of the 96-well plates.
25. Incubate plates at 39°C, 5% CO_2 for 7–10 days for stable transfectants to grow. There is no need to change the culture medium during this period.

3.3. Selecting Targeted Transformants

1. After 7–10 days selection, a subset of wells should contain large, mostly round clones that are normally visible to the naked eye from the underside of the 96-well plate (they can cover up to 20–40% of the area of the well). One can expect to find at least 2–6 clones per 96-well plate. It is important that only wells with single clones are selected for further analysis, so ignore all wells with multiple or suspected multiple clones. Single clones should be picked by gently pipetting culture medium present in wells until cells detach from the plate. It is crucial to use fresh pipette tips for the transfer and all subsequent manipulation of individual clones to minimise the risk of cross-contamination.
2. Transfer each clone together with its culture medium into a single well of a 24-well plate containing 0.5 ml fresh DT40 medium.
3. Monitor cell growth over the next few days. When the cells start to become dense, replica plate the 24-well plate into two additional 24-well plates using multipipettors when possible. From these three plates, one will remain the 'stock' plate, one will be frozen down at –80°C (*see* **Note 6**), and the third will be used for genomic DNA extraction.
4. Extract genomic DNA from each of the wells of the 24-well plate using Puregene cell and tissue DNA isolation kit.
5. To check for targeted insertion of your construct, use either long-range PCR or Southern blotting.

3.3.1. Long-Range PCR to Screen for Targeted Insertions

To confirm that the transfected construct has integrated into the correct genomic locus, PCR primers must be designed according to the strategy outlined in **Fig. 1.2**. It is important to validate the primers first on plasmids and wild-type genomic DNA.

1. PCR amplify the genomic DNA extracted from the clones using the conditions outlined in **Section 3.1**.
2. Check for the correct size PCR product by DNA electrophoresis.
3. Expand clones with targeted insertion.

3.3.2. Southern Blotting to Check for Targeted Insertions

1. Digest genomic DNA with appropriate restriction enzyme: 20 μl DNA in 30 μl final volume. Incubate overnight at 37°C.
2. Pour large 0.7% agarose gel in TBE (250 ml).
3. Load DNA samples onto gel and run gel overnight (35 V).
4. Incubate the gel in TBE + ethidium bromide for 20 min.
5. Take a picture of the ethidium bromide stained gel with a ruler placed on the side.
6. Incubate the gel in 500 ml of the depurination solution for 20 min exactly.
7. Discard the depurination solution and wash the gel with distilled water.

8. Incubate the gel in 500 ml of the denaturation buffer for 30 min.
9. Transfer overnight by assembling a Southern blot transfer sandwich.
10. Mark the well positions on the nylon membrane and crosslink DNA.
11. Soak nylon membrane in 5 × SSC and preheat Church and Gilbert solution to 62°C.
12. Place the nylon membrane in a roller bottle with DNA facing inside of the bottle and incubate for at least 30 min at 62°C with 20 ml of Church and Gilbert solution.
13. Prepare the radiolabelled probe (see manufacturer's instructions in NEBlot kit).
14. Add radioactive probe to the hybridization buffer and incubate overnight at 62°C.
15. Wash nylon membrane in following wash solutions at 62°C (1–5 ml per cm^2 of membrane):
    i. 2 × in low stringency wash solution for 5 min
    ii. 2 × in medium stringency wash solution for 15 min
    iii. 2 × in high stringency wash solution for 10 min
16. Remove blot from the last wash; drain and wrap in Saran Wrap for analysis on X-ray film.
17. Expand clones with correct targeted insertions.

3.4. Targeting the Second (or Subsequent) Alleles

Repeat **Sections 3.2** (Transfection of DT40) and **3.3** (Selecting stable transformants), on the heterozygous cell line. As a control, one can use wild-type DT40 to confirm that the second targeting construct can recombine successfully into the required locus. To target the second allele, use a different resistance cassette to that used for the first event or recycle the selection cassette as in **Section 3.6**. If introducing a different selection cassette, use double antibiotic selection to obtain stable transformants. This will eliminate positive clones in which the targeting event has occured in the same allele.

3.5. Inducible Tet-Off DT40 Cell Lines and Conditional DT40 Knockouts

If, after the second targeting event (**Section 3.4**), no colonies are obtained, but random integration occurs at normal levels (usually 20–30 clones per experiment but it can vary), then it is possible that disrupting the target gene is detrimental to the cell. To confirm this, it is important to test if the same construct integrates at the expected frequency when introduced into wild-type DT40 cells. If a gene is essential, either conditional knockouts can be generated (**Fig. 1.2B**) or a tetracycline-regulatable cDNA can be introduced into cells.

3.5.1. Inducible Tet-Off DT40 Cell Lines

1. Subclone the transgene for your targeted gene into the tetracycline-repressible vector pUHD10.3 *(21)*.

2. Linearise the vector with a restriction enzyme outside of the pCMV-polyA functional unit. PvuI is a good choice (only if it is absent from cDNA).
3. Purify linearised DNA and prepare for transfection as in **Section 3.2**.
4. Cotransfect the linearised DNA and the modified transactivator tTA3 *(22)* at a ratio of 10:1 (20 μg: 2 μg) by electroporation into DT40 cells with a single targeted allele (as in **Section 3.2**; *see* **Note** 7).
5. Select clones with random, but stably, integrated transactivator and cDNA by resistance to G418.
6. Screen clones expressing the cDNA of interest by Western blotting. To prepare protein extracts from cells, use the following method (adapted from *(24)*). Keep all reagents and the cell extract on ice for the duration of the experiment.
   i. Wash 5×10^6 cells in PBS and lyse in 100 μl hypotonic buffer by pushing cells through a 19G needle.
   ii. Collect the nuclei by centrifugation (2,700 × g, 10 s). The supernatant will consist of cytosolic proteins. If the protein of interest is cytosolic, transfer supernatant into fresh tube and proceed to iv. If the protein of interest is nuclear, then proceed to iii.
   iii. Wash the pelleted nuclei twice in hypotonic buffer and extract nuclear proteins in 100 μl of high-salt buffer.
   iv. Add an equal volume of 2 × Laemmli SDS sample buffer to the samples and boil for 10 min.
   v. Separate proteins by SDS-PAGE, and proceed with Western blotting to detect proteins of interest.
7. Select those clones that express the protein of interest 2–3-fold over that of the wild-type.
8. Test that the addition of doxycycline leads to repression of cDNA expression:
   i. Add 50 ng/ml of doxycycline to growing cells for 72 h. Include a control sample with no doxycycline added.
   ii. Take a sample every 12 h.
   iii. Prepare protein extracts (as above) for Western blotting to assess the degree of downregulation of the protein of interest.
9. Subsequent alleles can now be targeted; see **Section 3.2**.
10. After targeting the second allele, test if the addition of doxycycline leads to repression of cDNA expression and death of the cell population.
    i. Repeat step 8i, taking 2 samples every 12 h.
    ii. Use one of these samples to check that the protein levels are reduced by Western blotting.
    iii. Add Trypan blue solution to the second sample (1:5 ratio), mix well and incubate at room temperature (RT) for 5 min. Count cells on a haemocytometer to measure cell viability (*see* **Note 8**).

3.5.2. Conditional DT40 Knockouts

The following method describes the generation of conditional knockouts in DT40 cells. This method requires loxP flanked selection cassettes and the use of DT40 cells expressing tamoxifen-inducible Cre recombinase *(17)*. See an example for construct design in **Fig. 1.2B**. Care must be taken to include essential exon(s) between loxP sites. Use experimental techniques described in **Sections 3.1** through **3.3** to clone and target the construct into the first allele. Follow the protocol below, once a correct targeted insertion has been identified.

1. Treat cells for 24 h with 200 nM 4-hydroxy-tamoxifen to induce Cre-mediated recombination (*see* **Fig. 1.2B**).
2. Select those clones that have partial recombination by using Southern blotting or long-range PCR, as outlined in **Section 3.3**.
3. Expand partially recombined clones and target the second allele, using a straightforward knockout construct (**Fig. 1.2A**). Follow techniques described in **Sections 3.1–3.3**.
4. Expand positive clones.
5. Now a second 4-hydroxy-tamoxifen treatment (0.05 mM 4-hydroxy-tamoxifen) will allow the timely disruption of the gene of interest.

3.6. Recycling the Resistance Cassettes (see Notes 9 and 10)

1. Wash 3×10^6 targeted DT40 cells in 1 × PBS. Repeat.
2. Resuspend cells in 100 µl of reconstituted Nucleofector solution T (Amaxa).
3. Add 30 µg of a plasmid with Cre-recombinase.
4. Nucleofect the cells by placing in an Amaxa nucleofector, programme B-009.
5. Resuspend the cells in 2 ml of prewarmed DT40 medium and put into one well of a 6-well plate.
6. Place in the tissue-culture incubator at 39°C, 5% CO_2, for 48 h.
7. After 48 h, count the cells using a haemocytometer.
8. Perform serial dilution of the cells in DT40 medium such that there is one cell per 100 µl. Plate 100 µl of dilution into each well of a flat-bottomed 96-well plate.
9. Incubate this at 39°C, 5% CO_2 for one week without antibiotic selection.
10. After 7 days, eliminate wells with multiple clones and pick 24 single clones from the remaining wells as in **Section 3.3**. Transfer the cells into a 24-well plate with fresh DT40 medium. Allow cells to expand. When cells become dense, replica plate into a second 24-well plate.
11. To the second set of plates, add the antibiotics to which the original cells were resistant.
12. Leave for one week and assess which of the clones have lost their antibiotic resistance.

13. Expand these clones and extract genomic DNA using the Puregene Cell and Tissue DNA extraction kit.
14. Use PCR amplification to confirm the loss of the resistance cassette (see **Section 3.1** and **Fig. 1.2A**).

3.7. Analysing Targeted Clones by Immunofluorescence

1. Coat glass slides with poly-L-lysine:
   i. Add 2 mg/ml poly-L-lysine to glass slides and place at 37°C for 30 min.
   ii. Wash off poly-L-lysine with H_2O.
   iii. Wash slides a further 6 × 5 min to ensure all poly-L-lysine is removed.
   iv. Final wash in 70% ethanol and leave to air-dry.
2. Place coated glass slides in the Cytospin chamber (*see* **Note 11**).
3. Take 1 × 10^6 DT40 cells and place in a Cytospin chamber and spin in the Cytospin at 80 × g, 2 min.
4. Remove slides and check under the light microscope that cells have adhered. They should also be visible by eye as an opaque spot on the slide.
5. Use staining chambers to perform all wash steps. First wash in 1 × PBS for 1 min.
6. Fix in the fixative of choice for your antibody. For microtubules and centrosomal proteins, use 100% methanol, –20°C, 5 min, followed by a 5 min wash in 1 × PBS/0.1% Tween at RT. For kinetochore proteins, use 4% formaldehyde (freshly prepared), 37°C, 5 min, followed by a 5 min wash in 1 × PBS/0.2% Triton X-100 at 37°C.
7. Block nonspecific binding by incubation in blocking buffer, 5 min, RT. Dilute primary antibody in blocking buffer.
8. Add primary antibody to the opaque spot of cells on the slide. Ensure the spot of cells is covered by your primary antibody. Place the slide in humidity chamber (a 10 cm tissue culture dish with moist tissue will suffice) and put at 37°C for 2 h.
9. Wash the slides in 1 × PBS/0.1% Tween, 3 × 5 min, RT.
10. Add blocking buffer for 5 min, RT.
11. Add fluorescently labelled secondary antibodies, in the same way as for the primary. Incubate for one hour at 37°C.
12. Wash the slides in 1 × PBS/0.1% Tween, 3 × 5 min, RT.
13. Final wash in H_2O.
14. Add 5 μl of Prolong Antifade reagent, containing DAPI to the spot of cells and overlay with a glass coverslip.
15. Leave at RT overnight for the Prolong to set. For long-term storage, use 4°C.

3.8. Live Cell Imaging of Mitosis

1. Transfect the cells at least 24 h before imaging with pEGFP-Tubulin by Amaxa nucleofector (or electroporation), as stated in **Section 3.6**.
2. Coat chambered coverglass with poly-L-lysine, as in **Section 3.7 step 1**.
3. After transfecting cells, place them on poly-L-lysine-coated chambered coverglass (Lab-Tek) and incubate at 39°C, 5% CO_2 for 24 h.
4. Just before imaging the cells, add 1 μg/ml of Hoeschst to the medium to visualise DNA.
5. Follow cells undergoing mitosis using a confocal or epifluorescent microscope, ideally with a temperature and CO_2-regulated chamber. To enrich mitotic cells, *see* **Note 12**.

4. Notes

1. If Puregene kit is not available, follow this alternative method to extract genomic DNA (adapted from *(25)*). Lyse 10^7 cells overnight at 55°C in 300 μl of lysis buffer (100 mM Tris pH 8.5, 5 mM EDTA, 0.2% sodium dodecyl sulphate [SDS], 200 mM NaCl, 100 μg of proteinase K per ml). Add saturated NaCl solution (90 μl) and centrifuge at 13,000 × g, 10 min. Take off the supernatant and keep. Add 3 volumes of absolute ethanol and incubate at –20°C overnight to precipitate DNA Centrifuge at 13,000 × g, 15 min. Discard the supernatant. Wash the pellet in 70% ethanol. Centrifuge again at 13,000 × g, 15 min. Discard supernatant and leave the pellet to air-dry. Once dry, resuspend in H_2O.
2. When designing the targeting vector, take care when choosing the restriction enzymes used for subcloning. The enzymes used to subclone the left and right arms should be (a) absent from the target vector (i.e., pLoxNeo, pLoxPuro, pLoxBlasti), and (b) absent from the sequence contained within the left and right arms. In addition, in the final targeting construct, there must be a unique site outside of the left and right arms and not within the resistance cassette that can be used to linearise the vector before transfection.
3. If no product is obtained after PCR amplification, try the protocol with one of these modifications. First, decrease the combined annealing/extension temperature of 68°C to 67°C. In addition, increase the number of cycles in increments of two. Adding 2–10% DMSO may be necessary, especially if the template is GC-rich. It may also

be necessary to optimise the magnesium concentration: if no product is obtained, then vary the concentration of magnesium in the reaction. A further problem may be that the primers are designed against sequences that differ between the genomic DNAs of *Gallus gallus* and *Gallus gallus domesticus* (see Introduction). When primers target sequences within exons or 5'/3' untranslated regions, expressed short tag (EST) sequences from DT40 can be used to exclude polymorphisms. If, however, primers need to recognise intronic sequences, it is sometimes necessary to design 2–4 different primers at about 20–50 base pairs apart.

4. The resuspended pellet can be stored at –20°C at this point for use at a later date.
5. Only use flat-bottomed tissue culture-treated plates as the stably transfected colonies will then be visible from the underside.
6. To freeze down cells in a 24-well plate: when cells reach sufficient confluency, allow the cells to settle to the bottom of the well in 0.5 ml growth medium. Carefully remove half of the medium from the top of the wells. To each well add 0.25 ml DT40 medium, containing 20% DMSO, and mix well. Wrap the plate well in Parafilm. Place on ice for 15 min. Take a box of dry ice and put several layers of thick tissue over the top. Place the 24-well plate on top of this. Leave for 4 hours. After this time, the cells should have frozen and the plate can be transferred to the –80°C freezer for long-term storage.
7. The tTA3 transactivator is on a vector carrying G418 resistance and therefore the gene-targeting vectors used should not contain G418 resistance or should be recycled prior to their introduction.
8. Trypan blue is only taken up by dead cells and hence allows an assessment of cell viability by dye exclusion.
9. The method described here uses transient transfection of a Cre-expressing plasmid to remove the antibiotic resistance cassettes. For an alternative method using a DT40 cell line that has Cre stably integrated into the genome, see *(20)*.
10. If an Amaxa system is not available, an alternative method uses standard electroporation, as in **Section 3.2**.
11. If a Cytospin is not available, use twice the number of cells stated and place onto poly-L-lysine coated coverslips in a 24-well plate. Centrifuge in a bench-top centrifuge at 800 × g, 10 min.
12. To increase the ratio of mitotic cells, cells can be blocked in 200 ng/ml nocodazole for 8–10 h. Cells should be washed three times in prewarmed fresh medium prior to imaging. Alternatively, cells can be synchronised in early

S-phase by the addition of 1 μg/ml aphidicolin for 10 h. Mitotic index peaks at about 6–6.5 h following the removal of aphidicolin.

References

1. M. Zuccolo, A. Alves, V. Galy, S. Bolhy, E. Formstecher, V. Racine, J. B. Sibarita, et al. (2007) The human Nup107-160 nuclear pore subcomplex contributes to proper kinetochore functions. *Embo J.* **26,** 1853–1864.
2. E. Sonoda, T. Matsusaka, C. Morrison, P. Vagnarelli, O. Hoshi, T. Ushiki, K. Nojima, et al. (2001) Scc1/Rad21/Mcd1 is required for sister chromatid cohesion and kinetochore function in vertebrate cells. *Dev. Cell* **1,** 759–770.
3. A. Nishihashi, T. Haraguchi, Y. Hiraoka, T. Ikemura, V. Regnier, H. Dodson, W. C. Earnshaw, et al. (2002) CENP-I is essential for centromere function in vertebrate cells. *Dev. Cell* **2,** 463–476.
4. M. Okada, I. M. Cheeseman, T. Hori, K. Okawa, I. X. McLeod, J. R. Yates, 3rd, A. Desai, et al. (2006) The CENP-H-I complex is required for the efficient incorporation of newly synthesized CENP-A into centromeres. *Nature Cell Biol.* **8,** 446–457.
5. P. Vagnarelli, D. F. Hudson, S. A. Ribeiro, L. Trinkle-Mulcahy, J. M. Spence, F. Lai, C. J. Farr, et al. (2006) Condensin and Repo-Man-PP1 co-operate in the regulation of chromosome architecture during mitosis. *Nature Cell Biol.* **8,** 1133–1142.
6. G. Zachos, M. D. Rainey and D. A. Gillespie. (2003) Chk1-deficient tumour cells are viable but exhibit multiple checkpoint and survival defects. *Embo J.* **22,** 713–723.
7. G. Zachos, M. D. Rainey and D. A. Gillespie. (2005) Chk1-dependent S-M checkpoint delay in vertebrate cells is linked to maintenance of viable replication structures. *Mol. Cell. Biol.* **25,** 563–574.
8. H. Hochegger, D. Dejsuphong, E. Sonoda, A. Saberi, E. Rajendra, J. Kirk, T. Hunt, et al. (2007) An essential role for Cdk1 in S phase control is revealed via chemical genetics in vertebrate cells. *J. Cell Biol.* **178,** 257–268.
9. E. Sonoda, M. Takata, Y. M. Yamashita, C. Morrison and S. Takeda. (2001) Homologous DNA recombination in vertebrate cells. *Proc. Natl. Acad. Sci. USA* **98,** 8388–8394.
10. F. Bunz, C. Fauth, M. R. Speicher, A. Dutriaux, J. M. Sedivy, K. W. Kinzler, B. Vogelstein, et al. (2002) Targeted inactivation of p53 in human cells does not result in aneuploidy. *Cancer Res.* **62,** 1129–1133.
11. T. Waldman, K. W. Kinzler and B. Vogelstein. (1995) p21 is necessary for the p53-mediated G1 arrest in human cancer cells. *Cancer Res.* **55,** 5187–5190.
12. O. Topaloglu, P. J. Hurley, O. Yildirim, C. I. Civin and F. Bunz. (2005) Improved methods for the generation of human gene knockout and knockin cell lines. *Nucleic Acids Res.* **33,** e158.
13. K. D. Hanson and J. M. Sedivy. (1995) Analysis of biological selections for high-efficiency gene targeting. *Mol. Cell. Biol.* **15,** 45–51.
14. J. M. Buerstedde and S. Takeda. (1991) Increased ratio of targeted to random integration after transfection of chicken B cell lines. *Cell* **67,** 179–188.
15. J. M. Buerstedde, C. A. Reynaud, E. H. Humphries, W. Olson, D. L. Ewert and J. C. Weill. (1990) Light chain gene conversion continues at high rate in an ALV-induced cell line. *Embo J.* **9,** 921–927.
16. A. Hatanaka, M. Yamazoe, J. E. Sale, M. Takata, K. Yamamoto, H. Kitao, E. Sonoda, et al. (2005) Similar effects of Brca2 truncation and Rad51 paralog deficiency on immunoglobulin V gene diversification in DT40 cells support an early role for Rad51 paralogs in homologous recombination. *Mol. Cell. Biol.* **25,** 1124–1134.
17. H. Arakawa, D. Lodygin and J. M. Buerstedde. (2001) Mutant loxP vectors for selectable marker recycle and conditional knock-outs. *BMC Biotechnol* **1,** 7.
18. M. Gossen and H. Bujard. (1992) Tight control of gene expression in mammalian cells by tetracycline-responsive promoters. *Proc. Natl. Acad. Sci. USA* **89,** 5547–5551.
19. J. Wang, Y. Takagaki and J. L. Manley. (1996) Targeted disruption of an essential vertebrate gene: ASF/SF2 is required for cell viability. *Genes Dev.* **10,** 2588–2599.
20. H. Arakawa, D. Lodygin and J.-M. Buerstedde. (2001) Mutant loxP vectors for selectable marker recycle and conditional knock-outs. *BMC Biotechnol.* **1,** 7.
21. M. Gossen and H. Bujard. (1992) Tight control of gene expression in mammalian cells by tetracycline-responsive promoters. *Proc. Natl. Acad. Sci.* **89,** 5547–5551.
22. U. Baron, M. Gossen and H. Bujard. (1997) Tetracycline-controlled transcription

in eukaryotes: novel transactivators with graded transactivation potential. *Nucl. Acids Res.* **25,** 2723–2729.
23. J.-M. Buerstedde and S. Takeda. (1991) Increased ratio of targeted to random integration after transfection of chicken B cell lines. *Cell* **67,** 179–188.
24. G. Mosedale, W. Niedzwiedz, A. Alpi, F. Perrina, J. B. Pereira-Leal, M. Johnson, F. Langevin, et al. (2005) The vertebrate Hef ortholog is a component of the Fanconi anemia tumor-suppressor pathway. *Nature Struct. Mol. Biol.* **12,** 763–771.
25. R. Kwan, J. Burnside, T. Kurosaki and G. Cheng. (2001) MEKK1 is essential for DT40 cell apoptosis in response to microtubule disruption. *Mol. Cell. Biol.* **21,** 7183–7190.

Chapter 2

Functional Dissection of Mitotic Regulators Through Gene Targeting in Human Somatic Cells

Eli Berdougo, Marie-Emilie Terret and Prasad V. Jallepalli

Abstract

With the human genome fully sequenced *(1, 2)*, biologists continue to face the challenging task of evaluating the function of each of the ~25,000 genes contained within it. Gene targeting in human cells provides a powerful and unique experimental tool in this regard *(3–8)*. Although somewhat more involved than RNAi or pharmacological approaches, somatic cell gene targeting is a precise technique that avoids both incomplete knockdown and off-target effects, but is still much quicker than analogous manipulations in the mouse. Moreover, immortal knockout cell lines provide excellent platforms for both complementation analysis and biochemical purification of multiprotein complexes in native form. Here we present a detailed gene-targeting protocol that was recently applied to the mitotic regulator Polo-like kinase 1 (Plk1) *(9)*.

Key words: Gene targeting, rAAV, human somatic cells, mitosis, Plk1.

1. Introduction

Over the past several years, the methods used to generate knockout or knockin mutations in human somatic cells have greatly improved in efficiency *(3–8)*. The current method utilizes the high recombination potential of adeno-associated virus (AAV) vectors, requires less than 6 months for homozygous mutation of both alleles in a diploid cell *(10–12)*, and can be divided into several general stages. The first step is to design and assemble a targeting vector containing 5' and 3' arms that are homologous to the locus of interest. If desired, gene inactivation can be made conditional, simply by placing *loxP* sites on either side of a coding

Andrew D. McAinsh (ed.), *Mitosis: Methods and Protocols*, Methods in Molecular Biology *545*,
DOI 10.1007/978-1-60327-993-2_2, © Humana Press, a part of Springer Science+Business Media, LLC 2009

exon of interest. This allows the exon to be rapidly and irreversibly deleted at the desired time, via expression of Cre recombinase. This approach permits the study of genes that are essential for cell proliferation, as expected for many mitotic regulators. The second step is to cotransfect the targeting vector with helper plasmids into HEK293 cells, in order to generate infectious AAV particles. The third step is to infect the cell line of interest with these viruses and subsequently screen for clones that have undergone correct locus-specific recombination. The final step is to verify gene disruption by PCR and Southern blotting assays, and finally, to carry out the procedure a second time to obtain homozygous mutant clones.

Multiple human cell types are amenable to gene targeting with AAV vectors, including both transformed and nontransformed cell lines and primary cells isolated from patients *(7, 9, 13)*. Here we use telomerase-immortalized human retinal pigment epithelial cells (hTERT-RPE) to investigate Plk1 function in a nontransformed, nontumorigenic setting *(9)*. Unlike established cancer cell lines, hTERT-RPE cells tolerate severe (>90%) depletion of Plk1 via RNAi without effect *(14)*. In contrast, we find that homozygous deletion of the *PLK1* locus fully abrogates its function and recapitulates all known mitotic functions of this kinase *(9)*. Furthermore, these $PLK1^{\Delta/\Delta}$ cells could be reconstituted with a variant form of Plk1 that is uniquely susceptible to bulky ATP analogues, enabling chemical genetic dissection of Plk1's roles in late mitosis and cytokinesis *(9)*.

2. Materials

2.1. Cell Culture

1. HEK293 cells (ATCC CRL-1573)
2. Telomerase-immortalized human retinal pigment epithelial cells (hTERT-RPE; ATCC CRL-4000)
3. Medium suitable for propagation of HEK293 cells: D-MEM (Invitrogen) supplemented with 10% fetal bovine serum (FBS; Omega Scientific) and 0.1 mg/ml penicillin-streptomycin (Gemini Bio-Products)
4. Medium suitable for propagation of hTERT-RPE cells: D-MEM/F:12 medium with 15 mM HEPES, 2.5 mM L-glutamine, 2.4 g/L sodium bicarbonate (Invitrogen), supplemented with 10% FBS and 0.1 mg/ml penicillin-streptomycin
5. Hank's balanced salt solution (HBSS; Invitrogen)
6. 0.05% Trypsin-EDTA (Invitrogen)

7. Tissue culture-treated 96-, 48-, 24-, 12-, and 6-well plates (Corning)
8. Tissue culture treated T-25 and T-75 flasks (Corning)
9. G418 (Gemini Bio-Products)
10. Presterilized 8-port manifolds (CLPdirect)
11. Repeater Plus pipetter with sterile 50-ml Combitips (Eppendorf) or multichannel P200 and filter tips (Rainin)
12. 100 ml sterile basins (Fisher)

2.2. Production of rAAV Particles and Genomic DNA Extraction

1. pNY, Jallepalli lab *(9)*
2. QuikChange II XL site-directed mutagenesis kit (Stratagene)
3. AAV Helper-free system (Stratagene, 240071)
4. Lipofectamine Transfection reagent/Plus reagent (Invitrogen)
5. OptiMEM I medium (Invitrogen)
6. Disposable cell scrapers (Fisher)
7. Deep-well blocks (BD Biosciences)
8. Wizard SV 96 genomic DNA purification system (Promega, A2371)
9. Beckman-Coulter Allegra 25R with deep-well block rotor (S5700) or Vac-Man 96 Vacuum Manifold (Promega
10. Multichannel P200 and P1000 pipetters and filter tips (Rainin)
11. QIAampDNA Blood Mini Kit (Qiagen, 51106)

2.3. PCR Screening for Knockouts

1. Thin-wall 96-well plates (Simport) and sealing films (Fisher)
2. Tissue culture-grade water (Sigma)
3. DMSO (Sigma)
4. 10X PCR buffer: 166 mM ammonium sulfate, 670 mM Tris-HCl, pH 8.8, 67 mM $MgCl_2$, 100 mM beta-mercaptoethanol
5. 10 mM dNTPs (USB)
6. Primers (Integrated DNA Technologies)
7. Platinum Taq polymerase (Invitrogen)
8. Taq Extender (Stratagene)
9. Mineral oil (Sigma)
10. Bio-Rad Sub-Cell Model 96/192 gel system
11. Multichannel P10, P20, P200 pipetters and filter tips (Rainin)

2.4. FLP-Mediated Excision of the Neo Cassette

1. Fugene 6 (Roche)
2. pCAGGS-FLPe (Gene Bridges)
3. Puromycin (Gemini Bio-Products)
4. Purified adenovirus expressing Cre recombinase (Vector Development Laboratory Baylor College of Medicine)

3. Methods

The following method can be used to generate a conditional knockout of any desired locus in human somatic cells. The targeting strategy should be planned in its entirety before starting work (**Fig. 2.1**).

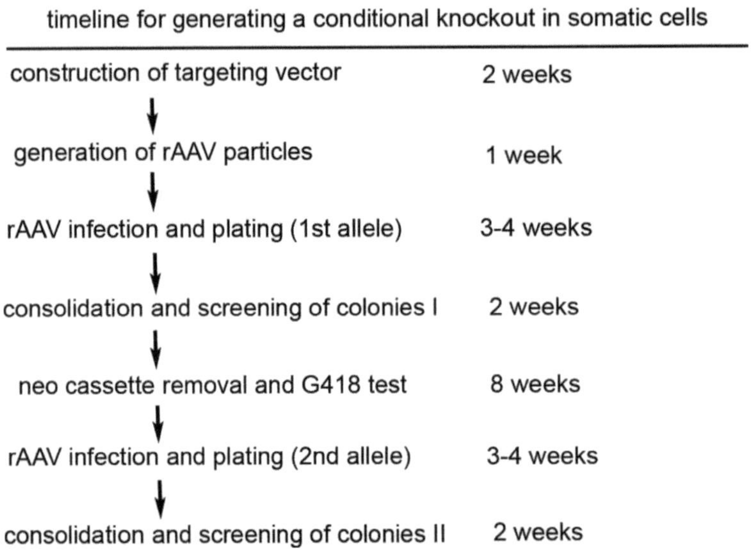

Fig. 2.1. Timeline for generating a conditional knockout in human somatic cells.

3.1. Characterization of the Genetic Locus and Primer Design

1. Choose the exon that you wish to conditionally delete. This exon will need to be flanked by tandemly oriented *loxP* sites, generating a so-called "floxed" allele. Use *Ensembl* to verify that length of this exon in basepairs is not a multiple of three (http://www.ensembl.org/index.html). This will ensure that the open reading frame undergoes a frameshift and premature termination after the exon is deleted.
2. Using RepeatMasker (http://www.repeatmasker.org/cgi-bin/WEBRepeatMasker), examine a 4-kb region centered on this exon for repetitive DNA content.
3. Choose 5' and 3' homology arms of about 1.0–1.5 kb in length, taking care to maximize the unique sequence content in each arm (*see* **Fig. 2.2A** and **Notes 1** and **2**).
4. Design oligonucleotides for amplifying and cloning the homology arms into the shuttle vector pNY, which contains a central *FRT*-neoR-*FRT*-*loxP* cassette (*see* **Fig. 2.2A**, primer pairs F1/R1 and F2/R2; *see* **Notes 3** and **4**).
5. Design several oligonucleotides for PCR screening and sequencing across both homology arms. The candidate

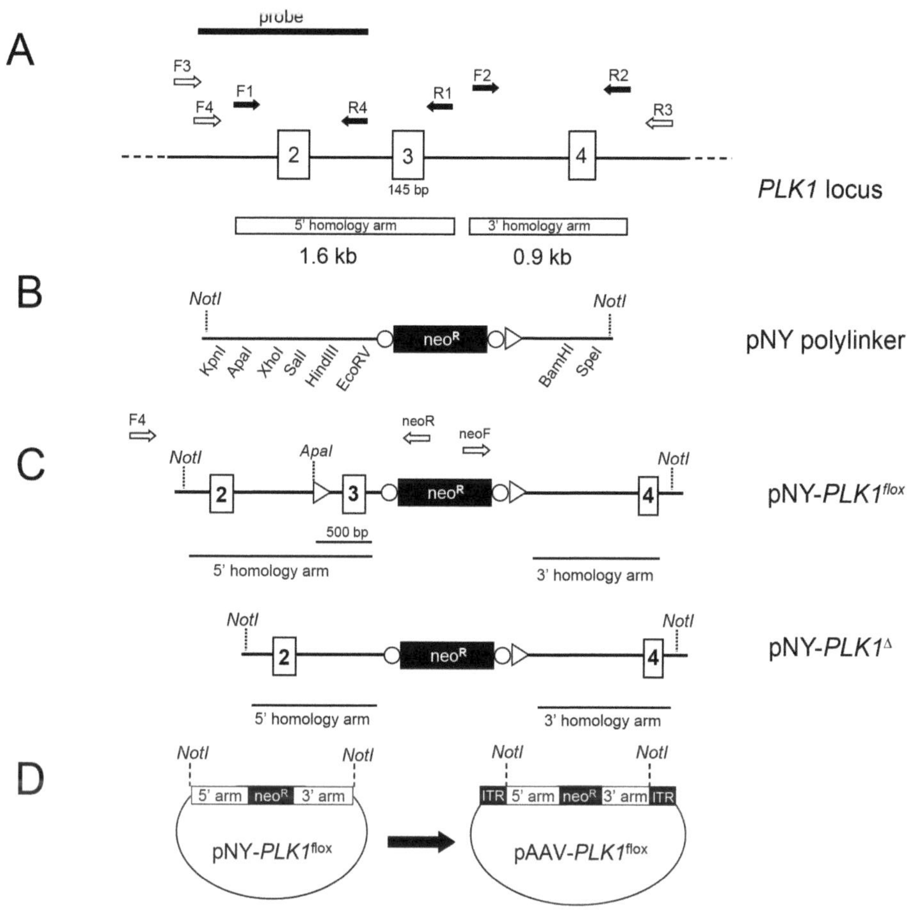

Fig. 2.2. Strategy for targeting exon 3 of *PLK1*. **(A)** Schematic of the 5' end of the *PLK1* locus. Primers used to amplify the homology arms for cloning are denoted by black arrows, and those used for PCR screening are denoted by white arrows. **(B)** Map of the pNY polylinker. *FRT* sites are shown as white circles and the *loxP* site is shown as a triangle. Unique restriction sites for cloning the 5' and 3' homology arms are also indicated. **(C)** Structure of the pNY-*PLK1^{flox}* and pNY-*PLK1$^{\Delta}$* shuttle plasmids. **(D)** Transfer of the NotI fragment from pNY-*PLK1^{flox}* to pAAV generates the final construct (pAAV-*PLK1^{flox}*) used for virus production. ITR, AAV-specific inverted tandem repeats.

screening primers should correspond to unique genomic sequences about 100–500 bp outside the region delimited by the 5' and 3' homology arms (*see* **Fig. 2.2A** screening primers F3, F4, and R3).

6. Harvest wild-type RPE-hTERT cells from a T-25 flask and prepare genomic DNA (gDNA) using the QIAamp DNA Blood Mini Kit. Determine the concentration of your DNA by measuring the OD 260 of your sample in a spectrophotometer.
7. Perform test PCR reactions using the conditions specified in **Section 3.6** with the candidate screening primers and locus-specific primers (F3/R1, F4/R1, F2/R3; *see* **Fig. 2.2A**),

using the gDNA prepared in step 6 as the template. Prepare 10-fold serial dilutions of your gDNA to identify those primers which have optimal sensitivity and specificity (*see* **Note 5**).

3.2. Construction of the Targeting Vector

1. Using a high-fidelity thermostable polymerase (e.g., PfuTurbo, Stratagene), amplify 5' and 3' homology arms from human gDNA (e.g., purified from the cell line of interest) or a human genomic BAC (bacterial artificial chromosome) clone identified by BLAST searching and obtained commercially (Invitrogen or BACPAC Resources).
2. Subclone homology arms into pNY and verify by sequencing (*see* **Fig. 2.2B**).
3. Introduce a *loxP* site into the appropriate homology arm of the targeting vector by site-directed mutagenesis (QuikChange II XL kit, Stratagene) or linker ligation. This *loxP* site should be in the same orientation as the existing *loxP* site in pNY. The introduced *loxP* site should also be marked with a novel restriction endonuclease site to facilitate downstream analyses. Verify the presence, orientation, and integrity of the inserted elements by sequencing and restriction digestion (*see* **Note 6**).
4. Restriction digest pNY containing the homology arms with NotI and gel-purify the insert (5' arm-[*FRT*-neoR-*FRT*-*loxP* cassette]-3' arm).
5. Restriction digest the recipient vector pAAV-*lacZ* with NotI. Treat this reaction with 1 μl of Calf Intestinal Phosphatase (CIP) and leave reaction at 37°C for an additional 15 min. Extract the DNA with phenol:chloroform, ethanol precipitate, and resuspend in 20 μl water. Gel-purify the NotI digested pAAV vector backbone (*see* **Note 7**).
6. Ligate the NotI digested insert from step 4 into the pAAV backbone from step 5 (*see* **Fig. 2.2D**).
7. Transform the ligations and screen the resulting colonies for recombinant AAV plasmids that contain your homology arms (see **Note 8**).
8. Prepare transfection-grade plasmid DNA using standard methods (silica-based kits or isopycnic centrifugation on cesium chloride gradients).

3.3. Generation of rAAV Particles

1. Thaw early-passage HEK293 cells in D-MEM supplemented with 10% FBS and 0.1 mg/ml penicillin-streptomycin (complete D-MEM). Grow cells at 37°C in a tissue culture incubator until 80–90% confluent.
2. Split HEK293 cells into two T-75 flasks and grow at 37°C in a tissue culture incubator until they are at 50–70% confluence.

3. Transfect each flask with 3 μg each of your targeting vector, pHELPER, and pAAV-RC (9 μg total) using Lipofectamine reagent and Plus reagent as follows. Mix DNAs in one sterile 1.5 ml tube with 750 μl OptiMEM I reduced serum medium and 36 μl of Plus reagent. In a second sterile 1.5 ml tube, mix 54 μl of Lipofectamine reagent with 750 μl OptiMEM I reduced serum medium. After 15 min at room temperature, drip the contents of the first tube into the second tube, and incubate for 30 min at room temperature to form DNA/lipid complexes. Wash the HEK293 cells twice with Hank's balanced salt solution (HBSS) and add 7.5 ml OptiMEM and incubate at 37°C for 15 min. Drip this DNA/lipid mixture onto your HEK 293 cells and incubate for 4 h at 37°C. Replace the medium with 15 ml of complete D-MEM and return the cells to the 37°C incubator.
4. Harvest the virus particles three days post-transfection. Begin by transferring the medium from each T-75 to a sterile 50 ml conical tube (*see* **Note 9**).
5. Add 5 ml of complete D-MEM to each flask, scrape off the remaining cells using a disposable cell scraper, and transfer each cell suspension to the 50 ml conical tube.
6. Set up a dry ice/methanol bath. Freeze the cell suspension in the bath. Transfer the tubes to a 37°C water bath and thaw. When the cell suspension is almost fully thawed, vortex the tubes at maximum speed for 1 min. Repeat this freeze/thaw cycle two more times (*see* **Note 10**).
7. Spin out the cell debris at 10,000 × g for 30 min at 4°C. Decontaminate the outside of each tube with 70% ethanol and place within a sterile tissue culture hood.
8. Carefully transfer the supernatant fraction to a new 50 ml conical tube. This is your working stock of infectious rAAV particles. Aliquot in 5–10 ml fractions and store at –80°C until ready to use.

3.4. Infection of Target Cells with rAAV Particles and Selection for Stable Integrants

1. Thaw an early-passage stock of your target cells of interest (hTERT-RPE cells in the example given here). Passage cells 1–2 days prior to infection so that they are ~30% confluent in a T-75 flask on the day of infection (*see* **Note 11**).
2. Wash the cells twice with 5 ml of HBSS. Add 6 ml of D-MEM/F:12 supplemented with 10% FBS and 0.1 mg/ml penicillin-streptomycin (complete D-MEM/F:12) + 6 ml of your rAAV preparation. Incubate at 37°C in a tissue culture incubator for 4 h.
3. Bring the volume up to 15 ml with complete D-MEM/F:12. Allow infection to continue for 48 h at 37°C in a tissue culture incubator.

4. Plate the rAAV-infected cells at sufficient dilution into 15 × 96 well plates to obtain no more than one clone per well. To do this, remove the medium from the flask and trypsinize the cells. Transfer the trypsinized cell suspension into a tissue culture vessel containing 300 ml of complete D-MEM/F:12+ 0.4 mg/ml G418. Mix well by gentle inversion or pipetting up and down and pour the cell suspension into a sterile basin. Dispense 200 μl/well into 15 × 96-well plates using a repeat pipetter with an 8-port manifold attachment or a multichannel pipette.
5. Wrap the stack of 96-well plates in plastic wrap and incubate at 37°C in a tissue culture incubator for 2–3 weeks until colonies form (*see* **Note 12**).

3.5. Consolidation of Colonies and gDNA Preparation

1. Two weeks after plating, inspect the 96-well plates for colonies using an inverted bright field microscope (*see* **Note 13**).
2. To consolidate colonies for PCR screening, carefully aspirate the media from the plates using an 8-port manifold attached to a vacuum line. Apply 50 μl of 0.05% Trypsin-EDTA using a P200 multichannel pipette and incubate plates for 10 min at 37°C in a tissue culture incubator.
3. Using an inverted brightfield microscope, verify that the colonies are fully detached from the well. Using a multichannel P200 pipette, disaggregate the cells and transfer 40 μl from each well to a deep-well block. We regularly pool 5–10 plates per deep-well block in order to reduce the number of PCR reactions required for the initial screen. However, single clones can also be analyzed if desired (*see* **Notes 14 and 15**).
4. Using the P200 multichannel pipette or a repeat pipetter with a 50 ml Combitip and 8-port manifold, refeed the 96-well plates with 190 μl of D-MEM/F:12 complete medium per well. Cover the plates in plastic wrap to minimize evaporation and return to the 37°C tissue culture incubator.
5. To prepare gDNA, add a sufficient quantity of HBSS to each well of the deep-well block to bring it to a final volume of 300 μl. Purify gDNA using the Wizard SV 96 genomic DNA purification kit (Promega). It is recommended that the eluted gDNA be used immediately for PCR screening, but if this is not feasible, it can be stored at –20°C (*see* **Notes 16** and **17**).

3.6. PCR Screen

1. All PCR reactions are done in thin-walled 96-well plates. The reaction conditions per reaction are as follows:

dd H20	6.3 μl
10 × PCR buffer	1.25 μl
10 mM dNTPs	1.5 μl
DMSO	0.75 μl
primerF (350 ng/ μl)	0.25 μl
primerR (350 ng/ μl)	0.25 μl
platinum Taq polymerase	0.125 μl
Taq extender	0.125 μl
genomic DNA	2.0 μl

Prepare a master mix (100 reactions per plate) and dispense 10.5 μl to each well. Add 2 μl of gDNA to each reaction using a P10 multichannel pipette and filter tips. Overlay each well with a drop of mineral oil and cover the plate with a piece of sealing film (*see* **Note 18**).

2. Place the plate in a thermal cycler and cycle as follows: 94°C × 30 s (1 cycle); 94°C × 15 s, 63°C × 30 s, 70°C × 2 min (4 cycles); 94°C × 15 s, 60°C × 30 s, 70°C × 2 min (4 cycles); 94°C × 15 s, 57°C × 30 s, 70°C × 2 min (40 cycles; *see* **Note 19**).
3. After completion of PCR reactions, remove the sealing film and add 12.5 μl of 2 × DNA loading dye to each reaction.
4. Using a P20 multichannel pipette, load 12.5 μl of each reaction onto a 0.8% agarose gel containing ethidium bromide cast with 104 wells (multichannel compatible). Run the gel at 150 volts for 30–45 min and photograph on a UV light box.
5. Identify the wells giving rise to correct PCR products (*see* **Fig. 2.3B,** white arrowhead).
6. Gel-purify positive PCR products from step 4 above and digest with the appropriate restriction enzyme (chosen during the design stage in **Section 3.2**) to assess where recombination occurred relative to the exogenously introduced *loxP* site (*see* **Fig. 2.3C** and **Note 20**).
7. Go back to your stack of 96-well plates and trypsinize those wells comprising each PCR-positive pool with 50 μl of 0.05% Trypsin-EDTA. Transfer each well to a new well of a 96-well tissue culture plate and add 150 μl of complete D-MEM/F:12. Return the plate to the 37°C tissue culture incubator and grow the cells until they reach confluence.
8. Prepare gDNA using the SV96 genomic DNA purification kit and rescreen individual clones by PCR. These clones are your targeted heterozygotes (i.e., cells with a *floxneo/+* genotype; *see* **Note 21**).
9. Expand your heterozygote clones to sufficient quantities and prepare gDNA using the Blood Mini Kit. Use this

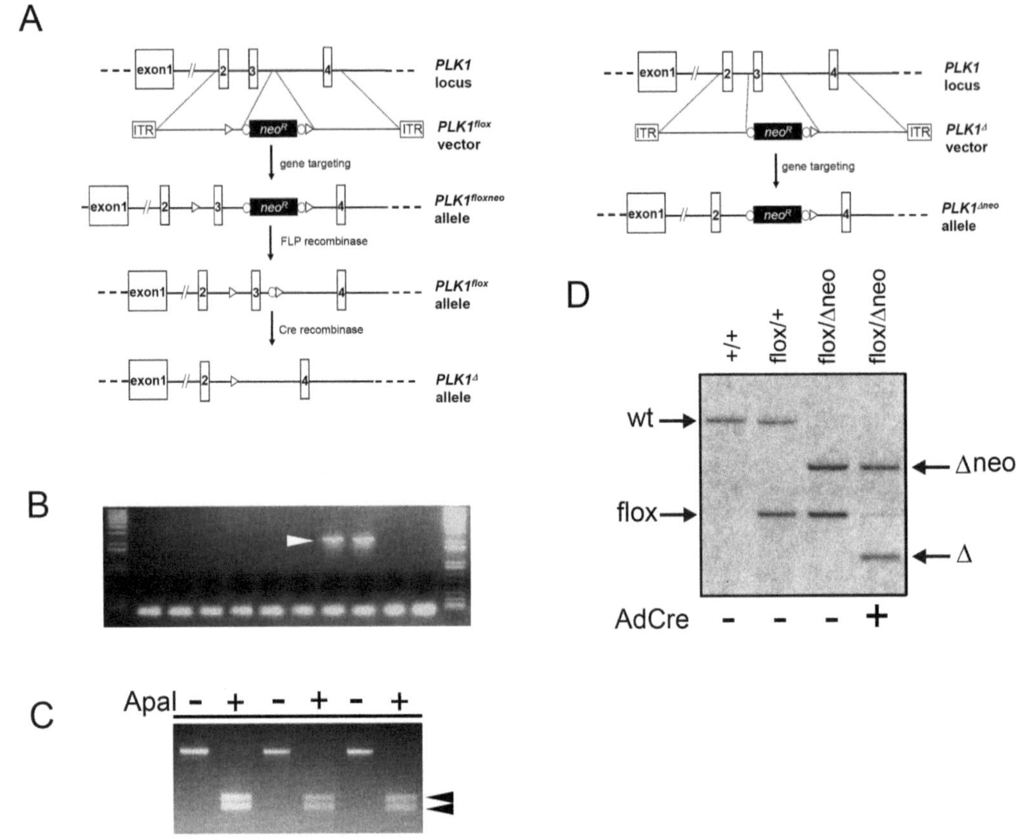

Fig. 2.3. Generation of *PLK1* conditional-knockout cells. **(A)** Structure of *PLK1* alleles generated after gene targeting, removal of the neomycin-resistance cassette, and deletion of exon 3. In this instance, the first allele of *PLK1* was targeted with a conditional-null vector (*left panel*), and the second allele was targeted with a constitutive-null vector (*right panel*). **(B)** Example of a positive "hit" from a genomic PCR screen using F4 and neoR primers. The expected product of 2266 bp is marked by a white arrowhead. **(C)** ApaI restriction digests demonstrate that recombination between the 5' homology arm and the chromosomal locus took place upstream of the *loxP* site between exons 2 and 3; in other words, a favorable crossover occurred. Fragments of 1107 bp and 1159 bp are highlighted by black arrowheads. **(D)** Verification of genotypes by Southern blotting. Genomic DNAs were digested with BamHI and SacI and hybridized with a [^{32}P]-labeled probe (*see* **Fig. 2.2A**). Wild-type (4.8 kb), *flox* (2.1 kb), *Δneo* (3.2 kb), and *Δ* (1.6 kb) alleles are indicated with black arrows. Note that conversion of the *PLK1^{flox}* allele to a *PLK1$^{\Delta}$* allele requires infection with an adenovirus expressing Cre recombinase (AdCre).

gDNA for additional PCR and Southern blot analyses to verify correct recombination at the locus *(15)*.

10. Expand heterozygote clones for long-term cryopreservation.

3.7. Removing the Neomycin Cassette by FLP Recombinase and Testing the Functionality of the "Floxed" Allele

1. Expand heterozygously targeted (*floxneo/+*) cells to 60–80% confluence in a T-25 flask (*see* **Note 22**).
2. Transfect heterozygously targeted cells with the pCAGGS-FLPe plasmid as follows. In a sterile 1.5 ml tube combine 9 μl Fugene with 500 μl of OptiMEM I and incubate at room temperature for 5 min. Add 3 μg of pCAGGS-FLPe,

mix, and incubate at room temperature for 15 min. Wash the target cells twice with HBSS and replace with 2.5 ml of complete D-MEM/F:12. Drip the DNA/Fugene mix onto the cells and return the flask to the 37°C incubator (*see* **Note 23**).

3. Trypsinize the cells 24 h post-transfection, and plate into a T-75 containing 3 μg/ml puromycin. Maintain the puromycin selection for 48 h, changing the medium after 24 h to remove the bulk of dead cells (*see* **Note 24**).
4. After selection, replace the medium with complete D-MEM/F:12 medium lacking puromycin and allow the cells to recover at 37°C in a tissue culture incubator for several days until they reach about 80% confluence.
5. Trypsinize the cells and prepare dilutions of the cell suspensions so that you can plate the cells using a multichannel pipette into 2–4 96-well plates at a density of 0.5 cells/well and 2.5 cells/well.
6. Wrap the plates with plastic wrap and allow them to grow for 2–3 weeks at 37°C in a tissue culture incubator until colonies form.
7. Check the plates for colonies using a brightfield microscope and identify wells that contain single colonies. Trypsinize these wells and transfer 48–96 clones to a new 96-well plate. Add complete medium and expand cells to confluence. These are your candidate *flox/+* cells.
8. Prepare gDNA from the entire 96-well plate as in **Section 3.5**.
9. Set up a PCR screen using a *neo*-specific primer (neoR; **Fig. 2.2C**) and a locus-specific primer to identify clones that have lost the *neo* cassette due to FLP-mediated excision (*see* **Note 25** and **Fig. 2.3A**).
10. Expand putative *flox/+* clones to sufficient quantities and prepare gDNA using the Blood Mini Kit. Verify the *flox/+* genotype by PCR and Southern blotting (*see* **Fig. 2.3D**).
11. Prior to the 2nd allele targeting, plate 10^6 *flox/+* cells into complete D-MEM/F:12 medium + 0.4 mg/ml G418 in a T-75 and allow cells to grow for 2 weeks. Check for complete G418 sensitivity by scoring for any colony growth (*see* **Note 26**).
12. Seed *flox/+* cells in a T-25 at a confluency of ~10%. The following day, when the cells reach ~20% confluence, infect with adenoviruses expressing Cre recombinase (AdCre (*see* **Section 2.4**)) at an MOI of 100–200.
13. Remove the AdCre-containing medium 24 h after infection and replace with complete D-MEM/F:12 medium.

14. Harvest the cells 48 h after infection and prepare gDNA using the Blood Mini Kit for Southern blotting to confirm the functionality of the *loxP* site (i.e., deletion of floxed sequences; *see* **Fig. 2.3A**, *PLK1*$^{\Delta}$ allele).

3.8. Targeting the Second Allele

1. Thaw and expand early-passage *flox/+* cells that have passed the G418 sensitivity test (described in **Section 3.7, step 11**) to a T-75 flask, such that they are ~30% confluent on the day of rAAV infection.
2. Thaw your rAAV virus preparation made in **Section 3.3.**
3. Infect *flox/+* cells with rAAV particles as in **Section 3.4** above (*see* **Notes 27–29**).
4. After 2–3 weeks, check for G418-resistant colonies using an inverted brightfield microscope.
5. Consolidate colonies and prepare gDNA as in **Section 3.5**.
6. PCR screen colonies and identify individual PCR positive wells as in **Section 3.6.**
7. To identify floxed homozygotes (*floxneo/flox*), perform a secondary PCR screen on the clones that scored positively in the first PCR screen, but in this case use locus-specific primers that span the *loxP* site, rather than a *neo*-specific PCR primer. Gel-purify PCR products and digest with the restriction enzyme used to mark the *loxP* site. The PCR product of a bi-allelic mutant (*floxneo/flox)* should cut completely, whereas the PCR product of a monoallelic (*floxneo/+)* mutant will cut only partially (less than 50%, due to random reannealing of the Watson and Crick strands during PCR).
8. Expand candidate *floxneo/flox* clones to verify their genotype by Southern blotting.
9. Expand candidate *floxneo/flox* clones for cryopreservation.
10. Generate *flox/flox* cells by removal of the neomycin cassette as in **Section 3.7.**
11. Expand early-passage *flox/flox* cells to 20–30% confluence and infect with AdCre.
12. 24 h after infection remove the AdCre containing medium and replace with complete D-MEM/F:12 medium.
13. Harvest cell pellets at 24 h intervals for several days post AdCre infection.
14. Prepare gDNA using the Blood Mini Kit to verify homozygous deletion of the targeted exon by Southern blotting (*see* **Fig. 2.3D**).

4. Notes

1. Where possible, minimize the repetitive DNA content of each homology arm to maximize the efficiency of locus-specific integration.
2. The packaging limit of AAV is 4.7 kb. Targeting vectors containing sequences larger than this will not package well, resulting in extremely low titers of transducing virus.
3. The NetPrimer program is invaluable in generating effective screening primers (http://www.premierbiosoft.com/netprimer/netprlaunch/netprlaunch.html).
4. The pNY vector is optimized for conditional knockouts, as marker removal is controlled by the Flp/FRT recombination system, whereas conditional exon removal is driven by Cre/lox recombination.
5. We stress the importance of testing the screening primers in order to identify those that give a strong and specific PCR product from very low levels of DNA (typically about 30 copies/reaction). For these tests, each screening primer is paired with a locus-specific primer that gives a PCR product of roughly the same size as that we wish to detect in the actual screen (2.5–3.5 kb). It may be helpful to vary the concentration of DMSO while testing your screening primers to determine optimal amplification conditions.
6. To avoid interference with splicing, we recommend placing the *loxP* site at least 100 bp upstream of the 5' end or downstream of the 3' end of the targeted exon. At the same time, the distance between the *loxP* site and the *neo*mycin-resistance cassette should be minimized in order to reduce the chances of a nonproductive crossover between these two elements.
7. The direct juxtaposition of two inverted tandem repeats (ITRs) results in plasmid instability. For this reason we use pAAV-*lacZ* as the source of the pAAV vector backbone.
8. The 5' and 3' ITRs in pAAV contain SmaI sites that can be used to confirm successful transfer of the NotI fragment from pNY to pAAV. The orientation of the insert does not seem to affect gene targeting efficiency as both the (+) and (–) strands are packaged.
9. Transfection of HEK293 cells with AAV producer plasmids may cause a cytopathic effect, but this is not strictly correlated with high-titer virus production.
10. Freezing the cells should take about 10 min in the dry ice/methanol bath. Remove the cells from the 37°C bath

when nearly all of the suspension is thawed. In our hands this takes about 13 min.

11. It is important to use early-passage target cells for rAAV infection.
12. Without removing the plastic wrap, check the stacks 3–5 days after plating to make sure the media is not yellow and the G418 selection is working.
13. The number of colonies can vary from infection to infection. We often get anywhere from 300 to 1500 colonies per experiment. Of these, typically 1–10% represent locus-specific recombinants. We find that HCT116 cells form compact colonies, whereas hTERT-RPE colonies are looser in structure because of the greater motility of this cell type.
14. The number of plates that you pool will depend on the total number of G418 colonies obtained in the experiment. If you have ~1 colony in every well, you will want to pool a maximum of 5 plates. However, if you have many fewer colonies, you may want to pool 10 plates. If you consolidate more than 5 plates in a deep-well block, be sure to adjust the volume of cells taken from each well such that the total volume in each well of the deep-well block is 300 μl (e.g., for 10 plates, take 30 μl from each well). Be sure to label each individual plate with a unique identifier so that you can accurately deconvolve your PCR positives from the pool stages.
15. The deep-well block can be stored at –20°C for a day or two or used immediately for gDNA preparation. We usually freeze the cells in the deep-well block for several hours prior to gDNA extraction to help with the lysis. If storing, cover the deep-well block with a piece of sealing film. Keep in mind that your candidate targeted heterozygotes (*floxneo/+* cells) will be confluent in several days and so it is critical to complete the PCR screening to identify positive clones in this time frame.
16. This protocol is optimized for a centrifuge outfitted with a swinging-bucket rotor that can accommodate deep-well blocks. Alternatively, one can use the Vac-Man 96 Vacuum Manifold with comparable results.
17. After preparing gDNA, we recommend setting up the PCR screen immediately as the gDNA may not be very stable and is susceptible to degradation with freeze/thaw cycles.
18. Using mineral oil avoids potential problems with reaction evaporation and cross-contamination.
19. Extension time may be varied according to the expected size of the PCR product (1 min per kb).
20. If you are only able to successfully PCR screen across the homology arm that does not contain the *loxP* site (i.e., you

cannot determine a favorable crossover), you will have to expand PCR positive clones and try to rescreen the *loxP*-containing homology arm using higher purity DNA prepared by the DNA Blood Mini Kit.

21. You should expect to identify a single PCR positive clone from each pool you deconvolve. If your initial plates contain multiple clones per well, a subsequent round of subcloning by limiting dilution will be necessary to reach clonality (and confirm heterozygosity) prior to the removal of the neomycin cassette.
22. We recommend expanding at least two independently derived *floxneo/+* clones for FLP mediated excision of the *neo* cassette in order to eliminate clonal bottlenecks in downstream analyses.
23. It is helpful to transfect an extra flask of *floxneo/+* cells with a GFP-expressing plasmid, which provides both an indication of transfection efficiency and a negative control for puromycin selection. For HCT116 cells we transfect pCAAGS-FLPe using the Lipofectamine Plus protocol as described in **Section 3.3,** as this gives a higher transfection efficiency than Fugene in this cell type.
24. We find that the puromycin selection is more effective for HCT116 cells than RPE cells. Because of this, we typically screen more FLP-transfected RPE clones (96) than HCT116 clones (48) in order to ensure recovery of *neo*-excised derivatives.
25. It can be helpful to perform a control PCR using locus-specific primers to confirm that your gDNA is present and of sufficient quality for screening.
26. Occasionally one may detect the presence of a small fraction of G418-resistant cells in a putatively *neo*-excised clone, presumably through low-level contamination by the parental clone or a *neo*-positive sibling. In our experience, such contamination can always be cured by an additional round of limiting dilution.
27. The amount of virus used in the second round of infection can be adjusted depending on the number of colonies obtained from the first round of infection. However, application of large amounts of virus (in excess of 6 ml per T-75) may also increase the risk of multiple integration events.
28. In general, homology arms do not have to be made from DNA that is isogenic to the target cell line. However, in rare cases where substantial polymorphisms exist, it is possible that these sequence differences may result in preferential and recurrent integration into one allele. In this unusual situation, we have found that reconstructing targeting vectors based on the specific polymorphisms present in the second (untargeted) allele can overcome this allele bias and

generate a homozygous knockout cell line (M.-E.T. and P.V.J., unpublished data).

29. It is also possible to target the second allele with a modified construct in which the targeted exon is deleted outright, rather than flanked by *loxP* sites (*see* **Fig. 2.3A**, right panel). Although this requires some additional cloning steps, it eliminates the need to consider where crossovers occur relative to the targeted exon. This can be beneficial in instances where the distribution of crossovers obtained with the first-allele targeting construct is found to be biased towards nonproductive integrants that failed to incorporate both *loxP* sites.

Acknowledgments

The authors thank Catherine Randall for generously providing the primary data used in **Fig. 2.3**. Work in the laboratory of P.V.J. is supported by grants from the National Institutes of Health (CA 107342) and the American Cancer Society (RSG-08-093-01-CCG). P.V.J. is a Pew Scholar in the Biomedical Sciences.

References

1. Venter J.C., et al. (2001) The sequence of the human genome. *Science* **291**, 1304–1351.
2. International Human Genome Sequencing Consortium. (2004) Finishing the euchromatic sequence of the human genome. *Nature* **431**, 931–945.
3. Hanson K.D. and Sedivy J.M. (1995) Analysis of biological selections for high-efficiency gene targeting. *Mol Cell Biol.* **15**, 45–51.
4. Jallepalli P.V., Waizenegger I.C., Bunz F., Langer S., Speicher M.R., Peters J., Kinzler K.W., Vogelstein B. and Lengauer C. (2001) Securin is required for chromosomal stability in human cells. *Cell* **105**, 445–457.
5. Wang X., Zou L., Zheng H., Wei Q., Elledge S.J., Li L. (2003) Genomic instability and endoreduplication triggered by RAD17 deletion. *Genes Dev.* **17**, 965–970.
6. Jallepalli P.V., Lengauer C., Vogelstein B., Bunz F. (2003) The Chk2 tumor suppressor is not required for p53 responses in human cancer cells. *J Biol Chem.* **278**, 20475–20479.
7. Papi M., Berdougo E., Randall C.L., Ganguly S., Jallepalli P.V. (2005) Multiple roles for separase auto-cleavage during the G2/M transition. *Nat Cell Biol* **7**, 1029–1035.
8. Larochelle S., Merrick K.A., Terret M.E., Wohlbold L., Barboza N.M., Zhang C., Shokat K.M., Jallepalli P.V. and Fisher R.P. (2007) Requirements for Cdk7 in the assembly of Cdk1/cyclin B and activation of Cdk2 revealed by chemical genetics in human cells. *Mol Cell* **25**, 839–850.
9. Burkard M.E., Randall C.L., Larochelle S., Zhang C., Shokat K.M., Fisher R.P. and Jallepalli P.V. (2007) Chemical genetics reveals the requirement for Polo-like kinase 1 activity in positioning RhoA and triggering cytokinesis in human cells. *Proc Natl Acad Sci USA* **104**, 4383–4388.
10. Hirata R., Chamberlain J., Dong R. and Russell D.W. (2002) Targeted transgene insertion into human chromosomes by adeno-associated virus vectors. *Nat Biotechnol.* **20**, 735–738.
11. Porteus M.H., Cathomen T., Weitzman M.D. and Baltimore D. (2003) Efficient gene targeting mediated by adeno-associated virus and DNA double-strand breaks. *Mol Cell Biol.* **23**, 3558–3565.

12. Kohli M., Rago C., Lengauer C., Kinzler K.W. and Vogelstein B. (2004) Facile methods for generating human somatic cell gene knockouts using recombinant adeno-associated viruses. *Nucleic Acids Res.* **32**, e3.
13. Chamberlain J.R., Schwarze U., Wang P.R., Hirata R.K., Hankenson K.D., Pace J.M., Underwood R.A., Song K.M., Sussman M., Byers P.H., Russell D.W. (2004) Gene targeting in stem cells from individuals with osteogenesis imperfecta. *Science* **303**, 1198–1201.
14. Liu X., Lei M., Erikson R.L. (2006) Normal cells, but not cancer cells, survive severe Plk1 depletion. *Mol Cell Biol.* **26**, 2093–2108.
15. Sambrook J. and Russell D.W. (2001) *Molecular Cloning: A Laboratory Manual*, 3rd Edition: Chapter 6, New York: CSHL Press, pp. 6.33–6.64.

Chapter 3

RNAi in *Drosophila* S2 Cells as a Tool for Studying Cell Cycle Progression

Mónica Bettencourt-Dias and Gohta Goshima

Abstract

Genetic studies on model organisms, particularly yeasts and *Drosophila melanogaster*, have proven powerful in identifying the cell cycle machinery and its regulatory mechanisms. In more recent years RNAi has been used in a variety of genome-wide screens and single molecule studies to elucidate the mechanisms of cell cycle progression. In *Drosophila* cultured cells, RNAi is extremely simple, and a strong effect can be observed by adding the dsRNA to the cultured cells, with few complications of off-target effects. Functions in cell cycle progression can be followed by a variety of assays. One of the advantages of these cells is that they allow high-resolution spatiotemporal observations to be made by microscopy, with no particular complexity in terms of media and temperature. Here we discuss protocols for RNAi in *Drosophila* S2 culture cells, followed by the study of mitotic progression, through immunocytochemistry, live imaging, and flow cytometry analysis.

Key words: Mitosis, cell cycle, RNAi, Drosophila, genomewide.

1. Introduction

Genetic studies on model organisms, particularly yeasts and *Drosophila melanogaster*, have proven powerful in identifying the cell cycle machinery and its regulatory mechanisms. The availability of the fully sequenced genome and the simplicity of culturing *Drosophila* cells, combined with the strong RNAi effect after the simple addition of long double-stranded (ds)RNAs *(1)*, led to the early development of RNAi in *Drosophila*. This permitted the implementation of a variety of genome-wide screens addressing questions such as cell growth, signalling, the regulation of cell

Andrew D. McAinsh (ed.), *Mitosis: Methods and Protocols*, Methods in Molecular Biology 545,
DOI 10.1007/978-1-60327-993-2_3, © Humana Press, a part of Springer Science+Business Media, LLC 2009

cycle progression, and mitosis *(2–6)*. These screens have driven the development of tools and resources for RNAi in *Drosophila*, such as algorithms that determine best primer combinations for designing dsRNAs with specificity, databases of primers, collections of primers, dsDNAs and dsRNAs, and finally collections of flies that harbour dsRNA transgenes (reviewed in Reference *(7)*; *see* **Links 1 and 12**).

Most available *Drosophila* cell lines, including haploid cell lines, have their origin in cultures generated through the dissociation of embryos *(8–10)*. In the field of mitosis, mostly S2 cells have been used, which were derived from embryos that were approximately one day old *(11)*. These cells are very easy to culture and manipulate. One of the advantages is that they allow high-resolution spatiotemporal observations to be made by microscopy, with no particular complexity in terms of media and temperature. A variety of other cell lines, such as more adherent or with other signalling properties, has been used for studying other biological problems *(12, 13)*. A complete list of cell lines, including cells carrying particular mutations can be found elsewhere *(7, 10, 14, 15; see* ***Links 1, 2****)*. Here we discuss protocols for RNAi in *Drosophila* S2 culture cells, followed by the study of mitotic progression.

2. Materials

2.1. Culturing Drosophila S2 Cells (see Note 1)

1. Incubator at 25°C (no need for refrigeration capability, if room temperature does not go above 22°C), tissue culture hood
2. DMSO for tissue culture, 70% ethanol
3. Medium for regular culture of S2: Schneider medium (Sigma), 10% heat-inactivated FBS (Sigma; *see* **Note 2**); optional 1/10 volume of antibiotic-antimycotic (e.g., Sigma A5955)
4. Cells: S2 (ATCC; CRL-1963)
5. Flasks (25 (T25) and 75 cm^2(T75); e.g., TPP9025 and TPP90075, respectively)
6. 6-Well plates (TPP92406); 24-well plates (TPP92424), and 96-well plates (TPP92696)
7. Cryogenic vials (e.g., Corning #430488)
8. Nalgene cryo freezing container (#5100-0001)
9. Hemacytometer (e.g., BrightLine 36219-00)

2.2. Making the dsRNA

1. Phenol/chloroform (1/1)
2. Nuclease free water
3. 7.5 M Ammonium acetate

4. Lysis buffer (100 mM NaCl, 10 mM Tris-HCl pH 8, 25 mM EDTA pH 8, 0.5% SDS (W/V), 0.1 mg/ml proteinase K (Qiagen Cat 19131))
5. T7 RNA polymerase, rNTPs, transcription buffer (e.g., Ambion MEGAscript RNAi Kit (AM1626); or Promega- T7 RiboMAX(TM) Express RNAi System-P1320)
6. Primers for controls (see text for choice and **Table 3.1**)

2.3. RNAi

1. Schneider medium (Sigma) with no supplementation
2. Serum (heat-inactivated)

2.4. Stable Cell Selection

1. Cellfectin (Invitrogen 10362010)
2. Hygromycin (50 mg/ml: Invitrogen)
3. pCOHYGRO (Invitrogen)
4. 20% serum-containing medium
5. pCOBLAST (Invitrogen)
6. Blasticydin (10 mg/ml: Invitrogen)

2.5. Assays for Studying Mitosis (Immunostaining, FACS, and Live Cell Imaging)

1. Mattek 35 mm Petri dish with 14 mm glass area (No. 1.5; ~0.17 μm; P35G-1.5-14-C).
2. DAPI (Invitrogen D3571).
3. Poly-L-lysine: 1 mg/mL in 50 mM Tris-HCl, pH 8.0).
4. 100 mM sterile CuSO4 solution.
5. Glass coverslips: VWR 13 mm (for 24-well plates; can use larger for 6-well plates), Thickness No 1.5.
6. 70% ethanol at –20°C.
7. Propidium iodide (Sigma P4170), stock 10 mg/ml in H_2O.
8. RNase free DNase (Qiagen; 1007885).
9. Fluorescence microscope (e.g., TE2000E (Nikon), Axovert 200 M (Zeiss)).
10. Fixative (*see* **Note 3**): Prepare 6.4% paraformaldehyde (PFA) solution by mixing 5 × HL3 buffer and 32% PFA solution. 5 ml is required per one 96-well plate (1 ml 5 × HL3, 1 ml 32% PFA, 3 ml water).
11. 5 × HL3 Buffer: 350 mM NaCl, 25 mM KCl, 100 mM $MgCl_2$, 50 mM $NaHCO_3$, 575 mM sucrose, 25 mM HEPES pH 7.2. Can be stored at 4°C for many months.
12. Paraformaldehyde (PFA; 32% solution: Electron Microscopy Sciences 15714-S).
13. PBST: PBS + 0.1% Triton.
14. SDS Buffer: 0.5% SDS in PBS.
15. BSA or goat serum (Sigma G9023).
16. DAKO mounting medium (Dakocytomation S3023) or 90% glycerol in PBS or Vectashield with or without DAPI.
17. Antibodies (*see* **Note 4** and **Table 3.2**).
18. Prepare antibody solutions using PBST + 5% goat serum. We need 4.2 ml for one 96-well plate.

Table 3.1
Primers used in this study

Gene/Primers*	Size**	Forward	Reverse
GFP	382	CTTCAGCCGCTACCCC	TGTCGGGCAGCAGC
Pavarotti/CG1258	604	AAATCCGTAACGAAACTAACCG	ACAACTGCTCTTGGCAGATACC
Nuf2/CG8902	283	ATTGAATGGCGTTATCAGTCG	GTGTACGTCTTGTTGGGAAAGC
Polo/CG12306	301	CGTTCTCCGCTTTGTGCTTGGTTTTCGTG	CGCTTGTAGGTTTTCCGCTGGTTGATGTCG
SAK/CG7186	230	ATACGGGAGGAATTTAAGCAAGTC	TTATAACGCGTCGGAAGCAGTCT
Cdc27	701	GATGGGACTCAAGAAACAATCG	TCTTCATGTAGAATTGCATGGC
Cdc16	659	TTCTATTAAAGCCGCAAAGTCC	AATGAGAGAACGTGGCTAGAGG

* add the T7 Polymerase binding site: TAATACGACTCACTATAGGG to the 5' end of each primer.
** Expected size using genomic DNA or GFP-encoding plasmid.

Table 3.2
Antibodies used in this study

Labelled Structure	Antigen	Antibody	Supplier	Dilution
Microtubules	α-tubulin+	Rat-YL1/2	Serotec YSRTMCA 78S	1/100**
Centrosomes	γ-tubulin+	Mouse-GTU-88	SIGMA T6557	1/100–1/400*,**
Mitotic Chromosomes	Phospho-histone H3+ (PH3)	Rabbit	upstate 06-570	1/250**
	Rhodamine Redex Anti-mouse++	Goat	Jackson 115-295-166	1/300
	FITCAnti-rat++	Goat	Sigma- F6258	1/300
	Alexa 647-Anti-rabbit++	Goat	Molecular Probes/Invitrogen A21244	1/300
	Alexafluor 488- anti-rabbit++	Goat	Molecular Probes/Invitrogen A11034	1/200

* If no treatment is performed with SDS use lower dilution. **For overnight staining, antibodies can be diluted further: YL 1/2: 1/150, GTU88 1/1000, pH3 1/1000.
+ Primary antibodies. ++ Secondary antibodies.

2.6. Genomewide RNAi

1. 96-well plastic plate
2. Glass-bottom 96-well plate with short skirt (e.g., Whatman 7706-2370, Iwaki 5866-096)
3. 12-channel pipetteman
4. Solution reservoir
5. Backing tape, black (PerkinElmer 6005189)

3. Methods

Depending on the question to be approached, it is first important to think of the best assay, the best timeframe for it, the best positive and negative controls, and ways of testing the efficiency of the RNAi. Here we describe three different assays: flow cytometry to monitor mitotic delay and changes in cell cycle progression, immunocytochemical assays where mitotic structures can be visualised, and live cell imaging, when time resolution is essential. A variety of different cell lines with fluorescent markers can be made or obtained to optimise and speed analysis. Cells maybe partially arrested in mitosis with dsRNA for cdc27 or cdc16 (see Primers in **Table 3.1**), or with the proteosome inhibitor MG132, if there is a desire to increase the percentage of mitotic cells available for analysis (*see* **Note 5**). Routinely we perform RNAi for 3–4 days, but in the case of very stable and/or abundant proteins, longer assays may be used (e.g., 8 days with retransfection after 4 days). We suggest the use of dsRNA for GFP (see Primers in **Table 3.1**) or for an intronic sequence as the negative control. In the absence of a good positive control for RNA interference suitable for the aspect of cell biology in which you are interested (a gene that shows a phenotype in the selected assay), we suggest using dsRNA for a kinesin called pavarotti/CG1258, depletion of which results in the appearance of large cells due to cytokinesis failure after 3 days (*(16); see* **Fig. 3.1**) or Nuf2/CG8902 *(2)* or SAK/CG7186 *(3, 17)* which serve as positive controls for kinetochore or centrosome disruption, respectively (**Fig. 3.1**). Finally, RNAi of polo kinase is a good control for mitotic arrest and consequent changes in FACS profile (*(3)*; *see* **Fig. 3.2** and **Table 3.1** for primers to make the dsRNA).

There are several ways of checking the efficiency of transfection. The strength of the phenotype observed in cells transfected with the positive control dsRNA should give an idea of the success of RNAi. Additionally, protein levels can be checked by Western blot analysis or, alternatively, the RNA depletion can be monitored by semi-quantitative or quantitative RT-PCR. In both cases, constitutively expressed genes, such as actin or eukaryotic initiation factor-4a, can be used as internal controls.

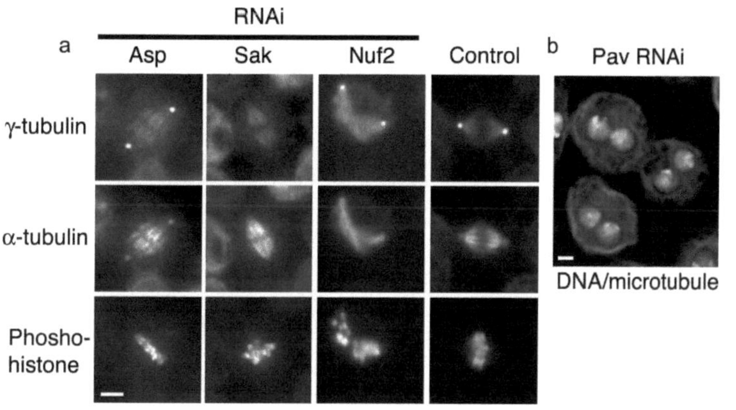

Fig. 3.1. Mitotic phenotypes observed after RNAi of Asp, Sak, Nuf2 and Pavarotti. γ-tubulin, α-tubulin and phospho-histone H3 are immunostained. **(a)** Spindle pole unfocusing (Asp), loss of centrosomes (Sak) and chromosome misalignment (Nuf2) are observed in the metaphase spindles. **(b)** Binucleate cells are frequently observed after Pavarotti RNAi as a consequence of cytokinesis failure. Adapted from *(2)*. Bars, 5 μm.

3.1. Culturing Drosophila Cells

Drosophila cell lines are ideally kept in a clean 25–27°C incubator. In contrast to mammalian cell cultures, there is no need for a controlled CO_2 atmosphere, because *Drosophila* cells tolerate wide changes of pH and most tissue culture media do not use bicarbonate buffering. Because of their ease of culturing, if needed, *Drosophila* cells can be kept in a drawer in the lab provided the room temperature is approximately 22–25°C. Additionally, their phenotype may change with time in culture, such as cell cycle profile or levels of expression of a transgene. As such, cells should be frozen periodically to allow for the use of cells with reproducible characteristics at any time point.

3.1.1. Subculturing

Drosophila cells double approximately every 24 h *(14)*. As such they should be routinely subcultured once to twice a week. Most of the cell lines grow as monolayers and detach by simple agitation, without trypsinization. Cells stick to the new flask but become loose over time. When culturing the cells, use an aseptic technique by wiping everything with 70% ethanol and working inside the tissue culture hood. Always check the cells on the microscope to make sure they look healthy and there is no contamination.

1. Resuspend cells from the bottom of the flask in their own media by pipetting up and down. Make sure the bottom of the flask looks transparent after cell resuspension.
2. Add 1:7 of the cells (if the cells have just been thawed use 1:3) to each of two fresh tissue culture flasks (total of 15 ml of media at 25°C in a 75 cm^2 flask).
3. Keep old flask as a backup until next time you passage the cells (*see* **Note 6**).

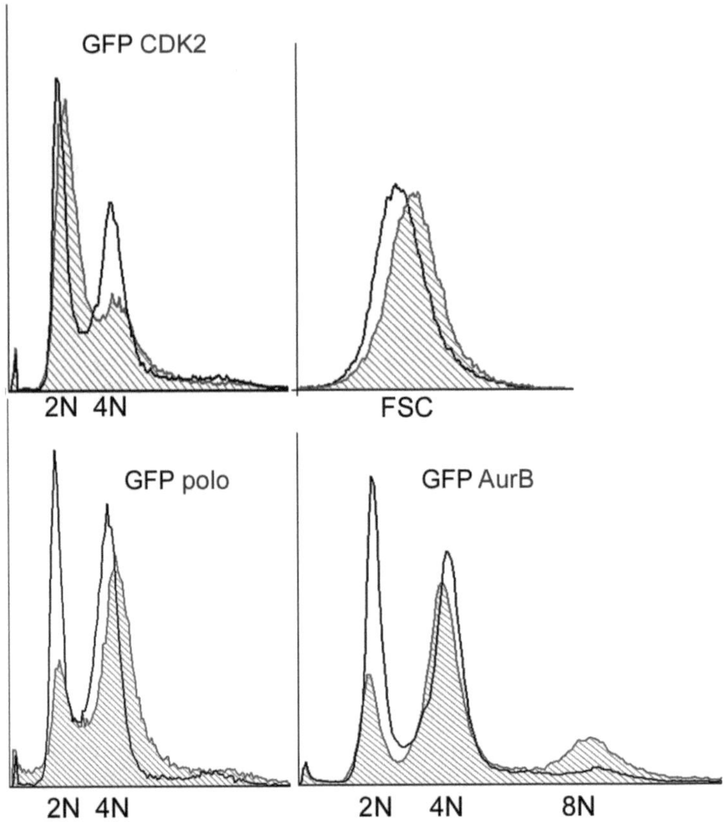

Fig. 3.2. Phenotypes observed after RNAi of CDK2, polo and AurB. Each panel shows a control FACS histogram (black, cells transfected with dsRNA for green fluorescent protein (GFP)) and a FACS histogram (lines) after RNAi of a kinase. The forward light scatter (FSC) profile reflects cell size. Note the increase in the population of bigger (FCS) cells delayed in G1/S after CDK2 RNAi. Increases in cell size are commonly observed when cells are delayed in G1 or G2 *(3)*. Note the increase in the proportion of cells with 4 N DNA content after polo RNAi. After RNAi of Aurora B, an 8 N peak becomes visible due to problems in cytokinesis. See *(3)* for other cell cycle phenotypes.

3.1.2. Freezing the Cells

Ten aliquots of cells can be frozen from one confluent T175 flask (40 ml) or three aliquots from a T75. Use aseptic techniques and work quickly because DMSO is toxic to cells whilst they are not frozen.

1. Label 10 cryo-tubes (e.g., cell type, passage number, name and date; use materials that are resistant to liquid nitrogen).
2. Prepare cryo freezing container (*see* **Note 7**).
3. Detach cells and centrifuge them at 300 g for 5 min and save 2.25 ml supernatant as a conditioned medium. Aspirate the rest.
4. Prepare a mix (cryogenic medium) of 2.25 ml conditioned medium (45%), 2.25 ml of fresh medium (45%), and 0.5 ml DMSO (10%).
5. Resuspend the pellet in 5 ml of cryogenic medium.

6. Place 500 μl of cell suspension into each cryo-tube, place on cryo freezing container and transfer to a –80°C freezer. The container will allow slow cooling of the cells (–1°C/min; *see* **Note 7**).
7. The following day transfer the tube on dry ice to a box in a <–80°C freezer or a liquid nitrogen container. Keep an organised map of your frozen vials (strain, date, passage number).

3.1.3. Recovery of Cells

1. Prepare the culture hood and a labeled sterile 15 ml centrifuge tube with 9 ml of prewarmed (25°C) medium without antibiotics.
2. Retrieve the desired vial of cells and transport it on dry ice.
3. Thaw the vial in a 37°C water bath, gently shaking the tube until the medium is almost completely thawed.
4. Spray the tube with 70% ethanol and transfer the cells to the prepared 15 ml tube inside the tissue culture hood.
5. Centrifuge at 300 g for 5 min, remove supernatant and resuspend in 10 ml medium and inoculate into a labeled small tissue culture flask (25 cm^2).
6. Check the cells every day; once they become confluent (1 day to 1 week) subculture them as described (we prefer 1:3 dilution at the first couple of passages; *see* **Note 8**).
7. Clear sample from database.

3.2. Making the dsRNA

3.2.1. Choosing the Target Sequence and Primers

Normally the dsRNA is produced by in vitro transcription of a PCR-generated DNA template containing the T7 promoter sequence on both ends (*see* **Note 9**). The template DNA can be amplified from cDNA (including ESTs from BDGP; *see* **Link 8**), whole cell cDNA (*see* **Note 10** for protocol), or genomic DNA templates, whichever is more convenient. We tend to use genomic DNA because the same template can be used for different targets. However, in some cases, when exons are very small and introns large, the use of cDNA may be required. The use of long dsRNAs to trigger RNAi was initially avoided in mammals because those molecules activate the interferon response; however, this response is not triggered in cultured *Drosophila melanogaster* cells (reviewed in *(7)*). The optimal length for maximum interference activity in *Drosophila* is 300–1000 bp *(18)*, but molecules of 150–3000 bp have been used with success. Usually most of the dsRNA corresponds to a putative exonic sequence, but dsRNA corresponding to an untranslated region (UTR) sequence can also be used *(19)*. Using the Apollo programme helps to visualise intron/exon boundaries and different transcripts (*see* **Link 13**). Expression of dsRNA specific for a determined region of an RNA species in *Drosophila* does not target regions upstream and downstream of the chosen targeted region. This allows the design of isoform specific targeted dsRNAs if desired *(20)*.

Specificity of the RNAi response in *Drosophila* has become an issue recently *(21, 22)*. Through the RNAi pathway, dsRNAs are cut into 21 bp sequences that then target the specific messenger RNA for degradation. Algorithms have been implemented that analyse all stretches of contiguous sequence identity of 21 bp or greater between the target gene and any other gene in the genome (*see* **Link** 7). These algorithms have been used in recent genomewide screens, and the target gene sequence can be chosen by inspection of primer databases from those studies (e.g., dsRNA sequences for ~15,000 genes are described in **Link 4**; select "RNA lib v2" in the "Database" scroll bar of "Online database") *(2)*. Alternatively DNA templates for dsRNA production can be purchased or the assay can be performed in an external facility (*see* **Note 11**). If you want to be particularly careful with specificity you can pick two or more different regions from which to make dsRNA and test both.

Once the target sequence has been chosen, we suggest using a programme for designing primers (as an example *see* **Link** 7). Complementary sequences should be 18–25 bp and the melting temperature should be 50–60°C. The T7 promoter sequence (TAATACGACTCACTATAGGG), needed for transcription, is added to the 5' end of each primer.

3.2.2. Making Genomic DNA

1. Grow cells to confluence in a 75 cm^2 flask. Remove supernatant and resuspend them in 10 ml fresh medium. Transfer the cell suspension to a 50 ml tube.
2. Count cells using a haemocytometer: after dispersing the cells in medium add 10 μl of the cell suspension to the haemocytometer; each large square of a standard haemocytometer usually contains 10^4 cells/ml.
3. Calculate the suspension volume containing 30×10^6 cells and centrifuge this volume at 300 g for 5 min at 22°C. Discard the medium and wash the cells twice in 25 ml PBS.
4. Resuspend the cells in 250 μl lysis buffer and transfer into an Eppendorf tube.
5. Incubate the samples at 50°C, overnight.
6. Remove proteins by phenol/chloroform (1:1) extraction. Add an equal volume (250 μl) of phenol/chloroform to cells, mix by inversion, and centrifuge for 5 min at maximum speed in a microcentrifuge.
7. Pipette the upper aqueous phase to a new tube and repeat the extraction once more.
8. Add an equal volume of chloroform and repeat the purification twice in the same way. Transfer the aqueous phase to a fresh tube.

9. Precipitate the DNA by adding ½ volume of 7.5 M ammonium acetate (or 1/10 sodium acetate 3 M pH 5.5) and 2 volumes of 96% (v/v) ethanol. Invert 6–8 times and incubate at –20°C for 15 min.
10. Centrifuge in a microcentrifuge at top speed for 10 min, pour off the supernatant and rinse the pellet with 1 ml ice-cold 70% (v/v) ethanol.
11. Dry the DNA at room temperature for 10–15 min and resuspend in 500 μl nuclease-free water. Test with a control primer pair. Aliquot and store at –20°C.

3.2.3. Synthesising the dsRNA

All work should be done in a sterile, RNase-free environment, wearing gloves and using only sterile, RNase-free solutions and materials. We have used the T7 RiboMAX TM Express RNAi system and the Megascript RNAi Kit (Ambion) to produce dsRNA. Here we describe the protocol for the former.

1. Mix the template DNA (50 ng of plasmid cDNA, 500 ng of isolated genomic or cDNA) with: 1 μl of Forward primer (20 μM), 1 μl of Reverse primer (20 μM), 5 μl 10 × PCR Buffer, 5 μl 25 mM $MgCl_2$, 1 μl 10 mM dNTP mix, 1 μl Taq polymerase (5U/μl), and water to a final volume of 50 μl. You can also use PCR master mixes (e.g., Promega M7502), which are more convenient.
2. We have successfully used the following PCR program

   94ºC 2 min

   94ºC 30 sec
   55ºC 30 sec } x 35 cycles
   72ºC 1 min

   72ºC 10 min
3. The PCR product can be purified by ethanol precipitation or with the PCR Clean UP kit (Qiagen). The quality and the quantity of the PCR product can be tested by gel electrophoresis (1.5% agarose-TAE buffer; *see* **Note 12**). Quantity can also be determined by measuring the absorbance at 260 nm (A_{260} × dilution factor × 50 = DNA concentration in μg/ml). The product should only have one band and it should run at the predicted size. There is no need to do this if you have characterised the primers before.
4. Use 2 μg of generated template DNA (with T7 polymerase binding sites) in 40 μl final volume of the transcription reaction.
5. Thaw the transcription reaction component at room temperature, except the enzyme mix (keep on ice).
6. Add the reaction components for a 40 μl final volume to an RNAse free 1.5 ml microcentrifuge tube: –20 μl RiboMAX™ 2 × buffer; –1–16 μl template DNA (2 μg); –0–15 μl nuclease free water; –4 μl T7 Express Enzyme Mix.

7. Mix gently and incubate at 37°C for 4 h. The recently made RNA strands will anneal during this time.
8. Remove the DNA template by digestion with DNase (*see* **Note 12**). This is performed by adding 2 μl DNase solution per 40 μl reaction volume and incubating for 30 min at 37°C.
9. Inactivate DNase by taking reaction to 65°C for 10 min and then cool down slowly (e.g., 10 min of each: 37°C, 32°C, 30°C, and 25°C).
10. Quantify the dsRNA by running 1/80th and 1/200th of the volume of the dsRNA on an agarose gel and check the quantity and the quality. The dsRNA should be a single band on the gel. Note that dsRNA migrates slightly more slowly than the dsDNA. Compare the amount of the dsRNA to two different amounts of known molecular markers (e.g., Low DNA Mass™ Ladder, Invitrogen (*see* **Note 13**)).
11. Store the dsRNA at –20°C at a concentration of 1 μg/ul in aliquots of 30 μg. The dsRNAs should be stable for at least 12 months at –20°C without detectable loss of efficiency.

3.3. RNAi

3.3.1. Silencing One Gene

1. Use exponentially growing cells just about to peak confluence. Remove the old medium from the flask and resuspend the cells in 10 ml fresh medium by pipetting.
2. Count and calculate the cell density using a haemocytometer.
3. In each well of a 6- or 24-well plate add 1×10^6 (in well of 6-well plate)/0.2×10^6cells (in well of 24-well plate) in 3/0.5 ml of medium. The cells should settle overnight. Double the amount of cells if you want to do the experiment on the same day (in that case use only 1/0.2 ml of media and wait for one hour after platting). If you are going to fix the cells directly on the wells, place them on top of sterilised coverslips (*see* **Note 14**).
4. The following day, prepare 1/0.2 ml of serum-free and antibiotic-free media, mixed with 30/6 ug of dsRNA in an Eppendorf.
5. Incubate for 15 min at room temperature.
6. Briefly remove the serum-containing medium from the cells.
7. Add the transfection medium to the cells and swirl the plate to produce uniform distribution of the liquid.
8. Incubate the plate at 25°C for 1 h.
9. Add 2/0.4 ml of medium, supplemented with 15% (v/v) FBS (*see* **Note 15**).

10. Place the cells in the incubator at 25°C until analysis, which is generally after 3–5 days.

3.3.2. Silencing Multiple Genes

More than one gene can be targeted by RNAi simultaneously, for large pooled screens *(23)* and studies of gene interaction *(16, 24)*. Because of the nature of the RNAi machinery, there is a limit of dsRNAs that can be added to the cells, after which the effect observed for each individual dsRNA may decrease *(25–28)*. In S2 cells, 3 dsRNAs have been successfully used (25 μg dsRNA/well (6-well plate) for each species at same time; *16, 23*).

3.4. Stable Cell Line Selection

A variety of vectors donated by several researchers is available nowadays for expression in Drosophila tissue culture cells from the Drosophila Genome Research Collection (*see* **Link 8, 9**). This is using the gateway system (Invitrogen) which allows much faster cloning, using recombination. A variety of different promoters has been used in these cells: the actin promoter (Act5c), metallothionein promoter (pMT, vectors available from Invitrogen), and heat shock protein 70 promoter (hsp70). Although the Act5 promoter is useful for nontoxic proteins and constitutive expression, vectors such as the pMT or hsp70 allow more flexibility, as induction of expression can be performed as desired (*see* **Note 16**).

1. Prepare happily growing cells in 24-well plate (~50% confluency, app. 0.75 million cells).
2. Prepare the following unique solution (mix well by pipetting) just before transfection:
   Plasmid(s) you want to introduce, each 0.5 μg.
   pCOHYGRO, 0.2 μg (*see* **Note 17**).
   Serum-free medium (SFM), 50 μl.
   Cellfectin (Invitrogen), 5 μl.
3. Leave at room temperature for 45 min.
4. Add 250 μl SFM and mix (total ~300 μl).
5. Replace the cell culture medium with the above ~300 μl solution.
6. Incubate at room temperature (or in incubator) for 3–5 h.
7. Add 300 μl of 20% serum-containing medium.
8. Incubate for 1–3 days (until they are nearly confluent).
9. Transfer to 6-well plate.
10. Add 1 ml of fresh medium (10% serum).
11. Incubate 1 day.
12. Add 10 μl of hygromycin (*see* **Note 17**).
13. Passage the cells when they reach confluency (usually 1/2–1/3). Add 7 μl hygromycin (50 mg/ml) per 1 ml fresh medium (*see* **Note 17**).
14. Once stable cell lines are obtained (~3–6 weeks), hygromycin addition is no longer necessary in the daily passage.

15. (Optional) Clonal isolation. Once a stable cell line is established, we can select clones that have the desired expression level of the transfected genes. To do this, dilute cells with fresh medium to make 6 ml of 10 cells/ml solution. Add 60 μl each to 96-well plate, and wait for 10 days. Check colonies under microscope (4× objective lens), and find wells that have single colonies. A few dozens of wells are expected to have 1–2 colonies. Pick up ~20 colonies, and check expression of the transfected genes by immunoblot or microscopy (in the case of GFP-tagged protein).

Note: When only transient expression is needed, there is no need to use pCOHYGRO. Cells can be directly assayed after step 8. Transient expression can be used in combination with RNAi. We have, for example, successfully started the RNAi and expressed a transgene one day afterwards (washes involved) and harvested the cells 3 days later *(29)*.

3.5. Assays for Studying Mitosis

Many different assays can be performed to study mitosis, depending on the problem of interest. FACS analysis can be used to look at defects in cell cycle progression or to quantitate levels of a given protein at different cell cycle stages. RNAi can be performed in cell lines where the microtubules and other structures, such as centrosomes or kinetochores are labelled with GFP or variants such as RFP or mCherry *(2)*, and cells visualised live or fixed by fluorescence. A variety of different lines has been published over the last years (*see* **Note 18**). We provide here a simple protocol to fix microtubules and visualize the mitotic spindle, centrosomes, and DNA, but other markers could be equally used (*see* **Note 4**). It is important to emphasize that it is common to observe mitotic defects in wild-type Drosophila cultured cells, such as multipolar spindles due to supranumerary centrosomes (approximately 30% of the cells have supranumerary centrosomes and most have supranumerary chromosomes *(2, 3, 16)*). As such, appropriate quantitation of the phenotypes is an imperative. See, for example, *(3)*.

3.5.1. FACS Analysis

3.5.1.1. Fixing

1. Spin the cells (more than 10 million) at 300 g for 5 min.
2. Resuspend well in 200 ul PBS.
3. Add 2 ml 70% ice-cold ethanol (in PBS or water) drop by drop and vortexing.
4. Put on ice for at least 30 min. Samples can be kept for long periods at 4°C by adding 0.1% sodium azide.

3.5.1.2. Analysis

1. Spin the cells-300 g for 5 min.
2. Drain out ethanol.

3. Wash 2 times with PBS by spinning and transfer to Eppendorf for immunostaining.
   If only analysis of DNA content is desired, then go to step 11.
4. Resuspend in 1 ml PBS/1% BSA/0.25% Triton X100 and incubate on ice for 15 min.
5. Pellet cells, resuspend in (dilute antibodies in PBS/1%BSA):
   100 ul of (1:100) rabbit anti-phospho histone H3.
6. Incubate 30 min at room temperature.
7. Wash 2 times in 1 ml PBS/1% BSA.
8. Resuspend in 100 μl anti-rabbit Alexafluor 488 (1:200).
9. Incubate in the dark 30 min at room temperature.
10. Wash cells 2 times in 1 ml PBS/1% BSA.
11. Ressuspend in 1 ml PBS with 100 μg/ml RNase (previously boiled for 5 min to kill DNase-stock 100 mg/ml, 1000 X) and with 100 μg/ml of PI (stock 10 mg/ml-100X).
12. Incubate in a 37°C waterbath for 30 min in dark.
13. Take for FACS. Before putting each sample into the FACS machine pass it through 25G needle/1 ml syringe. Carry out FACS analysis using FL1 (FITC/Alexafluor 488) and FL3 (propidium iodide). Acquire data from 30,000 cells for each sample. Results can be analysed using a variety of programmes, including using Summit from Dako Cytommation and Multicycle.

3.5.2. Immunostaining

If cells are already plated in 24-well plates on top of coverslips, then start at step 5.

1. Put coverslips (13 mm diameter, 1.5) inside 24-well plate.
2. Wash coverslips inside wells with media; dispose of media.
3. Resuspend cells from experiment (usually one well from a 6-well plate) and plate 250–300 μl cells in each well (add more if there is much cell death or small cell number).
4. Wait one hour to settle the cells at 25°C before fixation.
5. Prepare 6.4% paraformaldehyde (PFA) solution by mixing 5× HL3 buffer and 32% PFA solution. If 5 ml are required then use 1 ml 5× HL3, 1 ml 32% PFA, and 3 ml water (*see* **Note 3** for alternative fixative).
6. Carefully remove the media from the wells; do not wash, as this will remove mitotic cells.
7. Fix cells in the wells by adding 400 μl of fixative for 15 min at room temperature.
8. Wash 2 times in PBS (in the wells, 5 min).
9. Make fresh PBSTB (PBS, 0.1% triton, 1% BSA).
10. Permeabilise and block for 1 h with fresh PBSTB.
11. Make primaries solution in PBSTB.

12. Remove coverslips from wells with forceps and place on parafilm (cell side up) in big Petri dishes (this procedure saves a lot of antibody).
13. Add 40 μl of antibody solution per coverslip and incubate with primaries overnight at 4°C, in humid chamber (incubation time can be optimised for each antibody).
14. Wash 3 times in PBSTB in 6-well plate, 5 min per wash.
15. Make secondary antibody solution: 40 μl in PBSTB. Ensure diluted antibodies are spun down before use to avoid crystals.
16. Return to Petri dishes on parafilm and incubate with secondaries for 2 h at room temperature (foil covered).
17. Wash for 5 min in PBSTB (in 6-well plate).
18. Wash for 5 min in PBS (in 6-well plate).
19. Mount in vectashield with DAPI (if you wish to see DNA, otherwise other mounting media can be used).
20. Seal with nail varnish.
21. Store at 4°C.

3.5.3. Live Cell Imaging

These instructions assume the use of Mattek 35 mm Petri dish with 14 mm glass area. Several cell lines expressing GFP or mCherry-tagged mitotic markers are available in the field, such as cells expressing GFP-tubulin, H2B-GFP (histone), mCherry-tubulin/H2B-GFP, γ-tubulin-GFP/mCherry-tubulin, and Mis12-GFP (kinetochore)/mCherry-tubulin (*see* **Note 18**). Some markers have copper-inducible promoter. In such case, add 30–100 μM $CuSO_4$ at least a day before observation. Glass-bottom dishes can be coated with poly-lysine in order for cells to stick tightly. In this case, cytokinesis is not affected. Alternatively, cells can be spread on ConA (concanavalin A: Sigma C7275)-coated dish. This treatment is very convenient as it enables imaging of mitotic apparatus (e.g., spindle) in a single focal plane. However, this treatment inhibits cleavage furrow invagination during cytokinesis.

1. Add 50 μl of 1 mg/ml poly-lysine solution to the glass surface, wait for >1 h, and then rinse with water. Alternatively, poly-lysine-coated glass-bottom dishes are available at Mattek (P35GC-1.5-14-C). Alternatively, to coat the glass with Con A, prepare a 0.5 mg/mL solution of Con A in water. After filtration, Con A can be kept at –20°C for at least several months. Pipette 30 μl Con A solution to cover the glass surface of Mattek Petri dish (spread Con A with pipette tips, to cover entire glass surface). Allow the dish to dry at room temperature (several hours to overnight). Do not rinse the coverslips.
2. Seed cells in normal medium. If the culture is nearly confluent, plate 150 μl fresh medium and 25 μl cell suspension.

Add the cell solution only to the glass part (14 mm area) to keep cell density high, rather than to entire Petri dish. However, for long-term imaging, add ~1 ml medium later to avoid concentration of the medium due to evaporation.

3. Cells will attach to the glass within minutes of plating on poly-lysine or ConA and can be seen to adopt a spread, phase-dark morphology within an hour or so (ConA) by phase contrast microscopy. Cells are healthy for the next few hours (*see* **Note 19**).

3.6. Genomewide RNAi

Libraries of dsRNAs can be obtained from several vendors (e.g., Open Biosystems, Ambion; *see* **Note 11**). The following instructions assume the screening of mitotic phenotypes after immunofluorescence to visualize tubulin, γ-tubulin, and phospho-histone H3 using 96-well glass-bottom dishes (**Fig. 3.3**). For other assays such as FACS analysis, modify after step 7.

1. Add 1 μg of library dsRNA to each well of a 96-well plastic plate, and dry. The plates can be stored at 4°C for several weeks.
2. Transfer exponentially growing S2 cells to a centrifuge tube. Spin 5 min at 300 g.
3. Remove supernatant and resuspend cells with serum-free medium (SFM). Spin 5 min at 300 g.

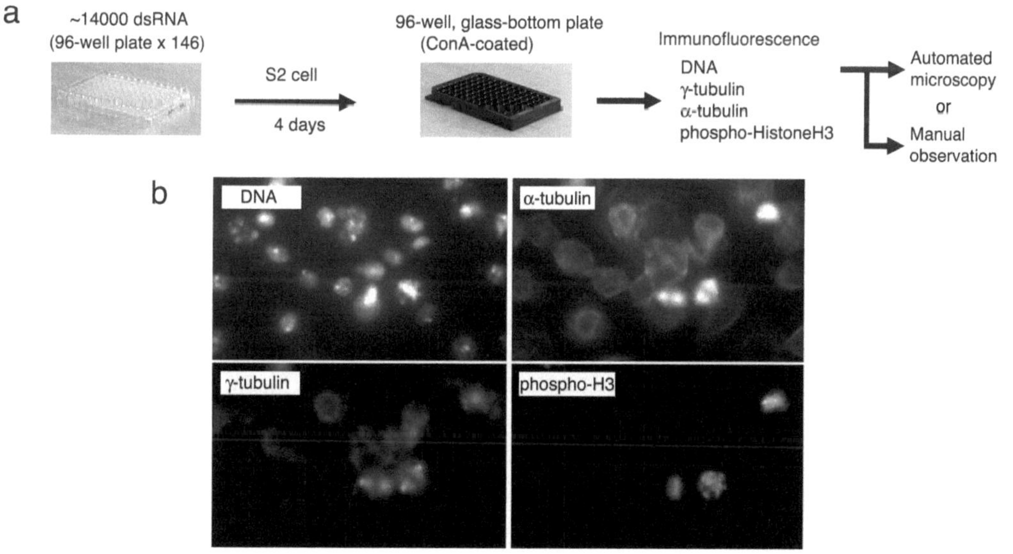

Fig. 3.3. Procedure of a genome-wide RNAi screen for mitotic spindle morphology in S2 cells. **(a)** S2 cells in serum-free medium are plated on 96-well plastic dishes in which dsRNAs are spotted in advance. After 4 days, cells are transferred to 96-well glass-bottom plate that is coated with concanavalin-A (Con A), followed by fixation and immunostaining. **(b)** Cell images are acquired by fluorescence microscopes. Note that phospho-histone H3 antibody stains only mitotic cells. Adapted from *(2)*.

4. Remove supernatant, count cells, and resuspend to a density of 2.0×10^6 cells ml^{-1} in SFM.
5. Pipette 35 μl of the cell suspension into each well of a 96-well tissue culture plate. We can use a 12-channel pipetteman to save time.
6. Incubate for 50 min and add 65 μl of 15% FBS containing medium.
7. Grow cells for 3–4 days at 25°C.
8. Two days before cell fixation, coat a fresh glass-bottom 96-well plate with concanavalin-A. Add 40 μl of a 50 μg/ml Con-A solution to each well, incubate at least 1 day at room temperature to dry. Do not use the plate before it is completely dry. The dried coated plates can be stored for up to 3 days in the fridge.
9. For transfer of cells to Con-A coated plates, add 100 μl of fresh medium into each well of RNAi-treated cells (total should now be 200 μl).
10. Resuspend cells in the plastic tissue culture plate by pipetting up and down ~20 times with a 12-channel pipette. Avoid foaming.
11. Transfer 60 μl of the cell suspension from each well into each well of the prepared glass-bottom plate coated with ConA.
12. Incubate 120–180 min at room temperature.
13. Discard medium (hold a plate, turn it upside down, and dump medium).
14. Immediately fix cells by addition of 45 μl of PFA solution (fixative).
15. Incubate for 15 min at room temperature.
16. Prepare 0.5% SDS solution (in PBS). We need 6 ml for one 96-well plate.
17. Discard PFA solution (check how with your institution). Add 50 μl SDS solution. Incubate for 10 min.
18. Prepare primary antibody solution: 4 ml PBST and 200 μl goat serum for one plate.
19. Discard SDS solution. Wash twice with PBST (each well 70 μl).
20. Discard PBST. Add 40 μl primary antibody solution. Incubate for ~180 min at room temperature.
21. Wash twice (incubate 1–5 min between each wash step) with PBST (70 μl).
22. Add 40 μl secondary antibody solution for each well. Incubate for >60 min at RT.
23. Wash 3 times (incubate 1–5 min between each wash step) with PBST (70 μl).
24. Add two drops of DAKO mounting medium after removing PBST by 12-well pipetteman. To reduce the cost of

reagents, 90% glycerol (in PBS) can substitute DAKO mounting medium (*see* **Note 20**).

25. Store the plates (well-labelled and sealed by black backing tape) at –20°C or 4°C for future observation of the phenotypes.

4. Notes

1. We have successfully used S2 cells and a derived clone that grows in serum-free media, DMEL. Although the use of S2 is most common, for some applications the latter may be easier to use as no serum is required in the media. Although we refer to the S2 throughout the text, most protocols are the same. The DMEL cells can be purchased from Invitrogen- DMELs (10831-014). The medium is Express Five-SFM (Gibco, 10486); antibiotics and antimicotics can be added as for the regular S2.
2. We have experienced that S2 cells become unhealthy with certain batches of Schneider's medium or serum. We therefore select good batches of serum and medium, before the ongoing batch runs short. We get a few medium and/or serum batches from Invitrogen or other vendors and we first culture S2 cells with the new medium through at least 3 passages and check if cells remain healthy under the microscope. Bad medium/serum will induce cell burst or aggregation. We then check if RNAi works well with the new medium/serum by treating cells with dsRNAs against Pavarotti (cytokinesis gene) and control GFP (see **Section 3**). After 4 days, we make sure that the majority of the cells treated with Pavarotti dsRNA are extremely large compared to the control (**Fig. 3.1b**).
3. We have used successfully a different fixative which also preserves well the microtubule cytoskeleton:

   Stock solutions: Formaldehyde from Polysciences 16%, PIPES 0.2 M pH 6.8, HEPES 0.2 M pH 7, EGTA O.5 M pH 6.8, $MgSO_4$ 1 M, H_2O. The stock solutions can be kept for long periods of time at room temperature after sterilization; the diluted solution is prepared fresh in each experiment. We keep the open formaldehyde vial at 4°C.

   To prepare 5 ml of diluted solution, we use formaldehyde (1.25 ml), PIPES (1.5 ml), HEPES (0.75 ml), EGTA (0.1 ml), MgSO4 (0.02 ml), and H_2O (1.5 ml). Fixation is performed at room temperature for 10 min.

If there is no need for very good preservation of microtubules, cells can be fixed with –20°C methanol for 3 min (very good for centrosomal antigens in general).

4. Other antibodies not shown in Table **3.2**, generated in individual labs, have been very useful in the *Drosophila* cell cycle field. That is the case of CID (CENPA-labels centromeres *(30)*, D-PLP (Drosophila pericentrinlike protein-labels interphase and mitotic centrosomes *(29, 31)* and CNN (labels mitotic centrosomes *(32)*.
5. Drosophila cultured cells are particularly difficult to synchronise. Most commonly researchers block them at different cell cycle stages for different analysis.
6. S2 cells can be alive for ~2 weeks in the same culture flask (without supplying new medium). As such, always keep the old flask. If you notice contamination of the current culture, you can go back to the old flask and transfer 1/3–1/7th of those cells to a new flask with fresh medium.
7. We have successfully frozen and recovered cells without this cryofreezing container.
8. When thawing cells, centrifuging them can be avoided by transferring thawed cells to a 6-well plate with 1.5 ml of media. After one hour make sure cells are attached and remove medium, adding fresh one (3 ml). If too dense, cells can be diluted at that stage.
9. If more convenient, it is equally possible to produce dsRNA using DNA templates containing two different promoter sequences at both ends (e.g., T7 and SP6).
10. cDNA from *Drosophila* can be prepared using TRIZOL reagent and SuperScript First-strand synthesis system (Invitrogen) according to manufacturer's instructions (*see* *(10)* for more details).
11. Always check flybase (*see* **Link 1**) for updates on RNAi resources. The Drosophila RNAi screening center at Harvard Medical School provides high-throughput RNAi screens (*see* **Link 2**). Open Biosystems provides the Drosophila RNAi Collection version 2.0, a collection of dsDNA constructs for generation of dsRNA (*see* **Link 10**). Finally, geneservice, in collaboration with Cyclacel Limited, offers a collection of 13,600 RNAi constructs in its Drosophila genome RNAi library (*see* **Link 11**).
12. In some laboratories, purification of PCR product is entirely skipped, and the crude PCR reaction mixture is used as the template of in vitro transcription. The resultant dsRNAs can also be used for RNAi without DNase treatment or purification.

13. We routinely quantitate the dsRNA using the free software for image analysis-imageJ (*see* **Link 14**) and performing titration curves with the markers.
14. Coverslips can be sterilised in a flow-hood by leaving them overnight in a Falcon/Petri dish with 70% ethanol. Take them out with clean forceps and allow them to dry before usage.
15. If you are using DMEL cells, add media without serum (supplemented or not with antibiotics) after the 1 hour incubation.
16. Inducible vectors are slightly leaky. As such some researchers use them without induction for low levels of expression. In addition, many media already contain copper. Check the media you are using as you may have some induction of the pMT promoter even without supplementing the media with $CuSO_4$.
17. Alternatively you can use pCoBlast (Invitrogen) and use Blasticydin as the selection antibiotic. For blasticydin, we start selection 2 days after incubation with cellfectin, by adding media with 30 μg/ml for 7 days. We then change the media to 50 ug/ml for 5 days, routinely culturing them afterwards with 20 μg/ml. If you want to select for two different plasmids (e.g., RFP, GFP), transfect them in the same amount and select them in the same way as described (e.g., selected only by hygromycin). The majority of the established stable cells possess both plasmids. Always use a well where no selection plasmid has been added to verify that your antibiotics are working fine and killing the undesired cells.
18. Besides the cell lines that you may encounter in papers, a variety of GFP-marked cell lines is listed in the RNAi database at **Link 4** (select "GFP localisation").
19. If the following phenomena are observed, cells are not in a good condition. (1) Cells get stuck in metaphase. (2) Microtubules start to bundle. (3) Very low proportion of cells in mitosis. In that case imaging condition should be changed (e.g., reduction of exposure time) or new cells should be prepared for imaging.
20. Phenotypes can be scored manually under microscope. However, if an automated microscope with higher NA objectives and laser-based autofocusing system is utilized (e.g., ImageXpress Micro (Molecular Devices)), autofocusing will work better when the samples are in PBS rather than in glycerol-based mounting medium. This is because a laser-based autofocusing system detects the difference of refractive index between samples and glass. After imaging is over, PBS can be substituted by mounting medium for long-term storage.

5. Links

Overall database for flies

1. Flybase: http://www.flybase.org/

Databases for RNAi protocols, targeted sequences, and phenotypes

2. *Drosophila* RNAi Screening Center at Harvard Medical School: http://www.flyrnai.org/
3. *Drosophila* RNAi Cell Microarray Methods: http://jura.wi.mit.edu//sabatini_public/fly_array
4. Database of screen for genes required for mitotic spindle assembly in *Drosophila* S2 cells http://rnai.ucsf.edu/mitospindlescreen/index.html
5. FLIGHT: Integrating genomic and high-throughput data http://flight.licr.org/
6. Genome RNAi: http://www.dkfz.de/signaling2/rnai/index.php
7. SnapDragon: dsRNA (Amplicon or Hairpin) design http://flyrnai.org/cgi-bin/RNAi_find_primers.pl
8. Drosophila Research Genome Collection http://www.fruitfly.org/index.html
9. Gateway vectors (https://dgrc.cgb.indiana.edu/vectors/store/vectors.html?product_category=3)
10. Open Biosystems http://www.openbiosystems.com/RNAi/Non%2DMammalian%20RNAi/Drosophila%20RNAi%20library/
11. Geneservice http://www.geneservice.co.uk/products/rnai/Dros_RNAi.jsp

Database of Fly Stocks Carrying RNAi Constructs

12. Viena *Drosophila* RNAi center http://stockcenter.vdrc.at/control/main

***Drosophila* Genome Browser**

13. http://www.fruitfly.org/annot/apollo/

Free Software for Image Analysis

14. imageJ http://rsb.info.nih.gov/ij/

Acknowledgments

We would like to thank the fly community for all the available and shared reagents that were cited here. MBD and GG would like to acknowledge the Glover, the Vale, and the MBD labs for help in developing protocols and Inês Bento, Joana Lamego, Ana Rodrigues Martins, and Zita Carvalho Santos for critical reading of this chapter. Work in the MBD lab is sponsored by an EMBO Installation Grant, Fundação Calouste Gulbenkian, Fundação para a Ciência e Tecnologia, Crioestaminal, and Oeiras City Council. Work in the GG lab is sponsored by Special Coordination Funds for Promoting Science and Technology commissioned by the MEXT of Japan.

References

1. Clemens, J.C., Worby, C.A., Simonson-Leff, N., Muda, M., Maehama, T., Hemmings, B.A., and Dixon, J.E. (2000). Use of double-stranded RNA interference in Drosophila cell lines to dissect signal transduction pathways. *Proc. Natl. Acad. Sci. USA* **97**, 6499–6503.
2. Goshima, G., Wollman, R., Goodwin, S.S., Zhang, N., Scholey, J.M., Vale, R.D., and Stuurman, N. (2007). Genes required for mitotic spindle assembly in drosophila S2 cells. *Science* **316,** 417–4121.
3. Bettencourt-Dias, M., Giet, R., Sinka, R., Mazumdar, A., Lock, W.G., Balloux, F., Zafiropoulos, P.J., Yamaguchi, S., Winter, S., Carthew, R.W., Cooper, M., Jones, D., Frenz, L., and Glover, D.M. (2004). Genome-wide survey of protein kinases required for cell cycle progression. *Nature* **432**, 980–987.
4. Kiger, A., Baum, B., Jones, S., Jones, M., Coulson, A., Echeverri, C., and Perrimon, N. (2003). A functional genomic analysis of cell morphology using RNA interference. *J. Biol.* **2**, 27.
5. Boutros, M., Kiger, A.A., Armknecht, S., Kerr, K., Hild, M., Koch, B., Haas, S.A., Consortium, H.F., Paro, R., and Perrimon, N. (2004). Genome-wide RNAi analysis of growth and viability in Drosophila cells. *Science* **303**, 832–835.
6. Bjorklund, M., Taipale, M., Varjosalo, M., Saharinen, J., Lahdenpera, J., and Taipale, J. (2006). Identification of pathways regulating cell size and cell-cycle progression by RNAi. *Nature* **439**, 1009–1013.
7. Echeverri, C.J., and Perrimon, N. (2006). High-throughput RNAi screening in cultured cells: a user's guide. *Nat. Rev. Genet.* **7**, 373–384.
8. Debec, A., and Abbadie, C. (1989). The acentriolar state of the Drosophila cell lines 1182. *Biol. Cell* **67**, 307–311.
9. Debec, A. (1978). Haploid cell cultures of Drosophila melanogaster. *Nature 274*, 255–256.
10. Bettencourt-Dias, M., Sinka, R., Frenz, L., and Glover, D.M. (2004). RNAi in Drosophila cell cultures. In: M. Sohail, ed. *Gene Silencing by RNA Interference: Technology and Application*, CRC Press, Boca Ratan.
11. Schneider, I. (1972). Cell lines derived from late embryonic stages of Drosophila melanogaster. *J. Embryol. Exp. Morphol.* **27**, 353–365.
12. Peel, D.J., Johnson, S.A., and Milner, M.J. (1990). The ultrastructure of imaginal disc cells in primary cultures and during cell aggregation in continuous cell lines. *Tissue Cell* **22**, 749–758.
13. Ui, K., Nishihara, S., Sakuma, M., Togashi, S., Ueda, R., Miyata, Y., and Miyake, T. (1994). Newly established cell lines from Drosophila larval CNS express neural specific characteristics. *In Vitro Cell Dev. Biol. Anim.* **30A**, 209–216.
14. Echalier, G. (1997). *Drosophila* Cells in Culture, Academic Press, New York, USA.
15. Ashburner, M. (1989). Drosophila. *A Laboratory Handbook*, Volume I, Cold Spring Harbor Laboratory Press, Cold Spring Harbor.
16. Goshima, G., and Vale, R.D. (2003). The roles of microtubule-based motor proteins in mitosis: comprehensive RNAi analysis in the Drosophila S2 cell line. *J. Cell Biol.* **162**, 1003–1016.
17. Bettencourt-Dias, M., Rodrigues-Martins, A., Carpenter, L., Riparbelli, M., Lehmann, L., Gatt, M.K., Carmo, N., Balloux, F., Callaini, G., and Glover, D.M. (2005). SAK/PLK4 is required for centriole duplication and flagella development. *Curr. Biol.* **15**, 2199–2207.
18. Yang, D., Lu, H., and Erickson, J.W. (2000). Evidence that processed small dsRNAs may mediate sequence-specific mRNA degradation during RNAi in Drosophila embryos. *Curr. Biol.* **10**, 1191–1200.

19. Goshima, G., and Vale, R.D. (2005). Cell cycle-dependent dynamics and regulation of mitotic kinesins in Drosophila S2 cells. *Mol. Biol. Cell* **16**, 3896–3907.
20. Roignant, J.Y., Carre, C., Mugat, B., Szymczak, D., Lepesant, J.A., and Antoniewski, C. (2003). Absence of transitive and systemic pathways allows cell-specific and isoform-specific RNAi in Drosophila. *RNA* **9**, 299–308.
21. Ma, Y., Creanga, A., Lum, L., and Beachy, P.A. (2006). Prevalence of off-target effects in Drosophila RNA interference screens. *Nature* **443**, 359–363.
22. Echeverri, C.J., Beachy, P.A., Baum, B., Boutros, M., Buchholz, F., Chanda, S.K., Downward, J., Ellenberg, J., Fraser, A.G., Hacohen, N., Hahn, W.C., Jackson, A.L., Kiger, A., Linsley, P.S., Lum, L., Ma, Y., Mathey-Prevot, B., Root, D.E., Sabatini, D.M., Taipale, J., Perrimon, N., and Bernards, R. (2006). Minimizing the risk of reporting false positives in large-scale RNAi screens. *Nat. Methods* **3**, 777–779.
23. Lum, L., Yao, S., Mozer, B., Rovescalli, A., Von Kessler, D., Nirenberg, M., and Beachy, P.A. (2003). Identification of Hedgehog pathway components by RNAi in Drosophila cultured cells. *Science* **299**, 2039–2045.
24. Rogers, S.L., Wiedemann, U., Stuurman, N., and Vale, R.D. (2003). Molecular requirements for actin-based lamella formation in Drosophila S2 cells. *J. Cell Biol.* **162**, 1079–1088.
25. Hannon, G.J. (2002). RNA interference. *Nature* **418**, 244–251.
26. Dykxhoorn, D.M., Novina, C.D., and Sharp, P.A. (2003). Killing the messenger: short RNAs that silence gene expression. *Nat. Rev. Mol. Cell Biol.* **4**, 457–467.
27. Hutvagner, G., and Zamore, P.D. (2002). RNAi: nature abhors a double-strand. *Curr. Opin. Genet. Dev.* **12**, 225–232.
28. Gonczy, P., Echeverri, C., Oegema, K., Coulson, A., Jones, S.J., Copley, R.R., Duperon, J., Oegema, J., Brehm, M., Cassin, E., Hannak, E., Kirkham, M., Pichler, S., Flohrs, K., Goessen, A., Leidel, S., Alleaume, A.M., Martin, C., Ozlu, N., Bork, P., and Hyman, A.A. (2000). Functional genomic analysis of cell division in C. elegans using RNAi of genes on chromosome III. *Nature* **408**, 331–336.
29. Rodrigues-Martins, A., Riparbelli, M., Callaini, G., Glover, D.M., and Bettencourt-Dias, M. (2007). Revisiting the role of the mother centriole in centriole biogenesis. *Science* **316**, 1046–1050.
30. Blower, M.D., and Karpen, G.H. (2001). The role of Drosophila CID in kinetochore formation, cell-cycle progression and heterochromatin interactions. *Nat. Cell Biol.* **3**, 730–739.
31. Martinez-Campos, M., Basto, R., Baker, J., Kernan, M., and Raff, J.W. (2004). The Drosophila pericentrin-like protein is essential for cilia/flagella function, but appears to be dispensable for mitosis. *J. Cell Biol.* **165**, 673–683.
32. Megraw, T.L., Kao, L.R., and Kaufman, T.C. (2001). Zygotic development without functional mitotic centrosomes. *Curr. Biol.* **11**, 116–120.

Chapter 4

Production of Mitotic Regulators Using an Autoselection System for Protein Expression in Budding Yeast

Marco Geymonat, Adonis Spanos and Steven Sedgwick

Abstract

A novel protein expression system in budding yeast is described which has been used to express many yeast mitotic regulators as well as a wide range of other recombinant proteins from several different species. The expression system relies on autoselection with essential genes to maintain high copy numbers of expression plasmids. Autoselection permits expression cells to be grown in rich medium with no need for plasmid selection with drugs or nutritional conditions. This optimizes growth and expression of recombinant proteins. The use of the expression system is illustrated by purifying budding yeast mitotic regulators, Cdc14 and Net1, and recapitulating their activities in vitro.

Key words: Protein expression, budding yeast, recombinant proteins, autoselection.

1. Introduction

The ability to purify biologically active proteins is central to any biochemical and biophysical studies of protein behaviour. Unfortunately, it is not a trivial task to identify an expression system capable of producing sufficient amounts of soluble and active protein. Often the search for a suitable expression system is a matter of trial and error that entails a considerable investment in time, effort, and expense. However, a novel yeast expression system recently has proved capable of producing several mitotic regulators from budding yeast in addition to a wide range of other proteins from other species *(1)*. Compared to many other eukaryotic expression systems, yeast is attractive because it is technically

Andrew D. McAinsh (ed.), *Mitosis: Methods and Protocols*, Methods in Molecular Biology *545*,
DOI 10.1007/978-1-60327-993-2_4, © Humana Press, a part of Springer Science+Business Media, LLC 2009

undemanding to grow and it can be evaluated rapidly and inexpensively for recombinant protein expression.

The key feature of this new yeast expression system is its ability to maintain high copy numbers of expression plasmids in rich medium without the conditional drug or nutritional selection regimes normally associated with plasmid selection. Under these conditions, growth rate is maximal so that gene expression and the activity of inducible promoters are optimised. Plasmids are maintained by 'autoselection'. Autoselecting plasmids encode an essential *S. cerevisiae* gene in host cells where the equivalent chromosomal copy of the essential gene is deleted. Thus, these expression plasmids are absolutely required for the viability of their host cells even in apparently nonselective rich media. Expression of single recombinant proteins uses plasmids carrying the essential *MOB1* gene in *Δmob1* mutant host cells. Two recombinant proteins are simultaneously expressed from pairs of expression plasmids carrying either *MOB1* or *CDC28* genes in doubly mutant *Δmob1 Δcdc28* host cells. An additional feature of this autoselection system is that copy number is increased in a compensatory response to the partial repression of the plasmids' essential gene expression prior to induction (*see* **Section 1.4**).

1.1. Expression Hosts

S. cerevisiae MGY70 host cells are used for expression of a single protein. MGY70 has a deletion of the essential *MOB1* gene and contains a *URA3* 'maintenance' plasmid encoding *MOB1* to provide the Mob1 protein needed for viability before a *MOB1* expression plasmid is introduced (**Fig. 4.1**). The genotype of MGY70 is *MAT a ura3-1 trp1-28 leu2Δ0 lys2 his7 mob1::kanR pep4::LEU2/pURA3-MOB1*. This means that MGY70 is phenotypically Trp$^-$, His$^-$, and Lys$^-$ if grown on minimal medium.

MGY140 is used to express two proteins. MGY140 has deletions of the essential *MOB1* and *CDC28* genes and contains a *URA3* 'maintenance' plasmid encoding both *MOB1* and *CDC28* to provide the Mob1 and Cdc28 proteins needed for viability before introduction of *MOB1* and *CDC28*-based expression plasmids (**Fig. 4.1**). The genotype of MGY140 is *MAT α ura3-1 trp1-289 his3 leu2 lys2Δ0 mob1::kan*^{R} *cdc28::LEU2 pep4::LYS2/pURA3-MOB1 CDC28*. MGY140 is phenotypically Trp$^-$ His$^-$ on minimal medium.

MGY70 and MGY140 both have a deletion of the major protease gene, *PEP4*, to increase product stability during purification.

1.2. Expression Plasmids

The expression plasmids carry either *MOB1* or *CDC28* for autoselection and either *TRP1* or *HIS3* markers for nutritional selection during expression strain construction (*see* **Section 3.1.2**). They also encode GST or 6His sequences for affinity tagging and protease sensitive sites for removal of affinity tags after purification (**Figs. 4.2, 4.3** and **Table 4.1**). Multiple cloning sites in the

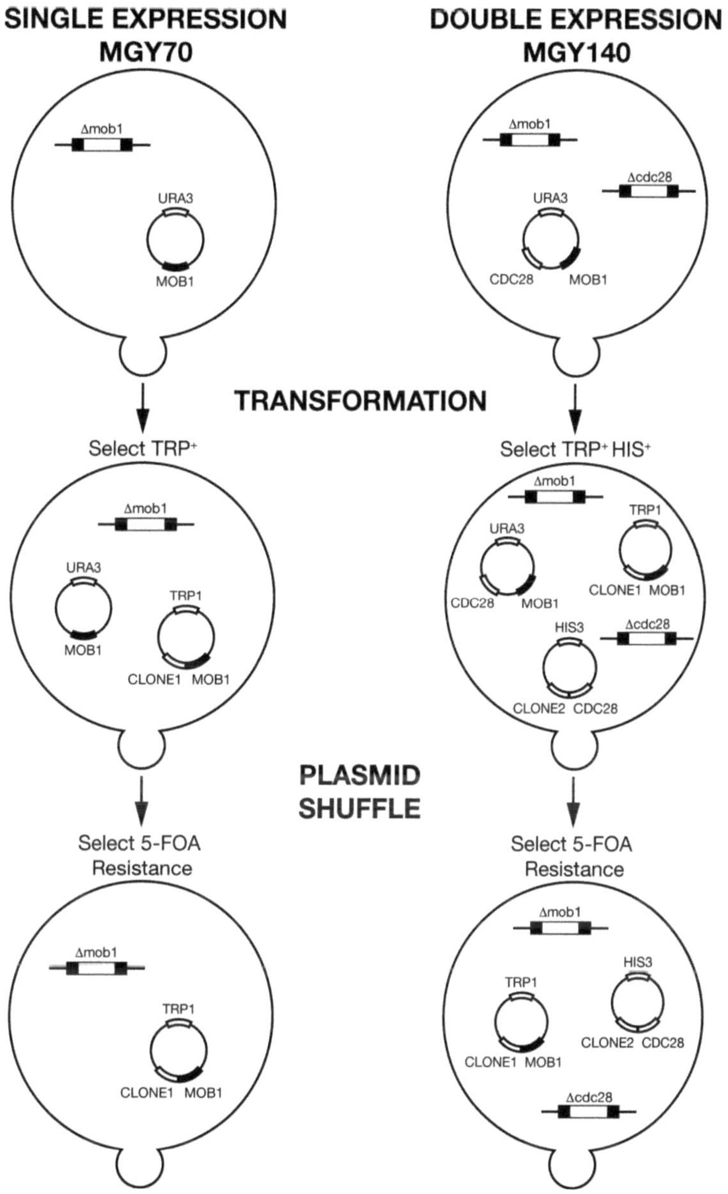

Fig. 4.1. Expression strain construction. The steps for constructing a single expression strain with MGY70 are presented to the left. Construction of double expression strain with MGY140 is shown on the right. Initially both cells at the top of the figure contain a *URA3* maintenance plasmid to maintain viability. After the transformation step the cells, centre figure, contain both maintenance and expression plasmids. Selection for 5-FOA resistance then selects for loss of the *URA3* maintenance plasmids so that the essential genes carried by the expression vectors become absolutely required for viability.

vectors are presented in **Figs. 4.2** and **4.3**. Recombinant protein expression is induced by galactose acting on the *GAL1-10* promoter of the vectors (**Figs. 4.2** and **4.3**).

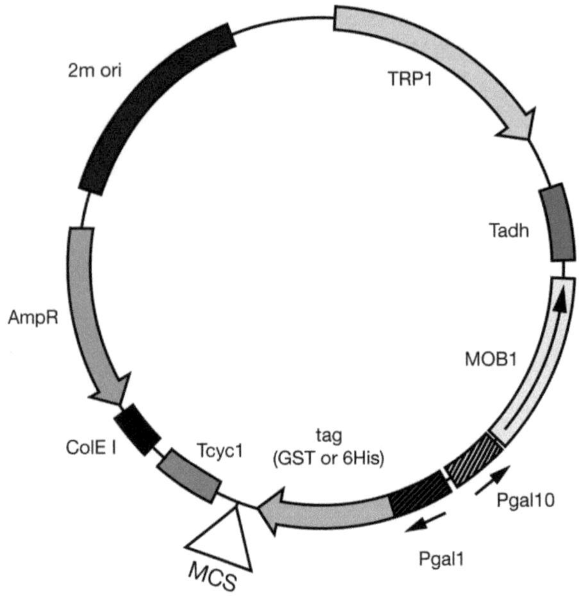

pMG1

BamHI Smal EcoRI

```
CTG GTT CCG CGT GGA TCC CCG GGA ATT TCC GGT GGT GGT GGT GGA ATT CTA
leu val pro arg gly ser pro gly ile ser gly gly gly gly gly ile leu
```

Thrombin cleavage site

Sall XhoI SacII

```
GAC TCC ATG GGT CGA CTC GAG TAA GCT TGG TAC CGC GGC TAG
asp ser met gly arg leu glu OCH ala trp tyr arg gly AMB
```

pMH919

rTEV cleavage site BamHI SmaI

```
GAA AAC CTG TAT TTT CAG GGC GCC ATG GGA TCC CCG GGA ATT TCC GGT GGT
glu asn leu tyr phe gln gly ala met gly ser pro gly ile ser gly gly
```

EcoRI Sall XhoI SacII

```
GGT GGT GGA ATT CTA GAC TCC ATG GGT CGA CTC GAG TAA GCT TGG TAC CGC GGC TAG
gly gly gly ile leu asp ser met gly arg leu glu OCH ala trp tyr arg gly AMB
```

pMH903

PreScission cleavage site BamHI EcoRI

```
CTG GAA GTT CTG TTC CAG GGG CCC CTG GGA TCC CCG GAA TTC CCG
leu glu val leu phe gln gly pro leu gly ser pro glu phe pro
```

SalI XhoI SalI

```
GGT CGA CTC GAG TAA GCT TGG TAC CGC GGC TAG
gly arg leu glu OCH ala trp tyr arg gly AMB
```

Fig. 4.2. Expression vectors for autoselection with *MOB1* and details of their multiple cloning sites.

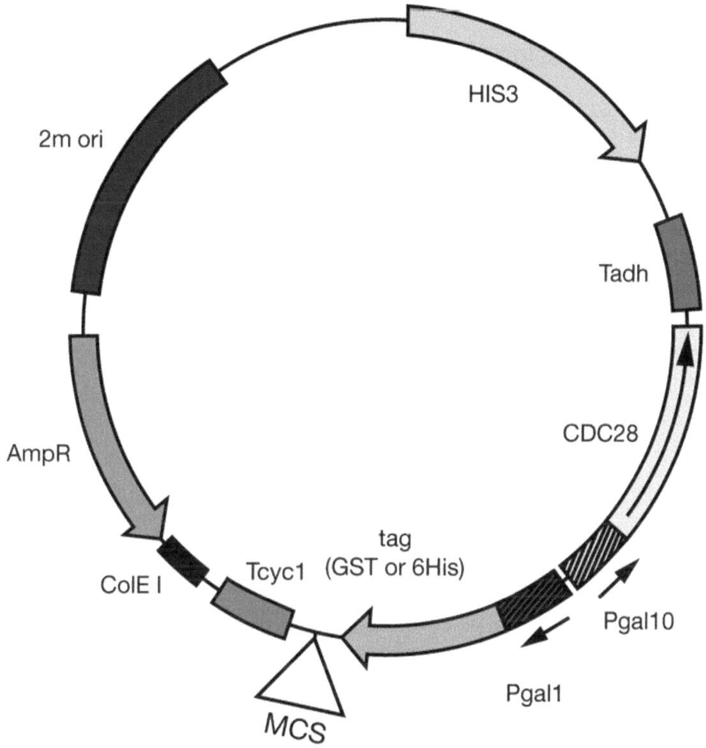

pMH925

BamHI SmaI

```
CTG GTT CCG CGT GGA TCC CCG GGA ATT TCC GGT GGT GGT GGT GGA ATT CTA GAC
leu val pro arg gly ser pro gly ile ser gly gly gly gly gly ile leu asp
```

Thrombin cleavage site

NcoI SalI XhoI SacII

```
TCC ATG GGT CGA CTC GAG TAA GCT TGG TAC CGC GGC TAG
ser met gly arg leu glu OCH ala trp tyr arg gly AMB
```

pMH940

rTEV cleavage site BamHI SmaI

```
GAA AAC CTG TAT TTT CAG GGC GCC ATG GGA TCC CCG GGA ATT TCC GGT GGT GGT
glu asn leu tyr phe gln gly ala met gly ser pro gly ile ser gly gly gly
```

SalI XhoI SacII

```
GGT GGA ATT CTA GAC TCC ATG GGT CGA CTC GAG TAA GCT TGG TAC CGC GGC TAG
gly gly ile leu asp ser met gly arg leu glu OCH ala trp tyr arg gly AMB
```

Fig. 4.3. Expression vectors for autoselection with *CDC28* and details of their multiple cloning sites.

Table 4.1
Expression plasmids

Expression Plasmid	Affinity Tag	Autoselection Marker	Nutritional Marker	Protease Site
pMG1	GST	*MOB1*	*TRP1*	Thrombin
pMH903	GST	*MOB1*	*TRP1*	PreScission
pMH919	6His	*MOB1*	*TRP1*	Tev
pMH925	GST	*CDC28*	*HIS3*	Thrombin
pMH940	6His	*CDC28*	*HIS3*	Tev

1.3. Using the System

The main steps for using the system are:

1. Cloning recombinant DNA coding sequences into *MOB1* or *CDC28* expression plasmids (**Table 4.1**).
2. Construction of an expression strain. The *URA3-MOB1* or *URA3-MOB1 CDC28* maintenance plasmids in the *mob1Δ* and *mob1Δ cdc28Δ* expression hosts are replaced with *MOB1* or *MOB1* and *CDC28* expression plasmids. This entails a plasmid transformation step to introduce the expression plasmids into the expression hosts (*see* **Section 3.1.2**) followed by a plasmid shuffling step (*see* **Section 3.1.3**) to select for loss of the maintenance plasmids (**Fig. 4.1**). Loss of the *URA3* maintenance plasmids is readily selected by resistance to 5-fluoroorotic acid (5-FOA; *see* **Note 1**).
3. Once the expression strain has been constructed, it is stable and can be stored at –80°C for future use with no need for further genetic manipulation.
4. Induction of recombinant protein expression.

1.4. Inducing Expression

Recombinant proteins are expressed from the *GAL1-10* promoter of the expression plasmids. The *GAL1-10* promoter is upregulated by galactose and is actively repressed by glucose, even when galactose is available. Thus, recombinant gene expression is repressed prior to induction and is induced only when galactose is added in the absence of glucose during the final period of growth. This avoids any toxic effect of overexpression on growth prior to induction. As the *GAL1-10* promoter is divergent it also regulates expression of the essential *MOB1* or *CDC28* genes (**Figs. 4.2** and **4.3**) so that repression of these selective genes before induction forces a compensatory increase in plasmid copy number (*see* **Note 2**).

2. Materials

2.1. Yeast Culture

1. Carbon sources. 50% stocks of glucose and sucrose are sterilised by autoclaving but 20% galactose stock solutions must be sterilised by filtration (*see* **Note 3**).
2. YEP 'rich' medium contains 1% Difco-Bacto yeast extract (BD cat no. 212750) and 2% Difco-Bacto peptone (BD cat. no. 211677). Solid YEP medium also contains 2% Bacto agar (BD cat. no. 214010). For convenience, prepare and sterilise multiple 500 ml bottles of medium without a carbon source. Carbon sources are added before use as follows: routine growth, 2% glucose; pre-induction, 1% sucrose; and induction, 1% galactose.
3. YNB Selective minimal agar for transformation. Make stocks of 10× 'Yeast Nitrogen Base without amino acids and with ammonium sulphate' (YNB Salts) (US Biological cat. no.Y2025). Filter sterilise and store in aliquots at 4°C. Make 400 ml stocks of 2.0% agar. Melt agar, cool to 55–60°C before adding one tenth volume 10× YNB salts and glucose to a final concentration of 2%. Add 20 mg/l each of uracil, lysine, and histidine for transformations with MGY70. For transformations with MGY140, add 20 mg/l each of uracil and lysine (*see* **Note 4**).
4. 5-FOA plates. Prepare and melt 3% Bacto agar (*see* **Note 5**). Add 1 mg/ml solid 5-fluoroorotic acid (5-FOA) (Toronto Research Chemicals Inc.; *see* **Note 6**). Cool to 55–60°C before adding one tenth volume 10× YNB salts and glucose to a final concentration of 2%. Add 20 mg/l each of uracil, lysine, and histidine for transformations with MGY70. For transformations with MGY140, add 20 mg/l each of uracil and lysine.
5. Freezer medium is made by including 20% glycerol in YEP-2% glucose medium.
6. A 10 mg/ml tetracycline stock solution is made in 70% ethanol. It is used at a final concentration of 20 μg/ml.

2.2. Yeast Transformation

1. 1 M filter-sterilised lithium acetate.
2. Polyethylene glycol (PEG), MW3350, (Sigma) is dissolved in water to give a 50% (w/v) solution (*see* **Note 7**). PEG solutions can be sterilised by filtration or by autoclaving.
3. 2 mg/ml denatured herring sperm DNA (Sigma, cat. no. D6898). The DNA is dissolved in 10 mM Tris-HCl pH 8.0, 1.0 mM EDTA (*see* **Note 8**). The DNA is denatured by boiling in a conical flask for 10–15 min followed by rapid cooling on ice. Aliquots of 0.5–1 ml are stored at –20°C. They are reheated and cooled prior to use.

2.3. Cell Lysis and Purification of GST-Tagged Proteins

1. The following recipe for a breakage buffer is generally applicable but may be varied for certain types of proteins (*see* **Note 9**). GST-breakage buffer for preparation of GST-tagged recombinant proteins comprises 50 mM Tris-HCl pH 7.5, 250 mM NaCl, 4 mM dithiothreitol, 5 mM ethylenediaminetetraacetic acid (EDTA), 10 mM NaF, 50 mM sodium glycerol 2-phosphate, 1 mM sodium orthovanadate (Sigma, cat. no. S6508; *see* **Note 10**), 5% glycerol, 0.5–1% NP 40 (Nonidet), 1 mM phenylmethylsulphonyl fluoride (PMSF; *see* **Note 11**), and 1 pill/50 ml Complete Protease Inhibitor without EDTA (Roche, cat. no. 04 693 132 001; *see* **Note 12**). The cell breakage buffer is made from stock solutions of the individual components on the day of use.
2. GST wash buffer: 50 mM Tris-HCl pH 7.5, 250 mM NaCl, 0.1% NP 40 (Nonidet), 1 mM dithiothreitol.
3. GST elution buffer: 50 mM Tris-HCl pH 7.5, 150 mM NaCl, 0.01% Triton, 20 mM reduced glutathione. A 1 M stock of reduced glutathione is made freshly before use in 1 M Tris-HCl pH 8.8.

2.4. Cell Lysis and Purification of 6His-Tagged Proteins

1. The following recipe for a breakage buffer is generally applicable but may be varied for certain types of proteins (*see* **Note 9**). 6His-breakage buffer for preparation of 6His-tagged recombinant proteins comprises 50 mM Tris-HCl pH 7.5, 500 mM NaCl, 0.1% NP40 (Nonidet), 10 mM NaF, 10 mM sodium glycerol 2-phosphate, 20 mM imadazole pH 8.0, 5 mM mercaptoethanol, 1 mM phenylmethylsulphonyl fluoride (PMSF; *see* **Note 11**), 5% glycerol, and 1 pill/ 50 ml Complete Protease Inhibitor without EDTA (Roche, Cat. No. 04 693 132 001; *see* **Note 12**). The cell breakage buffer is made from stock solutions of the individual components on the day of use.
2. 6His wash buffer contains 50 mM Tris-HCl pH 7.5, 500 mM NaCl, 0.1% NP40 (Nonidet), and 20 mM imadazole pH 8.0.
3. 6His elution buffer comprises 50 mM Tris-HCl pH 7.5, 150 mM NaCl, 0.01% Triton X100, and 250 mM imadazole pH 8.0.

2.5. Cell Lysis

1. 0.5 mm diameter glass beads for cell lysis can be obtained from Biospec Products, Bartlesville, Oklahoma, USA.
2. Small-scale cell lysis of 100–300 mg of cells requires a shaking device such as a Ribolyser machine (Hybaid) or a FastPrep® instrument (MP Biomedicals).
3. Up to 50 g of cells can be lysed in a Bead Beater (Biospec Products, Bartlesville, Oklahoma, USA). Alternatively, larger

amounts of cells are lysed in a French pressure cell or other large-scale lysis machines such as those made by Constant Cell Disruption Systems (Daventry, UK).

3. Methods

3.1. Expression Strain Construction

3.1.1. Plasmid Construction

The coding sequence for a recombinant gene product is cloned into the multiple cloning site of an expression vector to make an in-frame fusion with the GST or 6His affinity purification tag. Multiple cloning sites are presented in **Figs. 4.2** and **4.3**. Complete plasmid sequences may be obtained from the authors. Prepare a plasmid miniprep from 2 to 3 ml of *E. coli* using a standard commercially available plasmid purification kit.

3.1.2. Yeast Transformation

For single protein expression, plasmid constructs are transformed into *S. cerevisiae* MGY70 with selection on YNB minus tryptophan minimal medium for the *TRP1* marker of the vector. For co-expression, pairs of expression plasmids are co-transformed into MGY140 and plated on YNB minus both tryptophan and histidine to select for the *TRP1 MOB1* and the *HIS3 CDC28* plasmids. A simple lithium acetate transformation method is satisfactory, even for co-transformation of two different plasmids. The following is our simplified version that will give enough cells for at least 10 transformations (*see* **Note 13**).

1. On the evening before transformation, inoculate a 20 ml YEP-2%glucose culture with a single colony picked from a stock plate of MGY70 or MGY140 and grow overnight at 30°C with gentle shaking to obtain log phase cells. This is equivalent to approximately 2 × 10^7 cells/ml or OD_{600} ~1. Log phase cells should appear as budded cells under phase contrast microscopy. If the culture is overgrown, dilute to allow one or two more cell divisions or 2–3 h growth before use.
2. Meanwhile, prepare transformation mixture comprising 960 μl 50%PEG, 144 μl 1 M lithium acetate, and 200 μl 2 mg/ml single-stranded DNA.
3. Harvest the cells by centrifugation at 3000 × g for 5 min at room temperature. A bench centrifuge is suitable.
4. Resuspend pellet in 50 ml of water and centrifuge again.
5. Drain water and briefly centrifuge pellet again so that all remaining water can be removed with a pipettor. Resuspend

the pellet in the transformation mixture by drawing it in and out of a pipettor.

6. Add 5 μl miniprep DNA to each 100 μl aliquot competent cells. Put at 42°C in a water bath or hot block for 30 min.
7. With MGY70, plate the transformation mixture on YNB minus trytophan plates. MGY140 is plated on YNB lacking both tryptophan and histidine. Culture 2–3 days at 30°C to see growth of single transformant colonies approximately 2–3 mm diameter (*see* **Note 14**). Pick 2–3 colonies onto fresh minimal selective agar and streak to obtain single colonies (*see* **Note 15**). Cultivate for another 2–3 days at 30°C.

3.1.3. Plasmid Shuffling

The aim of the following procedure is to replace the initial maintenance plasmids of the expression strains with the expression plasmids as the sole source of essential gene product(s). After introduction of the expression plasmids, the host has a supply of essential gene products from both the original maintenance plasmids and the newly transformed expression plasmids. This means the *URA3* maintenance plasmids can now be lost spontaneously, provided that uracil is present in the growth medium. After loss of the *URA3* maintenance plasmids, the cells are phenotypically Ura$^-$ and can be selected by their resistance to 5-FOA (**Fig. 4.1**). The 5-FOA resistant cells therefore have the expression plasmid as the sole source of essential Mob1 protein. Thus, the expression plasmids have become essential for viability so that expression strains can now be grown on rich 'nonselective' media without plasmid loss.

1. Using sterile toothpicks, pick two or three isolated colonies from the YNB-minus tryptophan plate onto an 5-FOA agar plate marked into 2 or 3 sectors. Using fresh sterile toothpicks, streak out the transformed cells to obtain single colonies (*see* **Note 15**). Incubate at 30°C. 5-FOA resistant colonies 2–3 mm in diameter should appear after 3–4 days of growth. 5-FOA resistant cells can now be cultured on rich, YEP+2% glucose medium with no loss of expression plasmid (*see* **Note 14**).
2. If discrete individual colonies are not obtained, pick from the first 5-FOA plate onto a second one and incubate again at 30°C.
3. After testing for expression, freezer stocks of expression strains can be made for future use and require no further genetic manipulation.

3.1.4. Freezer Stocks

Use a sterile toothpick to spread a postage stamp-sized patch of cells onto a YEP + 2% glucose agar plate. Grow 1–2 days at 30°C. With a sterile toothpick, scrape off cells into 1–2 ml of freezer medium. Store at –80°C.

3.2. Recombinant Protein Expression

Initially, 50 ml cultures are sufficient for producing 100–300 mg of induced cells to test for recombinant protein expression. Afterwards, larger quantities of cells may need to be induced based on the yields produced in the initial protein 'minipreps'. Yeast grows best in conical flasks with moderate shaking (150–200 rpm). The culture should occupy no more than one third the total flask volume.

1. Day 1. Pick an expression strain colony into 10 ml of liquid YEP medium plus 1% glucose and grow overnight at 30°C to stationary phase (*see* **Note 16**).
2. Day 2. About 5 PM, make a 1:1000 dilution into fresh YEP medium with 1% sucrose as carbon source and grow for 16 h overnight (*see* **Note 17**).
3. Day 3. At about 9 AM the culture should be in log phase at approximately 2–5 × 10^7 ml or OD_{600} 0.5–1. Dilute threefold with YEP + 1.33% galactose to give a final concentration of 1% galactose, and grow a further 6–9 h to induce recombinant protein expression (*see* **Note 18**).
4. Cells are harvested by centrifugation at 2000 g for 5 min. After being washed once in sterile water, the cell pellets are snap-frozen and stored at –80°C. To save handling later, freeze samples for protein minipreps (*see* **Section 3.3**) in tubes suitable for the lysis machine employed.

3.3. Protein Minipreps

100–300 mg of cells can be obtained from 50 ml of induced culture and are used for 'protein minipreps'. Minipreps are usually used to test the two or three isolates of the initial expression strains constructed in **Section 2.1, step 3**. In this procedure, the binding and elution steps of the affinity-tagged proteins are essential for evaluation as it is not usual to see an induced protein in total cell lysates except by Western blotting. Sometimes protein minipreps can yield sufficient material for experimentation. In this case several samples can be processed in parallel to make a working stock of recombinant protein.

All solutions are precooled prior to use.

1. Up to 300 mg of cells from 25 to 50 ml cultures of induced cells are resuspended in 0.5 ml of chilled breakage buffer in a 2 ml screwcap tube. Glass beads, 0.5 mm diameter, are added to the meniscus.
2. Cells are lysed by four 10 s treatments with a shaking device such as a Ribolyser or FastPrep instrument. Cool the tubes on ice for 1 min or more between treatments to avoid overheating. In the absence of a Ribolyser machine, cell breakage can also be achieved by vortexing several minutes with pauses every minute for cooling. Breakage should be checked microscopically to determine a suitable period of

vortexing (*see* **Note 19**). Commercially available chemical lysis systems for yeast have been less successful with the strains used here.

3. To recover the lysates, the base of the Ribolyser tube is pierced with a red-hot needle. The tube is put into a 15 ml tapered plastic centrifuge tube which already contains a 1.5 ml screwcap microfuge tube without its cap. This assembly is centrifuged at approximately 2000 g for 15 s on a bench centrifuge to recover the lysate into the lower microfuge tube.
4. Particulates are removed by two 10 min centrifugations at 13,000 rpm in a refrigerated microfuge (*see* **Note 20**).
5. 100 μl glutathione-sepharose or nickel sepharose beads are required for each miniprep. Equilibrate the beads by two rounds of centrifugation at 2000 g for 15 s and resuspension in 1 ml of the appropriate wash buffer before resuspension with an equal volume of lysis buffer (*see* **Note 21**).
6. The cleared lysate from each miniprep is added to 100 μl equilibrated bead slurry in a 1.5 ml screwcap microfuge tube. The contents of the tube are mixed together for 1–2 h at 4°C on a roller mixer or rotor device.
7. The beads are recovered by centrifugation at 13,000 rpm for 30 s in a microfuge and are washed 5 times with 10 × volume of wash buffer (*see* **Note 21**).
8. The bead pellet is resuspended in a final 100–200 μl of wash buffer and transferred to a centrifugal filtration device such as a Costar® Centrifuge Tube Filter. The device is centrifuged for 30 s at 10,000 rpm. The lower collection tube containing wash buffer is discarded.
9. The beads are resuspended in 50–100 μl elution buffer in the upper portion of the filtration device which has been fitted with a new collection tube. The bead suspension is left for 5 min, and the eluate is collected by centrifugation.
10. Proteins produced are analysed by electrophoresis of a portion of the eluate through an SDS-polyacrylamide gel and are visualised by staining with Coomassie blue (*see* **Note 22**). An example of the co-expression of *S. cerevisiae* mitotic regulators, GST-Cdc14, and 6His-Net1 is shown in **Fig. 4.4**. This example shows how the recombinant products are sufficiently active to interact and recapitulate the in vivo inhibition of Cdc14 phosphatase by Net1 *(2–4)*. **Figure 4.4** also shows how chaperone proteins with an apparent molecular weight of ~70 kD often accompany the purified recombinant proteins.

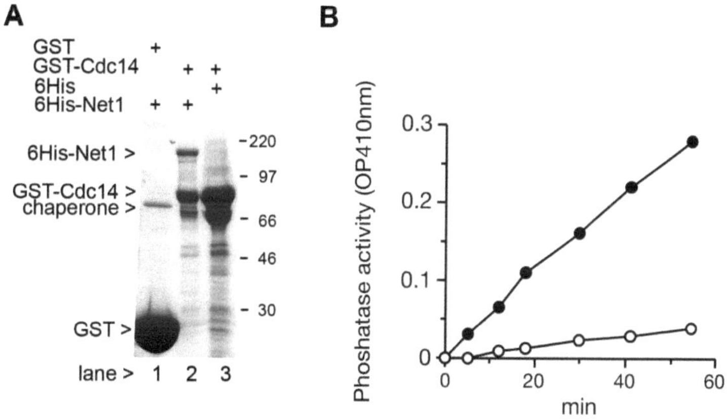

Fig. 4.4 In vitro recapitualtion of the inhibition of *S. cerevisiae* Cdc14 by Net1. Cdc14 phosphatase is a major effector of the FEAR and MEN pathways that complete mitosis once the replicated chromosomes have been equally partitioned into mother and daughter cell compartments at anaphase (*see (5)*). Before this stage of the cell cycle, Cdc14 is sequestered in the nucleolus in an inactive state by interaction with the Net1 protein *(2–4)*. To reproduce this interaction in vitro, GST-Cdc14 and 6His-Net1 were co-expressed in MGY140. *CDC14* was expressed in frame, 3' to the GST element of pMG1. The *NET1* coding sequence was cloned 3' to the 6His sequence of pMH940. Control cells contained either the GST-*CDC14*-pMG1 plasmid with empty pMH940 vector or 6His-Net1-pMH940 with empty pMG1. (**A**) Protein minipreps (**Section 3.3**) were prepared for affinity chromatography with glutathione sepharose. As would be expected, 6His-Net1 was not seen in lane 1 where the only product seen is the GST tag expressed from the empty pMG1 vector. However, 6His-Net1 was obtained when it was co-expressed with GST-Cdc14 (lane 2). Thus, 6His-Net1 is expressed and interacts with GST-Cdc14. Lane 3 was produced from a lysate from cells containing only GST-Cdc14 and the empty, 6His, pMH940 vector. (**B**) Phosphatase assays with GST-Cdc14 using the synthetic substrate p-nitrophenylphosphate *(6)*. GST-Cdc14 is active in vitro (•) but is inhibited by the presence of 6His-Net1/Cfi1 (○).

3.4. Larger-Scale Cell Lysis and Recombinant Protein Purification

3.4.1. Lysis with a Bead Beater

All manipulations use precooled solutions and equipment.

1. Cell pellets are thawed and resuspended on ice in a minimum of 5 ml breakage buffer/1 g cells.
2. Equal volumes of cell suspension and 0.5 mm diameter glass beads are used to fill the breakage chamber so that no free air space remains when the breakage rotor is inserted into the chamber.
3. The breakage chamber is immersed in an ice and salt mix. The cells are broken by 5 × 30 s treatments interspersed with 5 min cooling periods.
4. The cell lysate is separated from the glass beads by using a 500 ml disposable vacuum filtration tower from which the filtration membrane has been removed. Only a slight vacuum is applied to avoid foaming (*see* **Note 23**).

5. Cleared lysates are prepared by two 30 min clearing spins at 30,000 g.

3.4.2. Cell Lysis with a French Pressure Cell

All manipulations use precooled solutions and equipment.

1. Cell pellets are thawed and resuspended on ice in a minimum of 5 ml breakage buffer/1 g cells.
2. The cell suspension is passed twice through the pressure cell.
3. Cleared lysates are prepared by two 30 min clearing spins at 30,000 g.

3.4.3. Affinity Purification

1. Glutathione-sepharose or nickel sepharose beads are equilibrated in wash buffer. For this the bead slurry is pelleted by centrifugation at 2000 g for 15 s, and resuspended in 5–10 volumes of the appropriate wash buffer. This process is repeated before resuspension with an equal volume of lysis buffer. Note that 500 μl of bead slurry are required per gram of cell pellet.
2. The cleared lysate and sepharose beads are mixed together in a screwtop tube or bottle that is placed on a rotor or rolling table at 4°C for 2 h.
3. The affinity purification beads are recovered by centrifugation for 15 s at 2000 g and packed into a small column for washing and elution. The column is washed by gravity with 50–100 column volumes of wash buffer.
4. Proteins can be recovered by eluting the tagged fusion products from the affinity purification matrix using the elution buffers specified in **Sections 2.3, step 3** and **2.4, step 3**. Affinity-tagged material is eluted from the affinity purification column by stepwise elution. Elution buffer, approximately 50% the initial volume of bead slurry, is added to the top of the column, allowed to flow through under gravity and is recovered. Several individual elution steps are performed and assayed for protein content (*see* **Section 3.4.3, step 1**) to ensure complete recovery of eluted material. Alternatively, recombinant protein can be cleaved from the affinity purification tag by treatment with the appropriate proteases listed in **Table 4.1**. Protocols for protease treatment can be found at the protease suppliers' websites.
5. A BioRad protein assay or similar procedure is used to determine the protein content of the material released from the affinity-matrix. For the BioRad assay, the reagent is diluted to 20% strength and 1 ml aliquots are added to 1.5 ml microtubes. Absorbance at 595 nm is determined and compared with standards prepared with predetermined amounts of bovine serum albumin. A second sample is analysed by SDS-PAGE to check for product yield and integrity.

4. Notes

1. The *URA3* gene product is orotidine-5'-phosphate decarboxylase. Although it is normally deployed in uracil biosynthesis, it will also produce the toxic metabolite, 5-fluourouracil, from 5-FOA. Thus, cells with a *URA3* gene are killed by 5-FOA whereas cells without a functional *URA3* gene do not make 5-fluourouracil from 5-FOA and are therefore 5-FOA resistant.
2. The copy number of a single *MOB1*-based expression plasmid can approach 100 under autoselection in a rich medium. In the double expression system and in a rich medium, copy numbers for the two plasmids range between 70 and 100 each. In comparison, copy numbers under autoselection in a minimal medium are about 12 and with conventional auxotrophic selection in a minimal medium, plasmid copy numbers are about 3–5 (*see (1)*).
3. Galactose will dissolve more easily by using water preheated to 50°C. Galactose prices vary by more than fivefold and can be a major consideration. Galactose from Acros Organics (cat. no. 150615000) is one of the less expensive products and works well.
4. Rather than adding the specified supplements, an alternative is to use YNB dropout medium. For MGY70 transformations add Complete Supplement Mixture minus tryptophan powder (CSM-TRP, Bio 101Systems, cat. no. 4511-012) to solid agar stock and melt. For transformations with MGY140, add Complete Supplement Mixture minus tryptophan minus histidine (CSM-TRP-HIS, Bio 101Systems, cat. no. 4520-112) to solid agar stock and melt. Cool to 55–60°C before adding one tenth volume YNB salts plus 2% glucose. Dropout medium is slightly richer than plain minimal medium and so allows better growth but is more expensive. Uracil is included in both types of selection medium to encourage the loss of the maintenance plasmid that is required in the subsequent 5-FOA agar step.
5. To counteract the effect of 5-FOA in preventing agar from solidifying, 3% agar is essential.
6. It is important to add the 5-FOA powder when the agar is 75–80°C to avoid insolubility problems.
7. PEG dissolves more rapidly in water preheated to 50°C.
8. Undissolved herring sperm DNA is fibrous, tangled, and difficult to handle and weigh. It is easiest to dissolve the entire, factory-weighed contents of a bottle of solid DNA

in the supplier's bottle. Leave the DNA overnight at 4°C to dissolve before adjusting the final volume.

9. Lysis buffers generally contain both protease and phosphatase inhibitors. However, the amount and type of these and other reagents such as detergent and salt can be varied to optimise quantity and integrity of the tagged protein. Information on the type and quantities of such reagents can be found in the appropriate handbooks for glutathione sepharose 4B (Amersham Biosciences) or nickel sepharose (Qiagen). The presence of detergent, high salt concentrations, and protease inhibitors detailed here may therefore be modified if necessary to suit individual proteins. For instance, when purifying phosphatases or proteases, phosphatase inhibitors or protease inhibitors should be omitted.
10. To prepare a 200 mM sodium orthovanadate stock solution, dissolve 1.839 g of solid in 40 ml total water. Adjust to pH 10 with concentrated HCl. The solution is yellow. Heat the tube in a boiling water bath until the solution turns clear. Cool to room temperature and readjust pH to10 with concentrated HCl or NaOH as appropriate. Repeat the boiling–cooling cycle until the solution remains colourless at pH10 at room temperature. Adjust the volume to 50 ml with water. Store aliquots at –20°C.
11. A 100 mM stock of PMSF is prepared in isopropanol to avoid the degradation that occurs in water. Be aware that PMSF is toxic.
12. A ×25 stock of Complete Protease Inhibitor can be made by dissolving one pill in 2 ml of water and storing aliquots of the solution at –20°C.
13. More detailed accounts of yeast transformation can be found at http://home.cc.umanitoba.ca/~gietz/. However, high-efficiency methods are not needed for the single and double transformations carried out here. In general the protocol is quite resilient so that the conditions outlined are for general guidance rather than strict adherence.
14. If no colonies have grown by this time, recheck the selective medium, expression strain, or input DNA. Do not incubate further. Both MGY70 and MGY140 have been selected for vigourous growth so that colonies appearing later should be viewed with suspicion.
15. To obtain single colonies, pick with a sterile toothpick and make a small patch at one side of a fresh selective agar plate. With a fresh toothpick make one or two strokes out from the patch. It is important to streak in one direction only. With a fresh toothpick streak across the end of the first streak 2–3 times. Again, streak outwards only so that fewer

and fewer cells get drawn over the surface of the plate. Repeat with fresh toothpicks as space permits.

16. As an optional precaution against bacterial contamination, 20 μg/ml tetracycline can be added to the cultures on Day 1 and Day 2.
17. It is essential to use sucrose and not glucose in the Day 2 preinduction culture. Galactose will not induce expression when glucose is present. However, the small amount of glucose transferred from the Day 1 culture is entirely consumed before induction occurs and does not pose a problem.
18. Optimum times for induction may be determined empirically but, in general, yields were reduced when induction continued after these times. A second timing factor is the toxicity of some products which halts cell proliferation after induction. Strains expressing toxic proteins can be identified by failure to grow on YEP agar + 2% galactose. Yields of these toxic products may be improved on Day 3 by using larger starting inoculum for the final induction step in YEP-1% galactose medium and a shorter 4–6 h induction period.
19. Unbroken yeast appear bright and shiny under phase contrast microscopy whilst breakage produces granular and aggregated cell debris and cell wall 'ghosts' with a solid dark appearance.
20. After the clearing spins the colour of the pellets can also give an indication of breakage. Broken cell debris appears as a white band and a cream band below this layer contains unbroken cells. A thin ring of black debris is not unusual. Incomplete breakage usually arises with pellets greater than 250–300 μg. In this case yields can be improved by subjecting the pellet of the first clearing spin to a second bout of Ribolyser treatment to complete cell lysis.
21. After each centrifugation step the liquid layer is aspirated off using a vacuum line fitted with a fine tip to avoid accidental removal of the beads. A 10 μl plastic pipette tip is suitable.
22. If Western blotting is needed to detect a product then the protein is not well expressed in this system. In fact, it is sometimes possible to see purified protein bands in unstained gels. After removing one of the glass plates, look obliquely across the dry surface of the gel to see a band-shaped dimple in the surface of the polyacrylamide.
23. The glass beads can be reused but do become smaller and less effective after 2–3 cycles of use. They are rinsed extensively with water in the modified filtration tower described in **Section 3.4.1, step 4** and dried.

References

1. Geymonat, M., Spanos, A., and Sedgwick, S.G. (2007) A Saccharomyces cerevisiae autoselection system for optimised recombinant protein expression. *Gene* **399,** 120–128.
2. Visintin, R., Hwang, E.S., and Amon A. (1999) Cfi1 prevents premature exit from mitosis by anchoring Cdc14 phosphatase in the nucleolus. *Nature* **398,** 818–823.
3. Shou, W., Seol, J.H., Shevchenko, A., Baskerville, C., Moazed, D., Chen, Z.W., et al. (1999) Exit from mitosis is triggered by Tem1-dependent release of the protein phosphatase Cdc14 from nucleolar RENT complex. *Cell* **97,** 233–244.
4. Traverso, E.E., Baskerville, C., Liu, Y., Shou, W., James, P., Deshaies, R. J., et al. (2001) Characterization of the Net1 cell cycle-dependent regulator of the Cdc14 phosphatase from budding yeast. *J. Biol. Chem.* **276,** 21924–21931.
5. Bosl, W.J., and Li, R. (2005) Mitotic-exit control as an evolved complex system. *Cell* **121,** 325–333.
6. Sheng, Z., and Charbonneau, H. (1993) The baculovirus Autographa californica encodes a protein tyrosine phosphatase. *J. Biol. Chem.* **268,** 4728–4733.

Chapter 5

Hydrodynamic Analysis of Human Kinetochore Complexes During Mitosis

Sarah E. McClelland and Andrew D. McAinsh

Abstract

Hydrodynamic analysis is a powerful tool to dissect the molecular architecture of macromolecular protein assemblies. These techniques have been successfully used in yeast systems but are also well suited to the analysis of protein complexes from human cells. Furthermore, the combination of hydrodynamic analysis with siRNA mediated protein depletion provides an excellent system to probe the composition of protein complexes isolated from human cells. In this chapter we describe the use of these approaches in the analysis of macromolecular protein complexes during mitosis in human cells, using the kinetochore as an example.

Key words: Size-exclusion chromatography, sedimentation equilibrium, glycerol gradient, Svedberg coefficient, Stokes radius, kinetochore, mitosis.

1. Introduction

The human kinetochore is a large proteinaceous structure ~80–100 nm thick and 0.5–1.0 μm in diameter that assembles on centromeric DNA, biorients paired sister chromatids on spindle microtubules, and controls anaphase onset via the spindle checkpoint *(1)*. This complexity of function is reflected in the extraordinary structural complexity of the human kinetochore; to date, affinity purification and sequence search-based approaches have identified more than 80 kinetochore proteins. The combination of affinity purification and mass spectroscopy is an excellent approach to identify novel factors, and can reveal protein interaction networks. Such networks, however, do not distinguish

Andrew D. McAinsh (ed.), *Mitosis: Methods and Protocols*, Methods in Molecular Biology *545*,
DOI 10.1007/978-1-60327-993-2_5,

clearly between proteins that are present in stable complexes and those that interact in a more transient fashion. Biochemical analysis using size-exclusion chromatography and glycerol gradients has proved a powerful approach to discriminate between these possibilities *(2)*.

Hydrodynamic analysis relies on the combination of two classical techniques: Sedimentation equilibrium analysis (specifically, glycerol gradient ultracentrifugation) and size-exclusion chromatography. Both these methods separate proteins or complexes based on their molecular weight and shape. Nonglobular proteins or complexes behave differently to the globular standards used to calibrate both sedimentation equilibrium analysis and size-exclusion chromatography systems, giving rise to inaccurate estimations of native molecular weight if only one technique is employed. Using information from both techniques not only allows a more accurate measurement of the molecular weight of a protein or complex, but also provides information about the degree to which a complex is distorted from a globular state. These techniques are equally effective in the analysis of either native complexes from cell extracts or recombinant purified proteins. Although the methodology to analyse yeast extracts has been described *(3)*, here we provide detailed step-by-step instructions to allow any scientific researcher to perform the experiments and calculations necessary to probe the organisation of proteins and multiprotein assemblies in human mitotic cell extracts. Moreover, we show how combining this hydrodynamic analysis with siRNA-mediated protein depletion can be a powerful approach to probe the subunit composition of multiprotein complexes.

2. Materials

All chemicals are from Sigma-Aldrich, unless otherwise stated.

2.1. Cell Extract Preparation

1. H100 buffer (*see* **Note 1**; 50 mM HEPES pH 7.9, 1 mM EDTA, 100 mM KCl (*see* **Note 2**), 10% glycerol, 1 mM $MgCl_2$, 50 mM B-glycerolphosphate, 10 mM NaF, 0.25 mM $NaOVO_3$, 5 nM okadaic acid, 5 nM Calyculin A). Filter-sterilise and freeze in aliquots. On day of use add protease inhibitors (Complete (Roche)) and phosphatase inhibitors (Phophatase inhibitor cocktails A and B,(Sigma)).
2. Phosphate buffered saline (PBS).
3. Nocodazole (final concentration of 100 ng/ml; *see* **Note 3**).
4. Double distilled H_2O (ddH_2O).
5. 50 ml Falcon tubes.
6. Liquid nitrogen.

7. Porcelain mortar and pestle (Coors; Sigma).
8. High-speed ultracentrifuge, for example, Sorvall Ultra Pro 80.
9. Bradford Assay reagents or spectrophotometer system, for example, Nanodrop (Thermo Scientific).

2.2. Sedimentation Equilibrium Analysis; Glycerol Gradients

1. 2 × K150 buffer (100 mM HEPES pH 7.9, 2 mM EGTA pH 8, 2 mM $MgCl_2$, 600 mM KCl)
2. K150 + 5 % glycerol (5% (v/v) glycerol, 50% 2 × K150 buffer, 45% (v/v) ddH_2O)
3. K150 + 40 % glycerol (40% (v/v) glycerol, 50% 2 × K150 buffer, 10% (v/v) ddH_2O)
4. 13 × 51 mm polyallomer tubes (Beckman)
5. Optional: Gradient Master (Biocomp)

2.3. TCA Precipitation

1. Trichloroacetic acid (TCA)
2. Acetone
3. SDS-loading buffer (150 mM Tris.HCl pH 6.8, 1.2 % (w/v) SDS, 30% (v/v) glycerol, 15% (v/v) Beta-mercaptoethanol, 1.8 μg/ml Bromophenol blue)

2.4. Size-Exclusion Chromatography

1. K150 buffer (50 mM HEPES pH 7.9, 1 mM EGTA pH 8, 1 mM $MgCl_2$, 150 mM KCl; *see* **Note 4**).
2. Sephacryl size-exclusion chromatography column such as Pharmacia S-500 HR (Amersham).
3. High-pressure pump chromatography system such as Akta Purifier (Pharmacia).
4. Molecular weight globular protein standards (Amersham).

2.5. siRNA-Mediated Protein Depletion

1. Oligofectamine (Invitrogen).
2. siRNA oligos (Invitrogen/Qiagen).
3. Optimem (*see* **Note 5**; Invitrogen).
4. MEM media + 10 % FBS + 1% Penicillin/Streptomycin (all Invitrogen).
5. DMEM media + 10 % FBS + 1% Penicillin/Streptomycin (all Invitrogen).
6. 0.5 % Trypsin-EDTA (Invitrogen).

3. Methods

3.1. Making Mitotic Human Cell Extracts

In order to investigate the protein complexes involved in kinetochore function during mitosis we prepare our cell extract from human mitotic cells (*see* **Note 6**). Cells treated with the microtubule depolymerising drug, nocodazole, arrest in prometaphase due to lack of proper kinetochore–microtubule interactions leading to the sustained activation of the spindle checkpoint. Because mammalian cells 'round up' during mitosis, they are less strongly attached to the base of the dish and can be shaken loose. This allows us to further enrich for mitotic cells which can then be broken open to release the soluble protein contents. The main methods used to make concentrated soluble protein extracts from mammalian cells include freeze thawing, dounce homogenising, and liquid nitrogen grinding. We use the latter as it is a robust technique which produces protein extract of high concentration, typically around 30 mg/ml.

1. Grow cells in 15 cm dishes to 70–80% confluency (*see* **Note 7**).
2. Aspirate medium from cells and replace with medium containing nocodazole to a final concentration of 100 ng/ml (*see* **Note 3**).
3. 14–16 h later check the mitotic index using phase contrast microscopy to count the percentage of mitotic (rounded up) cells. There should be about 60–70% mitotic cells (*see* **Note 3**).
4. Perform mitotic shake-off: use a tip box to firmly tap all around the side of the dish. After 10–20 taps check that most mitotic cells have been detached using phase contrast microscopy. Repeat shake-off if necessary.
5. Using a 25 ml pipette take up the media, wash once over the dish, take up again, and collect in 50 ml Falcon tubes.
6. Spin cells down at 130 g for 5 min at 4°C, wash once with 50 ml PBS, spin cells down again, and remove medium.
7. Weigh the wet cell pellet (keep on ice) and resuspend in 1.5 volumes H100 buffer (+ protease inhibitor cocktail and phosphatase inhibitor cocktails added on day of use; *see* **Note 8**).

   Warning: The next steps are hazardous and should be performed in accordance with the local safety regulations (i.e., safety glasses and cold-proof gloves).

8. Chill mortar and pestle with three rounds of liquid nitrogen boiling. Freeze the mortar to the bench by adding some water around the base.
9. Drip cell suspension into liquid nitrogen-filled mortar and break pellets into a powder with pestle.
10. Grind pellets in a circular motion as the liquid nitrogen boils away, and really grind the powder once the liquid nitrogen has completely gone. Take care to scrape powder off the side of the wall with a precooled metal spatula.
11. Repeat liquid nitrogen boiling/grinding once for a total two rounds (*see* **Note 9**).
12. After last round of grinding, scoop powder out of the mortar with a precooled spatula and transfer to cooled 1.5 ml Eppendorf tubes.

13. Spin at 15.7 g in a bench-top centrifuge for 30 min at 4°C.
14. Remove supernatant, place in appropriate cooled tube (e.g., Beckman 1.5 ml tube for TLA 45 rotor) for the high-speed spin and add KCl to a final concentration of 150 mM (*see* **Note 10**), and NP40 to 0.05%, and spin at 125,000 g for 30 min at 4°C.
15. Remove supernatant = final extract. Determine the protein concentration using either Bradford assay or spectrophotometry then aliquot and freeze at –80°C. A typical yield should be in the range of 20–30 mg/ml protein.

3.2. Sedimentation Equilibrium Analysis; Glycerol Gradients

Sedimentation equilibrium analysis involves the centrifugal force-driven movement of protein complexes through a glycerol or sucrose gradient. Protein complexes migrate through the gradient at a rate dependent on two parameters: the molecular weight of the complex and the elongation of the complex. Elongated complexes do not move through the gradient as quickly as a globular counterpart of the same molecular weight and so will behave as a globular complex of smaller molecular weight. Glycerol gradients are a simple technique for estimating the sedimentation coefficient (Svedberg unit or S value) of proteins or protein complexes. The sample is centrifuged through a gradient of glycerol, which provides a near-constant rate of movement through the gradient. As the proteins move farther away from the centre of the rotor they experience increased centrifugal force, which is counteracted by increased viscosity of the gradient. The gradient can be varied to optimise resolution in the range in which you are interested. A 5–40% gradient provides a range of about 25–1000 kDa and is a good starting point. Once you have determined the region of interest then the gradient can be varied to optimise separation of your protein or complexes. This is critical as poor resolution leads to an inability to separate out complexes of similar size. There are several ways to prepare a glycerol gradient, from various commercially available gradient pouring equipment, to layering by hand. We use the Gradient Master system to prepare gradients simply and reproducibly.

3.2.1. Preparation of Glycerol Gradient Using Gradient Master System

1. Place a 13 × 51 mm polyallomer tube in the metal marker block, and mark tube using upper level.
2. Fill tube up to mark with K300 containing 5% glycerol and protease inhibitors.
3. Fill syringe with 40% glycerol and use filling syringe to place 40% glycerol at the bottom of the tube, carefully filling up to the mark once again (keeping the end of the syringe just below the interface between the 5% and 40% glycerol) so the whole tube is now full: 40% glycerol below the mark and 5% above it.

4. Carefully insert a black lid into the top of the tube, keeping the tube level, and allowing any air to escape through the small hole in the lid.
5. Select 'SW50 short glycerol gradient 5–40%' on gradient maker, place tubes in the tube holder, and run the program.
6. When ready to load the sample, remove lid carefully, and gently layer your sample (*see* **Note 11**) onto the top of the gradient.

3.2.2. Setting Up a Glycerol Gradient Centrifugation Run

1. Cool the AH641 rotor and the buckets in the cold room for ~30–60 min.
2. Chill ultracentrifuge to 4°C.
3. Load gradient, place carefully in bucket, screw down the lid so the numbers on the lid and bucket align, and place in rotor (always run all buckets even if some are empty). Spin at 6900 g for 14 h, 30 min at 4°C (*see* **Note 12**).

3.2.3. Glycerol Gradient Fractionation

Perform steps 1–3 in the cold room.

1. Label 25 Eppendorf tubes 1–25 and chill in freezer for a few minutes before use.
2. Carefully remove 200 μl from the top of the gradient using a P1000 with a cut tip, taking the sample from the centre of the gradient, not the side (a P200 can also be used) and place into tube 1.
3. Repeat until glycerol gradient is completely fractionated. If there is any left, collect it in a tube labelled 26 (*see* **Note 13**).
4. (Optional) TCA precipitate the samples (*see* **Section 3.2.4**) to concentrate before SDS-PAGE analysis.
5. Load 10–15 μl of each sample on to SDS-PAGE to analyse. (Alternatively you can freeze the samples until needed.)

3.2.4. TCA Precipitation

1. Add 35 μl of 100% TCA to each 200 μl fraction (or 175 μl for a 1 ml fraction) to give a final concentration of 15% TCA (v/v).
2. Vortex for 30 s.
3. Incubate for 30 min on ice.
4. Spin at 15.7 g in a bench-top centrifuge for 15 min at 4°C.
5. Remove supernatant (careful with pellet).
6. Add 500 μl acetone and vortex for 30 s.
7. Spin at 15.7 g in a bench-top centrifuge for 15 min at room temperature.
8. Remove supernatant and dry pellet until acetone cannot be smelt (*see* **Note 14**).
9. Resuspend in 1 × SDS Loading Buffer for SDS-PAGE analysis.

3.3. Size-Exclusion Chromatography

Size-exclusion chromatography involves introducing cell extract or purified protein to a size-exclusion column of suitable resolution, and collecting fractions. The column matrix consists of many tiny beads with pores of varying sizes. Large complexes cannot explore as much of the matrix as smaller complexes, so make a shorter path between the beads to the bottom of the column, resulting in a small elution volume. In addition to molecular weight, the shape of the protein complex also affects the elution volume; elongated complexes cannot pass through as many pores as their globular counterparts, and therefore behave as a globular complex of larger molecular weight. Size-exclusion chromatography can be used to estimate the Stokes radius of proteins or protein complexes. The choice of column used should be based on the size of the proteins or complexes being investigated. Ideally, the working range will be in the middle of the elution range. A good resolution is essential for accurate measurements, therefore we strongly recommend the Sephacryl S-500 column (Amersham). This column gives good resolution over a wide range. The HPLC system used should be able to provide constant pressure, such as the Akta purification system.

Perform All Steps in Cold Room/Cold Cabinet.

1. Determine column void volume. This is the elution volume of the smallest particle that cannot fit through any of the pores in the column matrix. This can be determined by running a 1 mg/ml solution of Dextran blue (included with the molecular weight standards from Amersham) through the column, and measuring the elution volume either by SDS-PAGE analysis or measuring the absorbance of each fraction at 380 nm using a spectrophotometer. The void volume is the volume at which Dextran blue begins to elute.
2. Run the globular protein standards in the first run (*see* **Section 3.4.1**), and repeat at least twice under the conditions you will use for your protein or extract (*see* **Note 15**).
3. Wash the column with 2 column volumes (CV) water; then equilibrate with 1 CV K150 or K300 buffer (*see* **Note 16**).
4. Load sample (retain an aliquot to serve as an input lane for analysis). It is best to keep the sample volume as low as possible; 1–2% of CV is a good guide. Run the column at 0.1 ml/min overnight (*see* **Note 17**) collecting 1 ml fractions (*see* **Note 18**).
5. TCA precipitate if necessary (*see* **Section 3.2.4**) and analyse using SDS-PAGE.

3.4. Hydrodynamic Calculations

3.4.1. Calibration Curves

It is essential to calibrate the glycerol gradient and size-exclusion chromatography systems for the exact conditions used in your experimental setup. Make up a solution containing protein standards covering the MW range of interest (we use a mixture of the Low and High molecular weight standards from Amersham). Weigh out 5 mg/ml (1 mg/ml for Ferritin) of each standard and dissolve in 1 ml water or K300/K150 buffer (+3% glycerol for use in glycerol gradients). Run 50 μl on a glycerol gradient and 0.5–1 ml on size-exclusion chromatography. At this concentration you should be able to analyse the fractions using SDS-PAGE without the need for TCA precipitation. Note that several of these standards are homomultimeric and so will separate into monomers of smaller molecular weight under denaturing SDS-PAGE conditions. Perform at least two independent standard runs for each technique. You will use the known Stokes radius and Svedberg coefficients for the standards to construct calibration curves.

Svedberg Coefficient (From Glycerol Gradient)
Plot a graph with the *x*-axis as Svedberg coefficient and the *y*-axis as the peak fraction from glycerol gradient centrifugation (use the mean from three independent runs). Add the data from the globular standards and plot a linear regression line through the points. Use the equation of the regression line to estimate the Svedberg coefficient of your complex of interest, again using the mean peak fraction from at least two independent runs.

Stokes Radius (From Size-Exclusion Chromatography)
Plot a graph with the *x*-axis as the Stokes radius and the *y*-axis as $-\log K_{av}$. The constant $-\log K_{av}$ is used because its relationship with the Stokes radius is linear, and is calculated from **Equation (5.1)**:

$$K_{av} = \frac{\text{elution volume } (V_e) - \text{Void volume } (V_0)}{\text{Total volume } (V_t) - V_0} \quad (5.1)$$

V_e = elution volume in ml.
V_0 = Void volume of the column.

Add the data from the standards run and plot a linear regression line through the points. Again, use the equation of the regression line to estimate the Stokes radius of your complex of interest from the experimentally determined K_{av}.

3.4.2. Native Molecular Weight Determination

Once you have estimated the Svedberg coefficient and Stokes radius of the protein or complex of interest, the native molecular weight can be determined from **Equation (5.2)** (*see* **Note 19**).

$$\text{MW(Da)} = \text{Stokes radius (cm)} \times \text{Svedberg coefficient } (10^{-13}\,\text{s}) \times \frac{6\pi\eta_0 N}{(1 - \upsilon p)} \quad (5.2)$$

η_0 = viscosity of H_2O ($\sim 1.005 \times 10^{-2}$ g/cm.s) *(4)*.
N = Avogadro's number (6.022×10^{23}/mol).

υ = partial specific volume ($\sim$ 0.724 cm^3/g, or can be more accurately determined using values for each amino acid *(5)*).

p = density of solvent ($\sim$ 0.998 g/cm^3).

3.4.3. Calculation of Axial Ratio

Once the Stokes radius and Svedberg coefficient have been determined and the native molecular weight calculated, you can calculate an axial ratio for your protein or complex, assuming a prolate ellipsoid shape. First calculate the frictional ratio, the ratio between the translational frictional coefficient (*f*) of the measured Stokes radius and the translational frictional coefficient (f_0) of a perfect sphere with the same molecular weight as your protein or complex, to give the ratio f/f_0:

$$(f/f_0) = \frac{\text{Stokes radius (cm)}}{[(3\upsilon\,\text{MW})/(4\pi\text{N})]^{1/3}} \qquad (5.3)$$

MW = native molecular weight determined above (in Daltons), υ and N as before *(4)*.

A frictional ratio of about 1 indicates that your protein or complex is likely to be globular, whereas values above 1 indicate a nonglobular state.

To obtain the axial ratio, the Perrin or shape function *P*, can be used. *P* is related to $f/f_0$and the hydration value (ω) of the protein:

$$P = (f/f_0)[(\omega/\upsilon\text{p}) + 1]^{-1/3} \qquad (5.4)$$

Where ω can be estimated to be 0.35–0.40 and υ and p are as above.

The calculated value for *P* can then be used to give an indication of the axial ratio (a:b) of your protein or complex, assuming a prolate ellipsoid (rodlike) shape. The relationship between P and axial ratio is given in **Table 5.1**. This estimation of axial ratio is usually sufficient to give a rough idea of the elongation of your protein or complex (see *(6)* for a method to more accurately determine the axial ratio from *P*).

3.4.4. Calculation of Standard Deviation

To obtain a reasonable estimate of the error, the Stokes radius and Svedberg coefficients should be determined experimentally at least three times, and the mean and standard deviation of these values carried into the molecular weight and axial ratio calculations. When multiplying or dividing two constants with independent errors, the following equation should be used:
For example:

$$Z = AB: \qquad (z/Z)^2 = (a/A)^2 + (\mathbf{b}/\mathbf{B})^2,$$

Table 5.1
Axial ratios for prolate ellipsoids for the hydrodynamic function P (from *(5)*)

P Value	Axial Ratio
1.0000	1.0
1.0009	1.1
1.0031	1.2
1.0063	1.3
1.0103	1.4
1.0149	1.5
1.0201	1.6
1.0256	1.7
1.0315	1.8
1.0377	1.9
1.0440	2.0
1.1130	3.0
1.1830	4.0
1.2500	5.0
1.3140	6.0
1.3750	7.0
1.4340	8.0
1.4900	9.0
1.5430	10
1.9960	20
2.3590	30
2.6710	40
2.9500	50
3.2050	60
3.4420	70
3.6640	80
3.8740	90
4.0740	100

Source: Reproduced from *(6)*.

where z is the standard deviation for Z, b is the standard deviation for B, and a is the standard deviation for A. Bear in mind this does not include additional error which is introduced by other assumptions made during the hydrodynamic analysis.

3.4.5. Example Using the NDC80 Complex

The NDC80 kinetochore complex is an evolutionarily conserved four-protein kinetochore complex that is essential for kinetochore–microtubule attachment in yeast and humans *(7–9)*. This complex has been shown by EM and rotary EM to form an elongated rodlike structure *(10, 11)*. We use the NDC80 complex as an example to illustrate the use of hydrodynamic analysis in determining the native molecular weight and elongation of a stable kinetochore complex.

First, following the instructions in **Section 3.4.1**, size-exclusion chromatography and glycerol gradient systems are calibrated. Next, mitotically arrested HeLa cell extract is fractionated on size-exclusion chromatography and glycerol gradients and the fractions are analysed using antibodies to the Ndc80 subunit of NDC80 (**Fig. 5.1A,B**). This profile reveals one discrete peak (N1) indicating that Ndc80 is part of a single Ndc80-containing complex. The Stokes radius and Svedberg coefficient can then be calculated:

Stokes radius:
Elution volume = 82 ml (**Fig. 5.1A**).

$$K_{av} = \frac{(V_e) - (V_0)}{(V_t) - V_0} = \frac{(82\text{–}33)}{(120\text{–}33)} = 0.5632$$

Regression line from calibration curve:

$$y = 0.002x + 0.0504$$
$$(y = -\log\ K_{av}, x = \text{Stokes radius (Å)})$$

$$\text{Therefore Stokes radius} = \frac{-\log 0.5632\text{–}0.0504}{0.002}$$

Stokes radius = 99.4Å

Repeat experiment three times, and obtain mean Stokes radius ± standard deviation:
Run 1: 82 ml
Run 2: 82 ml
Run 3: 84 ml
Mean Stokes radius = 94.67 ± 8.3 Å.

Svedberg coefficient:
Peak fraction = 8 (**Fig. 5.1D**).
Regression line from calibration curve:

$$y = 1.2608x + 1.9274$$
$$(y = \text{peak fraction}, x = \text{Svedberg coefficient}).$$

$$\text{Therefore the Svedberg coefficient} = \frac{8\text{–}1.9274}{1.2608}$$

Svedberg coefficient = 4.82

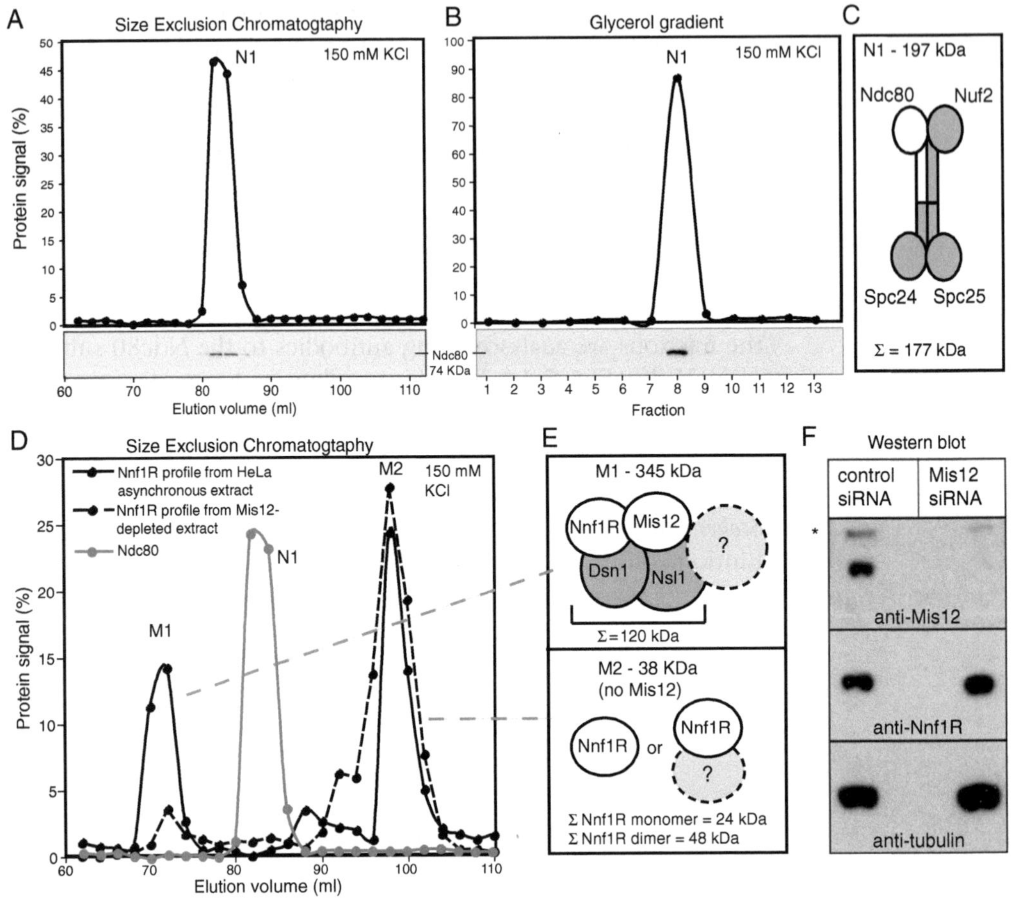

Fig. 5.1. Hydrodynamic analysis of the Ndc80 and MIND complexes from human cells. Whole cell extracts were prepared from nocodazole-arrested HeLa cells and fractionated on a HiPrep 16/60 Sephacryl S-500 HR size-exclusion chromatography column (**A**), and a 5–40% glycerol-density gradient (**B**), before immunoblotting with anti-Ndc80 antibodies *(13)*. (**C**) Cartoon of the Ndc80 complex indicating the predicted molecular weight calculated from the amino acid sequences of the subunits (Ndc80 = 74 kDa, Nuf2 = 54 kDa, Spc24 = 22.5 kDa, Spc25 = 26 kDa). (**D**) Whole cell extracts were prepared from asynchronous HeLa cells (solid black line) or from Mis12 siRNA-treated cells (dashed black line) and fractionated on a HiPrep 16/60 Sephacryl S-500 HR size-exclusion chromatography column before immunoblotting with antibodies to Nnf1R *(12)*. The profile of the NDC80 complex is overlaid in grey to illustrate that the NDC80 and MIND complexes can be clearly resolved using human mitotic cell extracts. (**E**) Cartoon representing the possible composition of the Nnf1R-containing complexes observed in (**D**) (Nnf1R = 24 kDa, Mis12 = 24 kDa, Dsn1 = 40 kDa, Nsl1 = 32 kDa). (**F**) Immunoblots of whole cell extracts from control and Mis12 siRNA *(15)* treated cells using antibodies to Mis12, Nnf1 *(12)*, and tubulin (Sigma) showing that Mis12 is efficiently depleted upon treatment with Mis12 siRNA whereas Nnf1R protein levels are unaffected (*represents a cross reacting band).

Repeat experiment three times, and obtain mean Svedberg coefficient ± standard deviation:
Run 1: fraction 8
Run 2: fraction 9
Run 3: fraction 8
Mean Svedberg coefficient = 5.08 ± 0.46 Å

Native Molecular weight:
From these experimentally determined values we can now calculate the native molecular weight of the Ndc80-containing complex using **Equation** (**5.2**):

$$\begin{aligned}\mathrm{MW(Da)} &= \text{Stokes radius (cm)} \times \text{Svedberg coefficient}(10^{-13}\,\mathrm{s}) \\ &\quad \times \frac{6\pi\eta_0 N}{(1-\upsilon\rho)} \\ &= (94.67 \times 10^{-8}) \times (5.08 \times 10^{-13}) \\ &\quad \times \frac{6\pi(1.005 \times 10^{-2})(6.022 \times 10^{23})}{(1-(0.724\text{–}0.998)} \\ &= 197897.7\,\mathrm{Da} \\ &= 197\,\mathrm{kDa}\end{aligned}$$

Standard deviation (s.d):

$$[\mathrm{s.d}/197]^2 = [8.3/94.67]^2 + [0.46/5.08]^2 = 24.8\,\mathrm{kDa}$$

Therefore the native molecular weight of the Ndc80 complex determined using hydrodynamic analysis from mitotically arrested HeLa cells is 197 ± 24.8 kDa, which is in good agreement with the theoretical molecular weight of the human NDC80 complex of 177 kDa.

Next, we can calculate the elongation of this complex using **Equations** (**5.3**) and (**5.4**):

$$\begin{aligned}(f/f_0) &= \frac{\text{Stokes radius (cm)}}{[(3\upsilon\mathrm{MW})/(4\pi\mathrm{N})]^{1/3}} \\ &= \frac{96.5 \times 10^{-8}}{[(3 \times 0.724 \times 191000)/(4\pi \times 6.022 \times 10^{23})]^{1/3}} = 2.5\end{aligned}$$

Because the f/f_0 value is well above 1, we already know that this complex is distorted from a globular state. The Perrin, or shape function, *P*, can now be determined:

$$\begin{aligned}\mathrm{P} &= (f/f_0)[(\omega/\upsilon\rho)+1]^{-1/3} \\ &= 2.5[(0.375/0.724 \times 0.998)+1]^{-1/3} = 2.17\end{aligned}$$

Table 5.1 relates a *P* value of 2.17 to an axial ratio of between 20:1 and 30:1 assuming a prolate ellipsoid shape.

This is in good agreement with structural studies of the human NDC80 complex *(10, 11)* and demonstrates the power of hydrodynamic analysis in determining the size and shape of multiprotein complexes from mammalian cell extracts.

3.5. siRNA as a Tool to Probe Multiprotein Complexes

The human MIND/Mis12 kinetochore complex contains four protein subunits; Nnf1R (Pmf1), Mis12, Nsl1, and Dsn1. Using antibodies to Nnf1R *(12)* and the approach described for the NDC80 complex we can analyse the hydrodynamic properties of the MIND complex. The MIND complex runs as two species on size-exclusion chromatography (**Fig. 5.1D**) and glycerol gradient ultracentrifugation *(13)* with molecular weights of ~38 kDa (M1) and ~345 kDa (M2). In order to investigate the composition of these complexes further, siRNA-mediated protein depletion can be used in combination with hydrodynamic analysis. Depletion of one subunit from a protein complex can result in the disintegration of the complex, degradation of the other components, or a shift in molecular weight of the complex. In the case of the MIND complex, depletion of the Mis12 subunit causes the degradation of the Dsn1 subunit, but not the Nnf1R subunit (*(14)*; **Fig. 5.1F**). (Currently the stability of the Nsl1 subunit in Mis12 depleted cells is unknown.) Therefore, as a simple example we can analyse the hydrodynamic properties of Nnf1R in asynchronous HeLa cell extracts depleted of Mis12 to determine whether the M1 and/or M2 MIND complexes contain Mis12 (*see* **Note 20**).

1. Design 3 siRNA oligos to target your gene (*see* **Note 21**). We used the following sequence to target Mis12: GGACAUUUUGAUAACCUUUTT *(15)*.
2. Validate the specificity and efficiency of the siRNA treatment using Western blotting or indirect immunofluorescence (*see* **Note 22**). **Figure 5.1F** shows the specific and efficient depletion of Mis12 in Mis12 siRNA but not control siRNA-treated cells.

DAY 1

1. Culture HeLa cells in a 10 cm dish until 80–90% confluency is reached.
2. Trypsinize cells: wash twice with PBS then add 1 ml 0.5% Trypsin-EDTA to the dish. Incubate at 37°C for 2–3 min until cells detach from the dish.
3. Take up HeLa cells in 10 ml fresh DMEM media supplemented with 10% FBS and 1% penicillin/streptomycin.
4. Add 240 μl of cell suspension into 10 ml fresh DMEM media supplemented with 10% FBS and 1% penicillin/streptomycin in another 10 cm dish

DAY 2

5. Change media to 6 ml MEM media supplemented with 10% FBS and 1% penicillin/streptomycin.
6. Mix the following components in Eppendorf tubes with filtered tips:
   Tube 1: 600 μl Optimem; 18 μl siRNA oligos (at 20 μM)
   Tube 2: 144 μl Optimem; 36 μl oligofectamine

7. Wait 8 min; then add tube 2 to tube 1 and mix by pipetting.
8. Wait 25 min;, then add the contents of tube 1 dropwise to cells.

DAY 3

9. Change media to fresh DMEM media supplemented with 10% FBS and 1% penicillin/streptomycin.

DAY 4

10. After an appropriate time for efficient depletion of your protein add nocodazole to the cells (*see* **Note 23**).
11. After 14 h nocodazole treatment, harvest cells and prepare mitotic extract as in **Section 3.1**.
12. Follow instructions in **Sections 3.2** and **3.3** to determine the hydrodynamic properties of your complex of interest in the absence of your chosen subunit.

The size-exclusion elution profiles of Mis12-depleted extract and control extract can be seen in **Fig. 5.1D**. The larger molecular weight species (M2) is completely abolished when Mis12 is depleted, whereas the lower molecular weight species (M1) is unaffected. This indicates that Mis12 is normally present in the M2 complex but not the M1 complex. Furthermore, because we know that Dsn1 is degraded in the absence of Mis12 *(14)*, we can conclude that Dsn1 is also not part of the smaller molecular weight complex. The possible compositions of this 38 kDa (M1) complex therefore include a monomer of Nnf1R, a dimer of Nnf1R, a heterodimer of Nnf1 and Nsl1, or a heterodimer of Nnf1 and an alternative protein. Follow-on experiments using depletion of the other MIND subunits will lead to a more complete understanding of the composition of these two species. Combining siRNA with hydrodynamic analysis in this way is therefore a powerful approach to dissect the components of multiprotein complexes from human cells.

4. Notes

1. It is best for the lysis buffer (H100 here) not to contain detergent as this may disrupt native complexes, however, your protein of interest may be insoluble (present in low-speed or high-speed pellet; check these if you cannot detect your protein in the supernatant) and may require the addition of small molecules or detergent to enhance solubility.
2. Either NaCl or KCl can be used.
3. This may vary for different cell lines, so test to see what achieves best arrest, aiming for 60–70% arrested (rounded-up) cells.

4. Use at least 150 mM NaCl or KCl to minimise nonspecific interactions between proteins and the column matrix.
5. Change Optimem bottle every month and close bottle well as pH changes over time.
6. We have used HeLa cells in the main for these experiments. Alternative human cell lines can also be used, and we have also used HEK293 cells with equal success, however because HEK293 cells do not attach very strongly to the base of the dish, mitotic shake-off cannot be performed inasmuch as most of the interphase cells also detach during this process. Mitotic enrichment using nocodazole can still be performed, however.
7. We find that 8–10 15 cm dishes give about 1 ml of 30 mg/ml extract.
8. To stop here drip the suspended cells into Falcon tubes containing a little liquid nitrogen, allow nitrogen to boil away completely, and then place in –80°C freezer. The samples can be stored at –80°C for several weeks if necessary.
9. It is worth testing to see if three rounds give a better yield, although we found no difference between two and three rounds.
10. Remember, the extract currently contains 100 mM KCl from the lysis buffer. We use a final extract concentration of 150 mM for most of our analyses because this is close to physiological salt concentration. However, you may want to consider increasing the salt to a maximum of 600 mM if your complex of interest appears to be aggregating in high molecular weight complexes, in order to investigate lower-order assemblies. With high salt extracts, we advise also increasing the salt concentration in the glycerol gradient and size-exclusion chromatography running buffers.
11. The sample must be in less than 60% of the lowest percentage of glycerol used in the gradient to avoid the sample sinking into the gradient before centrifugation, and up to 200 μl total volume.
12. Select the soft spin and brake off settings on the centrifuge if available, because this will minimise any perturbation of the gradient during acceleration and deceleration.
13. It is useful to know if anything is aggregating or very large and therefore reaching the bottom of the gradient.
14. Be careful not to overdry the pellet as this makes it difficult to redissolve the pellet in SDS loading buffer.
15. It is also good to periodically run a set of standards to check that the calibration is still accurate.
16. We do not usually add protease inhibitor cocktail to this buffer due to the large volumes required, but protease inhibitors could be added if protein degradation is a problem.

17. The column can be run faster, but this slower speed gives the best resolution, which is important when distinguishing between complexes of similar sizes.
18. Once the void volume has been determined there is no need to collect any fractions before the void volume has been eluted.
19. Note that this method only gives an estimate of MW and has a typical systematic error of about ±20% of the calculated value.
20. Depletion of Mis12 abrogates the spindle checkpoint so it is not possible to prepare mitotically arrested extract in this case.
21. You will need to use a program like BLOCK-iT^{TM} siRNA Designer on the Invitrogen website to find an appropriate sequence. If this is the first time you are investigating a gene it is best to order three siRNA oligos and test them all.
22. It is best to determine the efficiency of protein depletion using indirect immunofluorescence or Western blotting if antibodies are available to your protein. If antibodies are not available, reverse transcriptase PCR can be used as a measure of the siRNA efficiency. Test 24, 48, 72, and 96 h transfections and different concentrations of siRNA to determine the most suitable time and least amount for efficient protein depletion.
23. It is sometimes not possible to obtain mitotically arrested cells when depleting certain kinetochore proteins because in many cases their depletion abrogates the mitotic checkpoint. In this case, skip the addition of nocodazole and simply harvest cells at the end of the RNAi treatment to make an asynchronous extract.

Acknowledgments

The authors would like to thank the members of the McAinsh lab for helpful discussions and reading of the manuscript. We would also like to thank Hiro Yamano for equipment and expertise relating to the glycerol gradient ultracentrifugation experiments. This work was supported by Marie Curie Cancer Care.

References

1. Cleveland D.W., Mao Y., Sullivan K.F. (2003) Centromeres and kinetochores: from epigenetics to mitotic checkpoint signalling. *Cell.* **112**, 407–21.
2. De Wulf P., McAinsh A.D., Sorger P.K. (2003) Hierarchical assembly of the budding yeast kinetochore from multiple subcomplexes. *Genes Dev.* **17**, 2902–21.

3. Schuyler S.C., Pellman D. (2002) Analysis of the size and shape of protein complexes from yeast. *Methods Enzymol.* **351**, 150–68.
4. Siegel L.M., Monty K.J. (1966) Determination of molecular weights and frictional ratios of proteins in impure systems by use of gel filtration and density gradient centrifugation. Application to crude preparations of sulfite and hydroxylamine reductases. *Biochim Biophys Acta.* **112**, 346–62.
5. Cohn E.J., Edsall J.T. (1943) Proteins, Amino Acids and Peptides as Ions and Dipolar Ions, Reinhold, New York.
6. Harding S.E., Cölfen H. (1995) Inversion formulae for ellipsoid of revolution macromolecular shape functions. *Anal Biochem.* **228**, 131–42.
7. He X., Rines D.R., Espelin C.W., Sorger P.K. (2001) Molecular analysis of kinetochore-microtubule attachment in budding yeast. *Cell* **106**, 195–206.
8. Nabetani A., Koujin T., Tsutsumi C., Haraguchi T., Hiraoka Y. (2001) A conserved protein, Nuf2, is implicated in connecting the centromere to the spindle during chromosome segregation: a link between the kinetochore function and the spindle checkpoint. *Chromosoma* **110**, 322–34.
9. DeLuca J.G., Moree B., Hickey J.M., Kilmartin J.V., Salmon E.D. (2002) hNuf2 inhibition blocks stable kinetochore-microtubule attachment and induces mitotic cell death in HeLa cells. *J Cell Biol.* **159**, 549–55.
10. Wei R.R., Sorger P.K., Harrison S.C. (2005) Molecular organization of the Ndc80 complex, an essential kinetochore component. *Proc Natl Acad Sci USA.* **102**, 5363–7.
11. Ciferri C., De Luca J., Monzani S., Ferrari K.J., Ristic D., Wyman C., Stark H., Kilmartin J., Salmon E.D., Musacchio A. (2005) Architecture of the human ndc80-hec1 complex, a critical constituent of the outer kinetochore. *J Biol Chem.* **280**, 29088–95.
12. McAinsh A.D., Meraldi P., Draviam V.M., Toso A., Sorger P.K. (2006) The human kinetochore proteins Nnf1R and Mcm21R are required for accurate chromosome segregation. *EMBO J.* **25**, 4033–49.
13. McClelland S.E., Borusu S., Amaro A.C., Winter J.R., Belwal M., McAinsh A.D., Meraldi P. (2007) The CENP-A NAC/CAD kinetochore complex controls chromosome congression and spindle bipolarity. *EMBO J.* **26**, 5033–47.
14. Kline S.L., Cheeseman I.M., Hori T., Fukagawa T., Desai A. (2006) The human Mis12 complex is required for kinetochore assembly and proper chromosome segregation. *J Cell Biol.* **173**, 9–17.
15. Goshima G., Kiyomitsu T., Yoda K., Yanagida M. (2003) Human centromere chromatin protein hMis12, essential for equal segregation, is independent of CENP-A loading pathway. *J Cell Biol.* **160**, 25–39.

Chapter 6

Isolation of Protein Complexes Involved in Mitosis and Cytokinesis from *Drosophila* Cultured Cells

Pier Paolo D'Avino, Vincent Archambault, Marcin R. Przewloka, Wei Zhang, Ernest D. Laue and David M. Glover

Abstract

The identification of all the individual components that constitute the plethora of complexes in each cell type represents perhaps the most exciting challenge of postgenomic biology. This is particularly important in the study of events such as mitosis and cytokinesis, in which rapid and precise protein–protein interactions regulate both the direction and accuracy of these intricate processes. Here we describe an experimental strategy to isolate protein complexes involved in mitosis and cytokinesis in cultured *Drosophila* cells. This method involves the tagging of the bait protein with two IgG binding domains of Protein A and the isolation of the tagged bait along with its interacting partners by a single affinity purification step. These isolated complexes can then be analysed by several methods including mass spectrometry and Western blotting. Although this method has proven very successful in isolating mitotic and cytokinetic complexes, it can also be used to characterise protein complexes involved in many other cellular processes.

Key words: Protein A, affinity purification, cell division, protein complex.

1. Introduction

The complex processes of mitosis and cytokinesis require the coordinated function of several multiprotein complexes. These include microtubule organising centres (MTOCs); microtubule associated protein (MAP) complexes, including molecular motors and their cargoes, and protein complexes that enable mitotic chromosome condensation and the centromeric and kinetochore structures. Similarly, cytokinesis requires correct assembly of the late central spindle and formation of multiple annular complexes

Andrew D. McAinsh (ed.), *Mitosis: Methods and Protocols*, Methods in Molecular Biology 545,
DOI 10.1007/978-1-60327-993-2_6, © Humana Press, a part of Springer Science+Business Media, LLC 2009

of actin, myosin, and cytoskeletal proteins that mediate cell constriction and the addition of new plasma membrane. These events are orchestrated by a set of evolutionarily conserved protein complexes that mediate spatiotemporally coordinated protein phosphorylation and dephosphorylation, proteolytic degradation, and protein assembly.

The sequencing of whole genomes of humans and model organisms and the application of mass spectrometry to protein identification have opened the door to the systematic study of the protein complement, including the identification of protein interactions. Various proteomic approaches have contributed to the study of the cell cycle and cell division in model systems and diverse methods have been employed to examine protein–protein interactions in vivo, none of which being without some disadvantage *(1)*. Yeast two-hybrid is easily applied to genomewide screens and an informative two-hybrid screen has been carried out on the entire *Drosophila* transcriptome *(2)*. However, although a standardised set of procedures has been developed to screen against spurious positives, this approach is still prone to identify erroneous interactions such that other means need to be applied to confirm these in the natural situation. An alternative strategy developed in the budding yeast *S. cerevisiae* and successfully used for systematic genomewide characterisation of native protein complexes is the TAP (Tandem Affinity Purification) method. This method involves the fusion of a 'TAP' tag (see below) to target proteins and the expression of these tagged constructs into the host organism *(3, 4)*. The 'classical' TAP tag is composed of two immunoglobulin-binding domains of *Staphylococcus aureus* protein A (PtA) and a calmodulin binding peptide (CBP) separated by a cleavage site for the tobacco etch virus (TEV) protease. The isolation of the TAP-tagged bait protein and its partners is accomplished through two affinity purification steps. The first step involves the binding of the TAP-tagged complex to an immunoglobulin (IgG) resin and its subsequent elution by TEV protease digestion. The eluate is then purified a second time using calmodulin beads. For protein complex characterisation, proteins are usually fractionated by SDS-PAGE and identified by mass spectrometry, although the whole protein mixture can also be analysed directly by mass spectrometry.

Attempts to use the TAP method in *Drosophila* and mammalian cells have been somewhat promising *(5–8)*. However, in our experience this method suffers from low levels of complex recovery either due to the two-step purification procedure or in protein recovery following TEV protease treatment. These drawbacks make the TAP methodology very costly and time-consuming for large-scale protein complex purification. Furthermore, single-tag affinity purification approaches coupled with protein identification by mass spectrometry have been successfully used in yeast and in mammalian cells indicating that tandem affinity purification is not always necessary to obtain complexes of sufficient purity for meaningful protein complex analysis *(9–11)*. Therefore, we tested different tags and purification procedures in order to develop an efficient, single-step affinity purification methodology for the isolation and analysis of protein complexes in *Drosophila* cultured cells. This method utilises only two IgG binding domains of PtA for protein tagging and has proven very successful in isolating mitotic and cytokinetic complexes using more than 60 different baits that localise to diverse subcellular locations including the mitotic and central spindle, the cleavage furrow, kinetochores, and centrosomes *(12–15)*.

2. Materials

2.1. Construction of Plasmids for PtA-Tagged Protein Expression

1. BP Clonase II Enzyme Mix and LR Clonase II Enzyme Mix (Invitrogen) used exactly according to the manufacturer's recommendations
2. AccuPrime Pfx DNA Polymerase (Invitrogen) or any other DNA polymerase suitable for amplifying target cDNA sequences
3. Primers with *att*B overhangs for amplification of target cDNA sequences
4. Primers, which are to be used for sequencing of entry clones (M13F, M13R, and internal, specific for the cDNA, if necessary)
5. Gateway© donor vector(s), for example, pDONR221 (Invitrogen)
6. PCR or Gel Purification Kit (e.g., QIAquick Gel Extraction Kit from Qiagen)
7. Plasmid purification kit suitable for the isolation of high purity DNA to be used in cell transfection (e.g., Qiagen® Midi or Maxi Purification kits)

2.2. Generation of Cell Lines Stably Expressing PtA-tagged Proteins

1. Drosophila SFM (Invitrogen) or other medium for growing D-mel cells (*see* **Note 1**). D-mel cells are available from Invitrogen, Cat. No. 10831-014. Media are supplemented with L-glutamine and penicillin/streptomycin (Invitrogen or PAA Laboratories).
2. Plasticware, including 6-well plates, 25 cm^2, 75 cm^2, and 175 cm^2 flasks (with Vent Caps), cell scrapers, and disposable pipettes, is obtained from Corning or Greiner.
3. Cellfectin© Reagent (Invitrogen) for transfection of D-mel cells.

4. MG132 may be obtained from Calbiochem (10 mM In Solution MG-132, Cat. No. 474791) or Sigma (powder, Cat. No. C2211).
5. Blasticidin may be obtained from Invitrogen or PAA. A 10 mg/ml stock solution is made up in water and kept at –20°C for 6–8 weeks.
6. pCoBlast (Invitrogen) or any other vector bearing Blasticidin resistance gene suitable for the expression in insect cells.

2.3. Freezing and Thawing Cell Lines

1. DMSO (Sigma)
2. Cryo-vials for making stocks of newly established cell lines
3. Styrofoam box or slow-freezing box, for example, Cryo freezing container (Cat. No. 5100-0001, Nalgene).

2.4. Conjugation of Magnetic Beads to Rabbit IgG for PtA Affinity Purification

1. Dynabeads© M-270 Epoxy (Cat. No. 143.02D; Invitrogen Dynal AS).
2. Purified Rabbit IgG (Cat. No. 55944; ICN/CAPPEL).
3. 0.1 M sodium phosphate buffer pH 7.4 (*see* **Note 2**)
4. 3 M ammonium sulphate; 0.1 M glycine-HCl pH 2.5 (*see* **Note 2**).
5. 10 mM Tris-HCl pH 8.8 (*see* **Note 2**).
6. 0.1 M Triethylamine (*see* **Note 2**).
7. Phosphate buffered saline (PBS) (*see* **Note 2**).
8. PBS with 0.5% Triton X-100 (*see* **Note 2**).
9. Stock solution of 6% sodium azide (*see* **Note 2**).
10. Magnetic separation stands for Eppendorf tubes, 5 ml tubes, and 15 ml conical tubes can be obtained, for example, from Promega (MagneSphere Stand Cat. Nos. Z5342, Z5343 and PolyATract System Stand Cat. No. Z5410).
11. Eppendorf tubes, 15 and 50 ml conical tubes, and 5 ml snap-cap polystyrene round-bottom tubes (12 × 75 mm style).

2.5. Affinity Purification of PtA-Tagged Baits and Interacting Partners

1. Protease inhibitor mix may be made up from separately obtained inhibitors mixed in buffers to be used for extraction or washing. Alternatively one can obtain a premixed cocktail of inhibitors, for example, Complete, EDTA-free (Roche, Cat. No. 11873580001).
2. A suitable homogeniser for making animal cell extracts, for example, Power Gen 125 from Fisher Scientific.
3. Plastic disposable homogeniser tips for the homogenisation (*see* **Note 3**).
4. For DNA shearing, several different endonucleases may be used, for example, DNase I, Benzonase, or micrococcal nuclease (*see* **Note 4**).
5. Additional equipment needed during the procedure: preparative centrifuge, bench-top centrifuge, rotary shaker,

incubator set to 30°C, and vacuum centrifuge (Speed-Vac) suitable for concentration of water-based samples.

3. Methods

In order to facilitate the cloning and expression of PtA-tagged proteins, we have constructed a series of vectors (**Fig. 6.1**) based on Gateway® technology (Invitrogen). This technology provides a swift, precise, and efficient cloning system that allows the transfer of Open Reading Frames (ORFs) into multiple vectors for protein tagging and expression. Gateway technology uses the recombination sites of the bacteriophage λ to direct the precise cloning of DNA sequences of interest. The technology is extensively described at the Invitrogen website (http://www.invitrogen.com/) where free manuals are available. We therefore refer the reader to these manuals for the description of the system and for the protocols used to generate vectors based on Gateway technology.

Our vectors were designed to express N- or C-terminal PtA-tagged proteins in *Drosophila* cultured cells using either a constitutive (actin 5C) or inducible (metallothionein) promoter **Fig. 6.1**). A TEV protease site was also inserted between the PtA

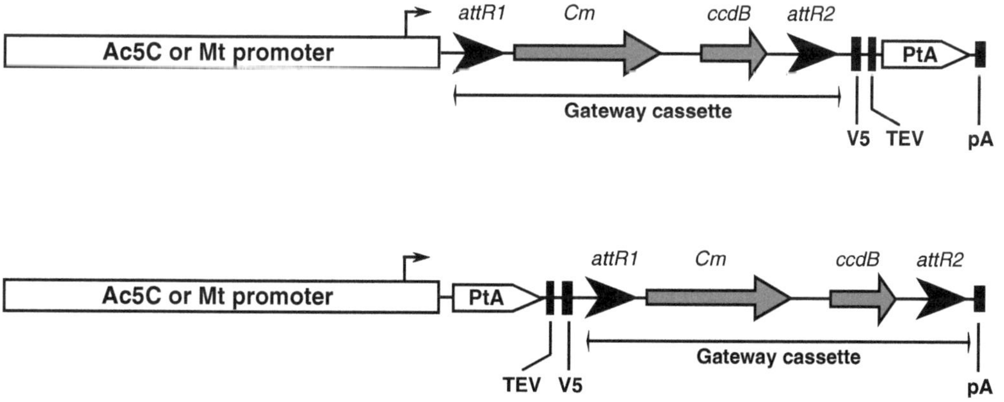

Fig. 6.1. Schematic representation of the vectors constructed to express PtA-tagged proteins in *Drosophila* cultured cells. At the top is shown a map of the vector for C-terminal PtA tagging and at the bottom the vector for N-terminal tagging. Both vectors were constructed with either a constitutive *actin5C* (Ac5) or inducible metallothionein (Mt) promoter. The vectors for C-terminal tagging contain a stop codon downstream of the PtA coding sequence and upstream of the polyadenylation sequence (pA), but lack a start (ATG) codon. Conversely, the vectors for N-terminal tagging contain a start codon and the translational start site of the *Drosophila polo* gene upstream of the PtA coding sequence, but lack a stop codon. In all vectors, a V5 tag and TEV protease site are inserted between the PtA sequence and the gateway cassette to allow elution of the native tagged protein and subsequent identification or further purification (see text). The gateway cassette contains the two *attR* recombination sites, the gene for the chloramphenicol resistance (*Cm*) and the *ccdB* toxic gene (for a full description please visit the Invitrogen web site).

tag and the Gateway cassette to allow elution of the native tagged protein. This eluted product contains a V5 tag for identification or further purification (**Fig. 6.1**).

3.1. Construction of Plasmids for PtA-Tagged Protein Expression

1. Open reading frames encoding the protein(s) of interest are amplified by PCR (Polymerase Chain Reaction) using high-fidelity polymerases and primers containing *att*B flanking sequences (see Gateway technology at http://www.invitrogen.com/). Parameters for the PCR reaction such as annealing and extension temperature and/or incubation time must be established according to the length and characteristics of the DNA template, the primer sequences, the model of the PCR machine, and the specific polymerase used. We generally use less than 30 cycles to limit the introduction of point mutations by the polymerase.
2. These PCR fragments are gel-purified and then introduced into a pDONR221 plasmid (Invitrogen) to generate 'entry vectors' using the BP recombination reaction. These clones are sequenced on both strands to check if mutations were introduced during the PCR reaction.
3. Mutation-free entry vectors are then employed in a LR recombination reaction using one of the plasmids for PtA tagging and expression described in **Fig. 6.1**.
4. The plasmids are purified using any plasmid purification kit suitable for the isolation of DNA pure enough to be used in cell transfection (Qiagen Midi/Maxi kit or similar).

3.2. Generation of Cell Lines Stably Expressing PtA-Tagged Proteins

1. Plate 3×10^6 D-mel cells per well in a 6-well plate in a final volume of 2 ml per well of medium without antibiotics (*see* **Note 5**). Let the cells adhere to bottom of the plate for a minimum of 2 h in a cell incubator at 25°C.
2. Prepare Solution A containing 5 μg plasmid DNA + 0.5 μg of 'helper' plasmid carrying a Blasticidin resistance (e.g., pCoBlast or similar) in 0.1 ml of medium without antibiotics and Solution B containing 15 μl Cellfectin in 0.1 ml of medium without antibiotics. Gently mix the two solutions by inverting the tubes 4–5 times.
3. Combine the two solutions, mix gently by flicking the tube, and incubate at room temperature for 30 min.
4. After incubation add 0.8 ml of medium without antibiotics to the DNA/Cellfectin mix.
5. Remove the 2 ml of original medium from the cells and replace it with the solution containing the DNA/Cellfectin mix. Incubate the cells overnight at 25°C.
6. The morning after remove the media containing the transfection mix and add 3 ml of fresh complete medium (with penicillin and streptomycin). Incubate at 25°C for 48 h.

7. Remove the old medium and add fresh complete medium containing 20 μg/ml Blasticidin. If the cells are in suspension, they must be centrifuged at 1000 *g* for 5 min before adding fresh medium.
8. In about 1–2 days the resistant cells should start dividing; at this point transfer the content of each well into a single 25-cm^2 flask in a final volume of 6 ml of complete medium containing 20 μg/ml Blasticidin.
9. When the cells become overconfluent, transfer them into a single 75-cm^2 flask in a final volume of 15 ml of complete medium with 20 μg/ml of Blasticidin.
10. Once the cells become overconfluent again, split the entire contents of the flask into three different 75-cm^2 flasks always in 15 ml of complete medium with 20 μg/ml of Blasticidin. At this point we usually mark the cells as passage one.
11. When the flasks become overconfluent, dilute the cells from one flask to a suitable concentration (about 2×10^6 cells/ml) in two or three 75-cm^2 flasks containing 15 ml of medium with 20 μg/ml of Blasticidin. Keep the other two flasks as backups until the new flasks are confluent again.
12. When the cells reach a concentration of about $0.8–1 \times 10^7$ cells/ml, they can be frozen in 1 ml aliquots (see below). Continue to propagate the rest of the cells at 25°C in complete medium with 20 μg/ml Blasticidin.

3.3. Freezing and Thawing Cell Lines

1. Harvest 1.5×10^8 cells and centrifuge in appropriate Falcon tubes for 5 min at 1000 g at room temperature.
2. Remove, but do not discard, the supernatant and resuspend the cell pellet in 10 ml of freezing medium composed of 10% DMSO, 45% fresh complete medium, and 45% conditioned medium (the supernatant from above).
3. Prepare 1 ml aliquots in suitable 1.8-ml cryogenic vials and freeze at –80°C in slow-freezing boxes.
4. After 24 h transfer the vials in liquid nitrogen or in a –140°C freezer for long-term storage.
5. To recover the frozen cells, gently thaw the cryogenic vial and transfer the contents into 5 ml of fresh medium in a 25-cm^2 flask. Incubate at 25°C until the cells have become adherent, but for no longer than 30 min.
6. Remove the medium without disturbing the cells and replace it with 5 ml of fresh medium.

3.4. Conjugation of Magnetic Beads to Rabbit IgG for PtA Affinity Purification

1. Resuspend an entire vial of Dynabeads M-270 Epoxy into 16 ml of 0.1 M phosphate buffer. Vortex bottle 30 s.
2. Split beads suspension into two 15-ml Falcon tubes.
3. Collect the rest of the beads in the glass vial with 1–2 ml of 0.1 M phosphate buffer. Combine with the rest.

4. Incubate for 10 min on a shaking platform or wheel at room temperature to allow potential bead aggregates to dissociate.
5. Wash briefly with 0.1 M phosphate buffer.
6. Meanwhile, prepare the Rabbit IgG. Dissolve the lyophilised powder (50 mg) in 3 ml of ultrapure H_2O (final concentration 17 mg/ml) (*see* **Note 6**).
7. Centrifuge 1 ml of Rabbit IgG solution for 10 min at maximal speed in a table-top centrifuge (around 14,000 rpm) and transfer the supernatant into a 50-ml Falcon tube. The remaining IgG solution can be stored at 4°C for several weeks.
8. Add 10 ml of 0.1 M phosphate buffer.
9. Add 5.5 ml of 3 M ammonium sulphate very slowly (drop by drop), shaking the tube (or vortexing at low speed) to avoid precipitation (*see* **Note 7**).
10. Filter the resulting solution with a syringe through a 0.2 μm filter (*see* **Note 8**). The resulting solution is the antibody mix.
11. Place the tubes containing the beads in magnetic holders. Wait until all are deposited on the side of the tube facing the magnet. Aspirate or carefully discard the supernatant while the tubes are in the magnetic holders.
12. Add 6 ml of antibody mix to each tube and vortex.
13. Rotate on a wheel at 30°C overnight to allow the conjugation reaction to proceed.
14. The next day, transfer the suspensions to 5-ml snap-cap Falcon tubes (around 3 ml per tube). Place the tubes in appropriate magnetic holders.
15. Wash once with 3 ml of 100 mM glycine-HCl (pH 2.5), vortex, and take it off promptly (*see* **Note 9**).
16. Wash once with 3 ml of 10 mM tris–HCl pH 8.8.
17. Wash once with 3 ml of fresh 100 mM triethylamine. Put it on, vortex, and take it off promptly (*see* **Note 9**).
18. Wash the coated beads 4 times, 5 min each, with PBS.
19. Wash once for 5 min with PBS + 0.5% Triton X-100.
20. Wash once 15 min with PBS + 0.5% Triton X-100.
21. Resuspend all the beads in a total volume of 2 ml of PBS + 0.02% NaN_3.
22. Store the coated beads at 4°C. The IgG-conjugated beads can be used for several weeks or even months, but use fresh for best results. See **Note 10** about reusing beads.

3.5. Affinity Purification of PtA-Tagged Baits and Interacting Partners

1. We found that the mitotic index of *Drosophila* tissue culture cells increases after treatment with the proteasome inhibitor MG132 without causing a complete bock in mitosis *(12, 14, 15)*. This results in the recovery of samples enriched in mitotic complexes (**Fig. 6.2**). Therefore in

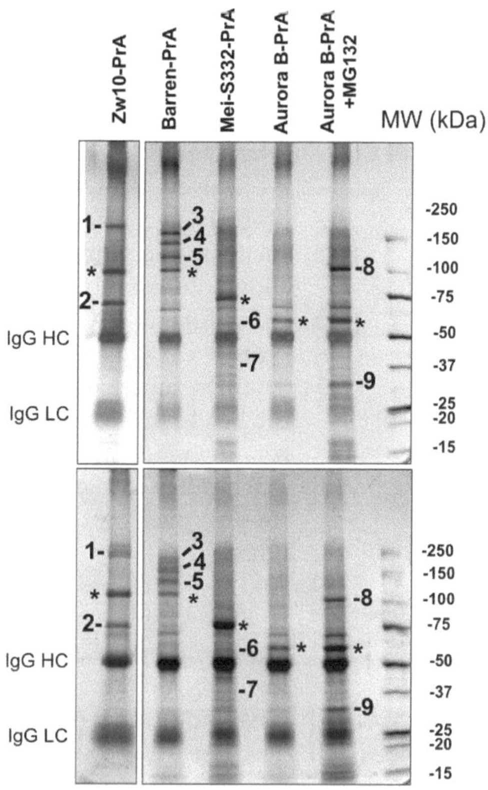

Fig. 6.2. Examples of protein complexes purified and resolved by SDS-PAGE. Fusion proteins indicated at the top of the lanes were purified from stably transfected D-Mel cell lines as described in the text. Mei-S332-PrA was expressed from the Mt promoter, and the other baits were expressed from the *actin 5C* promoter. Approximately 5×10^8 cells were used for each purification. Samples were resolved on 8–16% tris-glycine gels. The top gel was silver-stained and contained 4% of the total proteins purified. The bottom gel was Coomassie Blue-stained and contained 80% of the total proteins purified. The labelled bands were excised and proteins identified by mass spectrometry. The bands marked by an asterisk correspond to the baits. Zw10-PrA pulled down Rod (1) and Zwilch (2) (this complex is involved in regulating the spindle attachment checkpoint and in promoting cytokinesis). Barren-PrA pulled down Gluon (3), Cap-D2 (4), Cap-G (4), and SMC2 (5) (these proteins form the condensin I complex, promoting chromosome condensation in mitosis). Mei-S332-PrA (Mei-S332 promotes sister chromatid cohesion at the centromere) pulled down 3 subunits of the PP2A phosphatase complex: Pp2A-29B (6), Wdb (6), and Mts (7); Aurora B-PrA co-purified Incenp (8), and Borealin (9) (together, they form the passenger protein complex, which promotes kinetochore function and cytokinesis). In the case of Aurora B, treating the cells with a proteasome inhibitor (MG132, 25 μM, overnight) increased the yield of the complex.

most cases we treat cells with 25 μM MG132 for 4–5 h before harvesting.

2. Harvest cells by centrifugation and wash once with PBS containing protease inhibitors. Remove the supernatant. Cell pellets can be frozen in liquid nitrogen and stored at –80°C for weeks or even months.

3. Add the cold extraction buffer (EB) of choice (*see* **Note 11**). Use around 0.5 ml to 1 ml of EB per 1 × 10^8 cells (*see* **Note 12**). Resuspend by pipetting.
4. Lyse cells on ice with motorised homogeniser, set at high level, with 5 bursts of 5 s each, alternate by 10 s pauses to avoid overheating of the samples and excessive foam (*see* **Note 13**).
5. Incubate the samples at 4°C with agitation for 30 min to allow solubilisation of the cell debris (*see* **Note 14**).
6. Centrifuge at 4°C for 10–30 min at 1,000–25,000 *g* (*see* **Note 15**).
7. Transfer supernatant to a new tube (*see* **Note 16**). Add the beads, previously equilibrated in EB, and incubate on a rotating wheel at 4°C for 2–4 h (*see* **Note 17**).
8. Place tubes in a magnetic rack and wait for the beads to stick to the sides. Remove supernatant by aspiration or pouring.
9. Add 1–5 volumes of EB (relative to the original volume of EB used in step 3), vortex very briefly, and incubate on a rotating wheel for 5–10 min at 4°C. Put back on the magnetic holder and remove the supernatant. Repeat this wash 5 times.
10. The last wash solution should not contain glycerol, high salt, or high detergent (*see* **Note 18**). Leave a small volume of supernatant, resuspend beads, and transfer to a clean microfuge tube (contaminating proteins may stick to the walls of the original tube). Make sure to remove all the supernatant very carefully.
11. Add 500 μl of eluant (0.5 M NH_4OH, 0.5 mM EDTA, made fresh and filtered). Vortex and place on a rotating wheel at room temperature for 5 min. Place tube in magnetic rack; remove and transfer the eluate into a clean 1.5-ml microfuge tube. Repeat elution again and pool eluates (*see* **Note 19**).
12. Desiccate eluates using a Speedvac centrifuge (*see* **Note 20**).
13. Pellets can then be resuspended in a small volume (e.g., 25 μl) of 1X Laemmli SDS-PAGE sample buffer and resolved on gradient tris-glycine gels (8–16% gives a good resolution over a broad range of molecular mass) in a dust-free buffer and gel box.
14. Fix the gel for 20 min in 45% methanol, 1% acetic acid, and stain with Coomassie blue.
15. Bands are excised from the gel using a clean, sterile surgical blade for protein identification by mass spectrometry.

4. Notes

1. Drosophila Serum Free Medium (SFM, Invitrogen) is optimal for the growth of D-mel cells. However, Invitrogen has recently stopped producing this medium and other media can be used as replacements, such as Express Five SFM (Invitrogen) or HyQ SFX – INSECT (HyClone).
2. Information about the procedure and recipes for solutions used for the conjugation can be found in the Dynabeads information sheet.
3. Metal tips can also be used, but then they should be thoroughly washed before and after homogenisation of each batch of cells.
4. If DNase-I needs to be used (e.g., Cat. No. DNEP-5MG, Sigma), about 2000–3000 Kunitz units should be added to the extract made from 0.5 to 1.0×10^9 cells.
5. Unless otherwise stated all cell manipulations are carried out in a flow tissue culture hood.
6. It is crucial to dissolve the antibody thoroughly by pipetting up and down for a few minutes.
7. The salt helps the conjugation reaction, but local increases in salt concentration can lead to precipitation of the antibody.
8. This is necessary to remove any potential precipitates that may be invisible.
9. Leaving the beads in high or low pH for a long time may potentially lead to the dissociation of some IgG or IgG chains from the beads (although this has not been tested).
10. IgG-conjugated beads can generally be washed and reused (see protocols associated with the Dynabeads M-270 Epoxy). However, their binding capacity may be reduced and traces of proteins from previous purifications may remain, which is a major problem when purified proteins are identified by highly sensitive mass spectrometry.
11. The Extraction Buffer (EB) should be tailored for the protein or complex to be purified, either by experience or from empirical tests. The following EB has medium stringency/harshness and gave good results with a number of baits: 50 mM K-HEPES (or Na-HEPES) pH 7.5, 100 mM potassium acetate, 50 mM KCl, 2 mM $MgCl_2$, 0.1% NP-40, 5 mM DTT, 2 mM EGTA, 5% glycerol, Complete protease inhibitors (Roche). The extraction of membrane proteins requires a higher concentration of detergent. The extraction of chromatin-bound proteins will require higher salt concentration and/or addition of DNase. The extraction of spindle proteins may require higher salt

concentration. Very stringent extraction conditions often disrupt protein interactions and therefore a compromise needs to be reached. It is impossible to know what the best conditions are until tests are conducted. Small-scale tests can be performed using 1×10^8 cells in 1 ml of EB and analyzed on a Coomassie- or silver-stained gel. Because contaminants are generally purified using different baits, it may help to conduct simultaneous test purifications with at least two baits that don't assemble in the same complex.

12. For test purifications, small-scale experiments can be performed using 1×10^8 cells/samples. For mass spectrometry preparative purifications, about $1–2 \times 10^9$ cells/sample (depending on the levels of expression) are usually sufficient.
13. The lysis can be visualised under a microscope. Nuclei often do not solubilise in low-detergent buffers and the addition of detergents, salts, and/or DNase can be used to increase their solubilisation.
14. If the bait is likely to be associated with chromatin, DNase should be added (0.2 mg/ml). Samples should be first incubated at either room temperature or 37°C for 1–5 min to activate the DNase, followed by incubation at 4°C for 30 min (step 5). The incubation at 37°C should be avoided if the bait is easily degraded.
15. Use a table-top centrifuge for small samples in microfuge tubes and a SS-34 rotor for large volumes. The speed and duration of the centrifugation should be optimised for each purification. A long (30 min) and fast (25,000 g) centrifugation will result in a clear supernatant and may yield very clean purifications. However, in these conditions large complexes may pellet and therefore a shorter or slower centrifugation should be used instead.
16. A small volume of the supernatant and resuspended pellet can be taken for Western blot analysis. A side-by-side comparison of the pellet and supernatant will provide information about the amount of bait extracted.
17. The amount of beads required depends on the abundance of the PtA-tagged protein expressed. When expressing under the control of the *actin 5C* promoter, 50 μl of bead suspension are usually sufficient for 1×10^8 cells in 1 ml of EB. For larger-scale purifications, use a lower ratio of beads/cells. Roughly 200 μl of bead suspension for 1×10^9 cells in 10 ml may be used. The optimal incubation time may vary and is generally a compromise between maximising the binding of the bait to the beads and minimising the spontaneous disassociation of the protein complex being purified.

18. Make sure to use a buffer without glycerol to avoid interference with the high-pH elution. Glycerol, high salt, or high detergent should also be avoided in the last wash solution because they can be eluted along with the proteins and interfere with sample preparation for mass spectrometry.
19. If a few beads are still present in the eluate, put the tubes in a magnetic rack and transfer the supernatant to a new tube.
20. Do not desiccate completely when preparing samples for gel-free mass spectrometry analysis but only reduce the volume to 100 μl. Add 1 ml of ultrapure H_2o and again reduce the volume to 100 μl. Again add 1 ml of H_2o and reduce to 100 μl. This treatment is necessary to allow evaporation of the ammonia. Store at –20°C until ready for mass spectrometry. Some mass spectrometry users or facilities may prefer to start with a pellet, which can be obtained using a standard methanol protein precipitation procedure.

Acknowledgments

This work was supported by the BBSRC and the CR-UK. VA was supported by the EMBO and the HFSP.

References

1. Archambault V. (2005) Cell cycle: proteomics gives it a spin. *Expert Rev Proteomics* **2**, 615–625.
2. Giot L., Bader J.S., Brouwer C., Chaudhuri A., Kuang B., Li Y., Hao Y.L. et al. (2003) A protein interaction map of Drosophila melanogaster. *Science* **302**, 1727–1736.
3. Gavin A.C., Bosche M., Krause R., Grandi P., Marzioch M., Bauer A. et al. (2002) Functional organization of the yeast proteome by systematic analysis of protein complexes. *Nature* **415**, 141–147.
4. Puig O., Caspary F., Rigaut G., Rutz B., Bouveret E., Bragado-Nilsson E., Wilm M. and Seraphin B. (2001) The tandem affinity purification (TAP) method: a general procedure of protein complex purification. *Methods* **24**, 218–229.
5. Yang P., Sampson H.M. and Krause H.M. (2006) A modified tandem affinity purification strategy identifies cofactors of the *Drosophila* nuclear receptor dHNF4. *Proteomics* **6**, 927–935.
6. Veraksa A., Bauer A. and Artavanis-Tsakonas S. (2005) Analyzing protein complexes in *Drosophila* with tandem affinity purification-mass spectrometry. *Dev Dyn* **232**, 827–834.
7. Forler D., Kocher T., Rode M., Gentzel M., Izaurralde E. and Wilm M. (2003) An efficient protein complex purification method for functional proteomics in higher eukaryotes. *Nat Biotechnol* **21**, 89–92.
8. Knuesel M., Wan Y., Xiao Z., Holinger E., Lowe N., Wang W. and Liu X. (2003) Identification of novel protein-protein interactions using a versatile Mammalian tandem affinity purification expression system. *Mol Cell Proteomics* **2**, 1225–1233.
9. Gingras A.C., Gstaiger M., Raught B. and Aebersold R. (2007) Analysis of protein complexes using mass spectrometry. *Nat Rev Mol Cell Biol* **8**, 645–654.
10. Archambault V., Chang E.J., Drapkin B.J., Cross F.R., Chait B.T. and Rout M.P. (2004) Targeted proteomic study of the cyclin-Cdk module. *Mol Cell* **14**, 699–711.
11. Ho Y., Gruhler A., Heilbut A., Bader G.D., Moore L., Adams S.L. et al. (2002) Systematic identification of protein complexes in Saccharomyces cerevisiae by mass spectrometry. *Nature* **415**, 180–183.

12. D'Avino P.P., Takeda T., Capalbo L., Zhang W., Lilley K., Laue E. and Glover D.M. (2008) Interaction between Anillin and RacGAP50C connects the actomyosin contractile ring with spindle microtubules at the cell division site. *J Cell Sci* **121**, 1151–1158.
13. Chen F., Archambault V., Kar A., Lio P., D'Avino P.P., Sinka R., Lilley K., Laue E.D., Deak P., Capalbo L. and Glover D.M. (2007) Multiple protein phosphatases are required for mitosis in Drosophila. *Curr Biol* **17**, 293–303.
14. D'Avino P.P., Archambault V., Przewloka M.R., Zhang W., Lilley K.S., Laue E. and Glover D.M. (2007) Recruitment of Polo kinase to the spindle midzone during cytokinesis requires the Feo/Klp3A complex. *PLoS ONE* **2**, e572.
15. Przewloka M.R., Zhang W., Costa P., Archambault V., D'Avino P.P., Lilley K.S., Laue E.D., McAinsh A.D. and Glover D.M. (2007) Molecular analysis of core kinetochore composition and assembly in *Drosophila Melanogaster*. *PLoS ONE* **2**, e478.

Chapter 7

Automated Live Microscopy to Study Mitotic Gene Function in Fluorescent Reporter Cell Lines

Michael H.A. Schmitz and Daniel W. Gerlich

Abstract

Fluorescence live microscopy is a powerful technique to study complex cellular dynamics such as cell division. The availability of fluorescent markers based on GFP fusion proteins for virtually any cellular structure allows efficient visualization of specific processes, and the combination of different fluorophores can be used to study their coordination. In this chapter, we present methods for automated live cell microscopy to study mitotic gene function systematically and in high throughput. In particular, we provide protocols for efficient generation of fluorescent reporter cell lines stably expressing combinations of cellular markers, and provide detailed guidelines for optimizing imaging protocols for automated long-term live microscopy.

Key words: Automated live-cell imaging, multidimensional microscopy, laser scanning confocal microscopy, widefield epifluorescence microscopy, mitosis, Green Fluorescent Protein (GFP), stably expressing reporter cell lines.

1. Introduction

Cell division is a highly dynamic process that is difficult to study by traditional fixed cell or biochemical approaches alone. The inherent stochastic nature of mitotic events such as chromosome capture by spindle microtubules often makes it difficult to draw definitive conclusions about specific mitotic transitions from fixed cell approaches. In addition, the fixation process can perturb cell morphology. To overcome these limitations, multidimensional live cell imaging has become an important tool *(1, 2)*. The availability of fully motorized microscopes enables automated imaging

Andrew D. McAinsh (ed.), *Mitosis: Methods and Protocols*, Methods in Molecular Biology *545*,
DOI 10.1007/978-1-60327-993-2_7, © Humana Press, a part of Springer Science+Business Media, LLC 2009

protocols that allow for systematic studies with high experimental reproducibility. Fluorescent reporter cell lines stably expressing combinations of cellular markers provide a powerful tool for high-content assays of mitotic gene function, for example, in RNA interference-based screening. Here, we provide detailed guidelines for the generation of double-stable fluorescent reporter cell lines and for automated live cell experiments using widefield or confocal microscopes.

2. Materials

2.1. Generating Stable Cell Lines

1. Cell line(s) of interest (wild-type strain)
2. pIRES vectors (Clontech)
3. FuGENE6 (Roche Applied Science)
4. Optimem (Gibco, Invitrogen)
5. Column purified DNA, dissolved in H_2O
6. Multiwell plates (6-, 12-, 24-, 48-, and 96-well; Nalgene, Nunc)
7. 10 cm and 15 cm cell culture dishes
8. Sterile Phosphate Buffered Saline (PBS)
9. Cell culture medium, for example, D-MEM
10. 0.5 mg/mL Genticin/G418 (Gibco, Invitrogen)
11. 0.5 μg/mL Puromycin Dihydrochloride (Calbiochem)
12. 5 μg/mL Puromycin Dihydrochloride (for hTERT-RPE1 cells)

2.2. Live Cell Imaging

1. Laser scanning confocal or widefield microscope (e.g., Zeiss LSM510 confocal, Molecular Devices ImageExpress-Micro).
2. Microscope incubation chamber providing constant temperature, and CO_2 (e.g., EMBL-EM, technology transfer, Germany/Light Imaging Services, Switzerland).
3. Digital thermometer.
4. Piezoelectric focus device (e.g., Piezosystem Jena Inc.).
5. High-speed shutter and filter wheels (e.g., Uniblitz, widefield epifluorescence microscope only).
6. Neutral density filters.
7. Plastic beaker (fitting into the microscope incubation chamber).
8. Piece of copper (to place into plastic beaker).
9. Chambered coverslips (e.g., LabTek, Nalgene Nunc; ibidi slides, integrated BioDiagnostics, and *see* Table 19.3 in *(3)*).
10. Imaging chamber stage adaptor(s) if needed (custom built).
11. Parafilm.

12. Imaging medium. Ideally with the same formulation as medium used for cell culture, but without phenolred and without riboflavin.
13. Silicon grease (only if CO_2-independent media are used) + 15 mL plastic syringe.

3. Methods

Using live imaging for high-throughput applications depends on three crucial requirements. First, reporter cell lines stably expressing fluorescently tagged markers ensure optimal reproducibility within a cell population. This becomes particularly important if assays rely on additional perturbation steps such as siRNA transfection. Second, the microscope has to be equipped with an environmental control for unperturbed cell proliferation. Third, imaging conditions need to be established that do not compromise cell viability and at the same time provide sufficient image quality. Finally, optimizing microscope software and hardware configurations can enhance throughput and spatiotemporal resolution.

3.1. Cell Lines for Live Cell Imaging

Selecting a cell type for live cell imaging often is a compromise between experimental amenability and potential degeneration from a wild-type somatic cell. Human cancer cell lines are particularly robust for establishing stable cell lines, and for performing imaging and RNAi experiments. However, they contain a high number of chromosomal aberrations and lack some features of cell cycle control. Therefore, conclusions obtained from cancer cells are not always valid for normal somatic cells. Diploid noncancer cell lines have recently become available by immortalization through telomerase overexpression *(4)*. Although these cell lines are also suitable for live imaging approaches and RNAi, we found it much more difficult to obtain cell lines stably expressing fluorescent markers. Our lab has decided to use a strain of the human cervical carcinoma cell line termed HeLa "Kyoto" (obtained from S. Narumiya, Kyoto University) as an experimental standard, and in addition the widely used retinal pigment epithelial cell line, which was immortalized by telomerase overexpression (hTERT1-RPE1, Clontech) as a noncancer cell line to validate observations from HeLa experiments.

The HeLa Kyoto cell line has several features that are particularly useful for live cell imaging:

1. It is used as a standardized background in several systematic studies, including genomewide screening *(5)*, proteomics, and GFP-localization.
2. Grows adherently on most glass and plastic surfaces and has low cell motility.

3. High efficiency for stable transfection.
4. Efficient RNAi.
5. Short cell cycle duration.
6. Relative insensitivity to light exposure. Compatible with long-term fluorescent imaging up to several days.

3.2. Generating Stable Reporter Cell Lines

It is essential for systematic imaging-based studies to have a reliable and reproducible labeling of the structure of interest. In our experience, transient transfection of fluorescently tagged marker proteins is limited by high variability in expression levels and intracellular localization, which can correlate with different degrees of potential overexpression toxicity. This is particularly critical in experiments that make use of additional perturbations such as siRNA transfection. We therefore generally use monoclonal reporter cell lines that stably express fluorescent markers. In many applications it is important to have a reliable reference marker that reports on the cell cycle state. As an informative marker for mitotic stages, we use fluorescent core histone 2b *(6)*. If this particularly bright marker is tagged in red, a co-expressed green marker can be used to study the coordination of a process of interest with mitotic chromosome dynamics (**Fig. 7.1A** and **B**). **Table 7.1** provides an overview of common markers for subcellular structures. We note that red/green fluorophore combinations are preferable above cyan/yellow, because this reduces toxic short wavelength illumination. To minimize artifacts by interfluorophore dimerization, it is important to use truly monomeric fluorescent proteins. For EGFP derivatives, a single point mutation reduces weak dimerizing activity, making this mEGFP variant the standard marker in our lab *(7)*. For red fluorophores, mCherry *(8)* provides a particularly bright signal, but to our experience has a higher tendency to induce cellular artifacts. We mostly use mRFP *(9)* as a red marker. Because mRFP is less bright than mEGFP, we typically visualize the more prominent cellular structure (e.g., H2B) with the red tag. A comprehensive list of available fluorophores is available at (www.fluorophore.org).

Generation of stable cell lines takes advantage of the finding that transiently transfected plasmids can integrate into the genome and become stably propagated. To select for these rare events a resistance marker is commonly placed on the expression plasmid for the gene of interest. However, selection against the resistance marker will frequently lead to selection of cells which only integrated this marker into the genome, and not the gene of interest. This problem is reduced to a minimum when the resistance gene and the gene of interest are expressed from a bicistronic transcript. We therefore strongly recommend bicistronic pIRES expression vectors (Clontech). Our experience showed that this increases efficiency of clone generation more than twentyfold (typically 95–99% of resistant colonies also

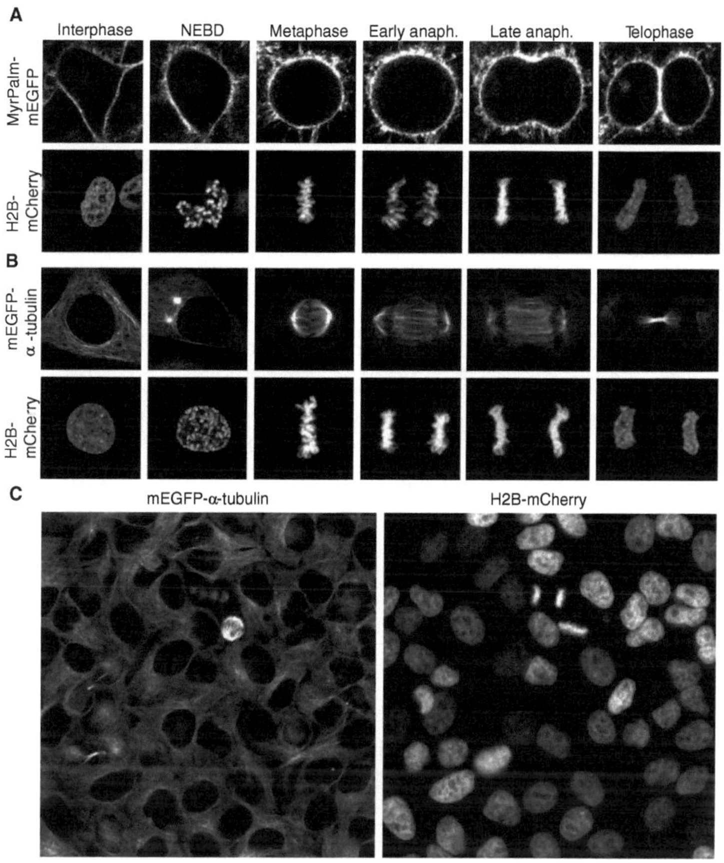

Fig. 7.1. Example fluorescent reporter cell lines expressing combinations of tagged proteins. (**A**) HeLa cell expressing a chromatin marker (H2B-mCherry) and a plasma membrane marker (MyrPalm-mEGFP). (**B**) HeLa cell expressing H2B-mCherry and mEGFP-α-tubulin. Both cell lines express H2B (mCherry) as a reference marker for mitotic progression. (**C**) Field of cells expressing the markers shown in (B).

positive for the gene of interest), and protein expression levels within a stably expressing clone are often more homogeneous (**Fig. 7.1c**).

The commonly used CMV promoter allows expression of the tagged gene at high levels, which yields bright fluorescence, but may be toxic for some marker proteins. However, the clonal expression level in stably transfected cells is typically quite variable, which allows selection for the right expression level after isolating the colonies. If lower expression levels are needed, the promoter can also be exchanged with a weaker expressing, a tetracycline inducible, or the protein's endogenous promoter.

3.2.1. Transfection of Cells

The following protocol allows the generation of stable HeLa and RPE reporter cell lines, using pIRES vectors with

Table 7.1
Collection of standard markers for subcellular structures

Structure/Localization	Tagged Gene	Reference
– Microtubules	– alpha-Tubulin	(Clontech)
– Microtubule tips	– EB3	*(10)*
– Centrioles	– Centrin	*(11)*
– Centromeres	– CenpA	*(12)*
– Actin	– Actin	(Clontech)
– Plasma membrane	– MyrPalm (myristoylated and palmitoylated)	*(13)*
– DNA replication factories	– PCNA	*(14)*
– Golgi	– Galactosyltransferase (GalT)	*(15)*
– Endoplasmic reticulum	– KDEL-sequence	*(16)*
– Mitochondria	– mt	*(17)*
– Nuclear envelope	– Lap2beta	*(18)*

Geneticin/G418 or Puromycin resistance markers and FuGENE6 (Roche Applied Science) as transfection reagent.

1. Subclone the fluorescently tagged gene of interest into a pIRES vector backbone.
2. Plate 1×10^5 cells/well into a 6-well dish 24 h prior to transfection. Culture in standard tissue culture medium (e.g., D-MEM supplemented with 10% fetal bovine serum (FBS) and 1% penicillin/streptomycin).
3. Dilute 3 μL FuGENE6 (Roche Applied Science) in 47 μL serum-free media (e.g., Optimem, Invitrogen); mix by pipetting.
4. Dilute 1.5 μg DNA in a total volume of 50 μL serum-free media; mix by pipetting.
5. Incubate both solutions for 5 min. Add the diluted FuGENE6 to the diluted DNA solution. Mix by pipetting and incubate for at least 30 min at room temperature.
6. Add the FuGENE6/DNA mix dropwise to one well of the 6-well dish.
7. Culture cells for 48 h to allow expression of the resistance marker.
8. Trypsinize cells; resuspend vigorously with 1 mL pre-warmed D-MEM to avoid cell clumping. Seed 1 × 100 μL and 1 × 900 μL cell suspension into two separate 15 cm dishes containing 15 mL selective media (*see* **Note 2**). In case too many colonies grow (visible after about 14 days), the second plate can be discarded later. Typically,

a total of 30–500 colonies per 15 cm dish is optimal. For HeLa Kyoto we use 0.5 mg/mL Genticin/G418 (Gibco, Invitrogen) and/or 0.5 μg/mL Puromycin Dihydrochloride (Calbiochem). For hTERT-RPE1 we use 0.5 mg/mL Geneticin/G418 and/or 5.0 μg/mL Puromycin.

9. Exchange the selection medium without further trypsinization every 1–2 days to remove dead cells and to maintain sufficient levels of selection pressure.
10. After 2–3 weeks, single colonies will appear as small "white dots" on the surface of the dish with a diameter of 2–3 mm (depending on cell type).
11. *See* **Note 1** if too many or no colonies are observed.

3.2.2. Isolating Colonies

It is critical to isolate colonies 3–4 weeks after seeding, when they appear as big white dots. Colonies should still be clearly separated from each other to avoid mixing with cells from adjacent colonies. On the other hand, harvesting cells from too small colonies might impair survival after transfer into the multiwell dish. Colonies can be seen best when the medium is removed and the dish tilted under the light of the cell culture hood. In cases of very bright fluorescent markers, positive colonies can already be detected with a standard inverted epifluorescence microscope through the plastic of the cell culture dish. However, most markers can only be detected with high NA objectives that are incompatible with imaging through thick plastic. In most cases, we recommend picking of about 10–20 colonies, with subsequent validation for expression and correct localization of a fluorescent marker on a high-resolution microscope. Be aware that some markers are expressed at much lower levels in stable cell lines as compared to transient transfections. In contrast to HeLa Kyoto, hTERT-RPE1 cells have a higher cell motility which results in larger and less confined colonies. Such colonies are more difficult to spot by eye and the dishes have to be checked more regularly to find the best time of picking colonies (*see* **Note 3**). Avoid mixing of individual clones to establish a purely monoclonal cell line. For this, always use new tips for each clone isolation and subsequent cell expansion step.

1. Remove medium of the 15 cm dish containing colonies.
2. Wash the cells 2× with sterile phosphate buffered saline (PBS).
3. Remove the PBS.
4. Quickly mark 12–24 colonies (select different sizes of colonies) with permanent ink on the outside bottom surface of the dish.
5. Immediately add 20 mL PBS to the 15 cm dish to prevent drying out.

6. Preload 100 μL of prewarmed (37°C) trypsin into 12–24 wells of a 48-well plate.
7. Use a 100 μL Gilson pipette set to 100 μL. Touch with the tip the marked colony and at the same time aspirate very slowly 100 μL of PBS/cells. Try to aspirate the whole colony by scratching the tip slightly around the marked region while aspirating (*see* **Note 3**).
8. Transfer the 100 μL cell/PBS solution to the 48-multiwell dish.
9. Resuspend by pipetting and incubate 5 min to allow the cells to detach from each other.
10. Add 300 μL of prewarmed D-MEM containing selection drug and resuspend again by pipetting.
11. Incubate for 24 h to allow cells to reattach.
12. Exchange the medium with fresh D-MEM containing selection drug (critical for cell survival).
13. When the cells have reached 80% confluence, transfer to a 24-well plate.
14. When the cells have again reached 80% confluence, transfer 25% of the cells into an 8-well chambered coverslip (LabTek, Nalgene, Nunc) to test expression and correct localization of fluorescent marker. Plate the remaining 75% of cells to a 12-well plate for further expansion.
15. Identify positive clones by immunofluorescence microscopy (for details: see also **Section 3.2.3**).
16. Further expand several positive clones to larger cell culture dishes and finally freeze multiple aliquots per clone.
    See **Note 3** for trouble-shooting.

3.2.3. Clone Characterization

Once expanded, clones should be carefully re-examined to ascertain normal proliferation and correct localization of the marker.

- Measure the mitotic index of the different clones; it should be comparable to the parental cell line.
- Measure population doubling times.
- Determine protein localization in live and/or fixed cells.
- To validate full-length expression, perform a Western blot. Thereby, the overexpression levels of the reporter gene relative to the endogenous protein can be determined.

3.2.4. Generating Double-Stable Reporter Cell Lines

For the generation of double-stable reporter cell lines, follow the protocols outlined above two times, but use a different resistance marker of the pIRES vector in the second round of transfection (**Fig. 7.1**). If one of the reporter genes is used as a reference marker in many cell lines (e.g., histone 2B for chromosome labeling as mitotic reference marker), generate this cell line first. If the integration of the second marker proves difficult, the second reporter gene should first be transfected and thereafter the reference marker. Typically, smaller tagged proteins are easier to express stably.

Cells should always be cultured in the presence of the respective selection drugs to prevent loss of the marker. However, we recommend using nonselective medium for RNAi transfection assays.

3.3. Imaging Chambers for Live Cell Microscopy

Chambered coverslips are well suited to image cells under physiological conditions on an inverted microscope. The advantage of these chambers is that the cells can be seeded directly into the imaging chamber, which can be mounted straight to the microscope stage thereby maintaining sterility. Chambered coverslips are commercially available in different shapes and sizes, in single or multiwell formats, and as open or closed systems (*see* Table 19.3 in *(3)*). Imaging chambers should have a glass bottom thickness of ~170 μM to match the optical correction of most microscope objectives. If cells attach weakly to glass, use poly-L-lysine coated dishes. Choosing the right imaging chamber not only depends on the assay to be performed (e.g., single or multiwell) but also on the stage insert of the microscope that holds the specimen. Some microscopes may require custom-built stage adaptors (**Fig. 7.2**).

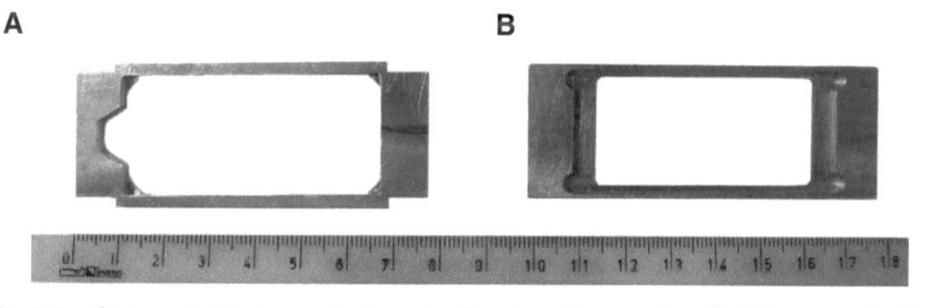

Fig. 7.2. Custom-built stage adaptors for chambered coverslips. (**A**) Stage adaptor for LabTekII. Note that the inner rims holding the LabTekII chamber have been minimized. (**B**) Stage adaptor for LabTekI. Both adaptors have standard coverslip outside dimensions to fit with most motorized microscope stages.

3.4. Microscope Environmental Control and Imaging Media

Appropriate incubation is absolutely critical for reproducible long-term live imaging experiments. Several aspects have to be considered, including a stable temperature, optimal pH, and minimized evaporation of the imaging media. Small changes in temperature, pH, or salt concentration can have severe effects on cell proliferation and viability. It is important to assess each of the above parameters independently. In addition, temperature changes outside the imaging chamber lead to focus drifts, such that it is essential to maintain a constant room temperature for long-term imaging. An optimal solution for a stable physiological environment is an incubator box around the stage and large parts of the microscope stand. Such systems are commercially available (EMBL-EM, technology transfer, Germany/Light Imaging

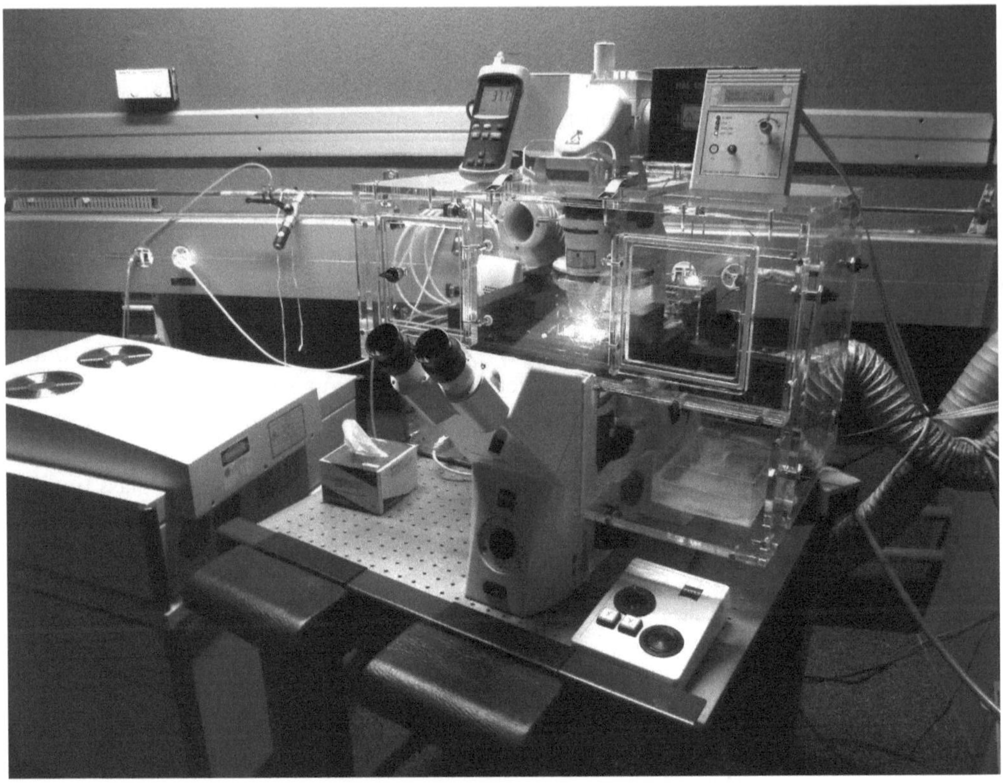

Fig. 7.3. Zeiss LSM510 confocal microscope equipped with incubation box (EMBL-EM, technology transfer, Germany). Temperature and CO_2 can be individually controlled. Humidity is provided by a water-filled plastic beaker.

Services, Switzerland) and are in general custom-built to the specific requirements of the microscope. These incubator boxes provide temperature, CO_2, and humidity control (**Fig. 7.3**).

3.4.1. Temperature

The major advantage of incubator boxes is that large parts of the microscope—including the stage and the objective revolver—are equilibrated to the same temperature as the specimen. This assures unperturbed cell proliferation and at the same time minimizes focus drifts. It is critical to heat the incubation box to the required temperature several hours before the experiment to allow full temperature equilibration of the whole microscope. Temperature deviations of a few degrees can dramatically delay mitotic entry and progression *(19, 20)*. If the microscope is not located close to the tissue culture room, the imaging chamber should be transferred on a prewarmed heat block (37°) to the microscope stage to avoid cooling of the cells which can already induce delays in mitotic entry by up to several hours. We further recommend permanently maintaining the incubator box at 37°C even when the microscope is not in use, and adjusting room heating or air conditioning to constant settings during day and night time. This is essential to prevent focus drifts during long-term experiments. If this is not possible, focus drifts need to be compensated by autofocus methods (*see* **Section 3.5.4**).

3.4.2. pH and Imaging Media

Most cell types are cultured optimally in a sodium bicarbonate-, CO_2-dependent medium. Even though CO_2-independent cell culture media can be used for live imaging with some cell types, our experience showed that CO_2-buffered media provide more reproducible results in long-term imaging experiments. Most culture media contain phenol red as a pH indicator. Although this can be helpful to determine proper function of the microscope incubator CO_2 control, it can cause perturbing background fluorescence (**Fig. 7.4**). This is particularly problematic when using

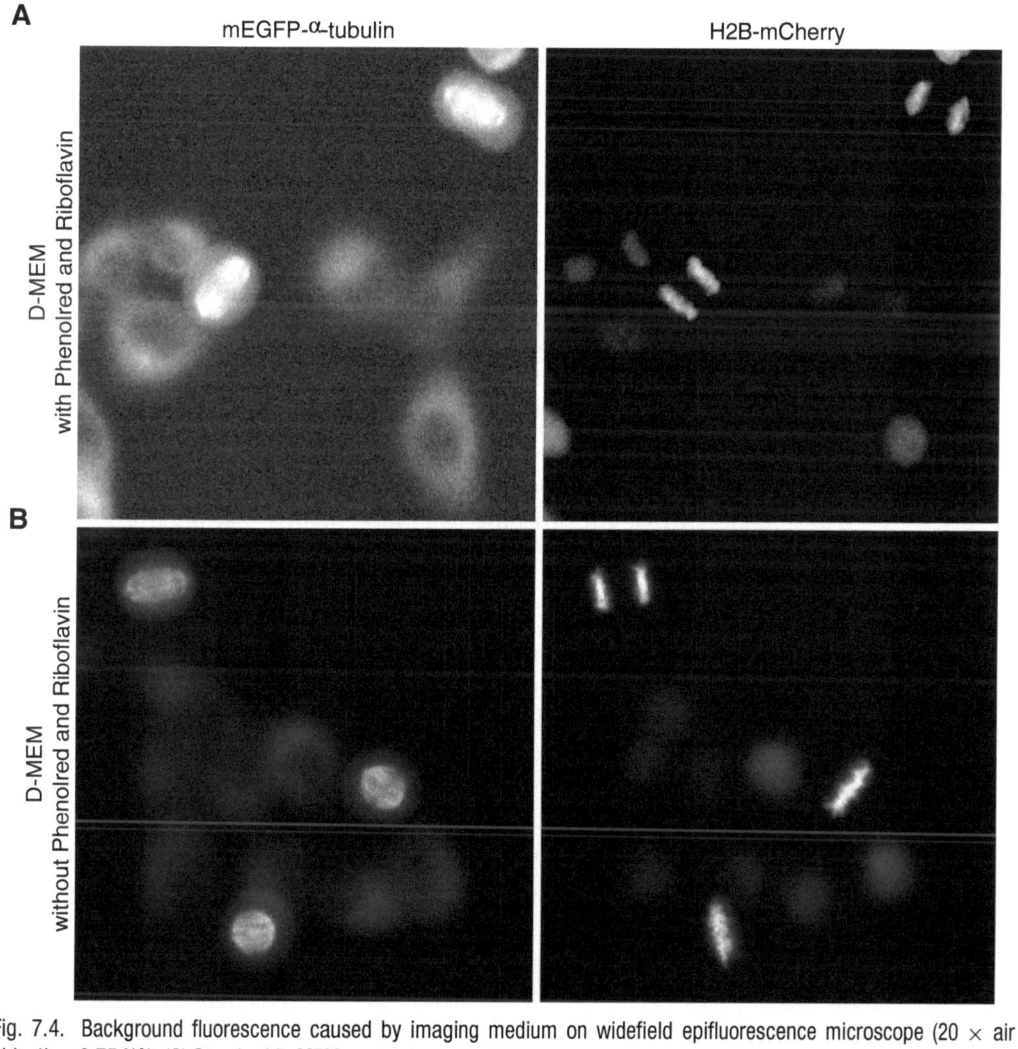

Fig. 7.4. Background fluorescence caused by imaging medium on widefield epifluorescence microscope (20 × air objective, 0.75 NA). (**A**) Standard D-MEM containing 10% FBS, phenolred, and riboflavin causes strong background fluorescence in the EGFP channel. (**B**) D-MEM containing 10% FBS, but without phenolred and without riboflavin minimizes background fluorescence in the EGFP channel. (**A**) and (**B**) were recorded and contrast adjusted with identical settings.

widefield epifluorescence microscopes and excitation wavelengths of 400–500 nm. Another source of high background fluorescence is riboflavin, particularly in widefield applications for GFP imaging. We therefore use D-MEM medium without phenol red and without riboflavin (custom order by Gibco, Invitrogen).

However, background fluorescence from the imaging medium is much lower on confocal microscopes, and it is therefore not as important to use a medium without phenolred and riboflavin as on widefield systems. In any case, the imaging medium used for microscopy should be first tested by growing cells in the chambered coverslip in a tissue culture incubator, and also inside the stage incubator of the microscope without imaging. Compare imaging media and standard tissue culture media in parallel (e.g., by using a multiwell chambered LabTek), and check if mitotic index and doubling rates of the cells are similar. Always prewarm the imaging media to 37°C before replacing the standard medium in the imaging chamber to avoid mitotic entry delays.

3.4.3. Evaporation of Culture Medium

Evaporation of the medium in the imaging chambers during long-term experiments (>2 h) can lead to significantly increased salt concentrations, and thereby may perturb cell viability. Filling the imaging chambers with as much imaging medium as possible minimizes these effects (*see* **Note 4**). Furthermore, the microscope incubator box should be humidified, for example, by placing a large beaker filled with water into the incubator box (a copper plate on the bottom of the beaker prevents growth of fungi and bacteria). Alternatively, the imaging chamber may be sealed. This is achieved by smearing silicon grease on the outside rim of the imaging chamber before placing the lid on it (*see* **Note 5**) or by adding a layer of mineral oil on top of the imaging media *(21)*. It is important that sealing should only be applied if CO_2-independent media are used. If the particular assay requires drug addition during the time-course of imaging, the plastic lid of the imaging chamber should be replaced with a sheet of parafilm. This reduces evaporation, but facilitates opening without mistakenly displacing the chamber to different stage positions.

3.5. Optimizing Image Acquisition

Key to successful live cell imaging experiments is an optimal balance between image quality and cell viability. In the following, we discuss guidelines towards this goal.

3.5.1. Confocal Versus Widefield Imaging

Widefield epifluorescence microscopes and confocal laser scanning microscopes are two robust and widely used imaging systems, which have specific advantages for different assay types. The following is a list of features that we consider important for an automated live cell microscopy setup:

- Full motorization of the microscope, and software support for three-dimensional multicolor time-lapse imaging.
- Software and hardware running stable for several days of continuous time-lapse acquisition at maximum speed.
- Compatibility with environmental incubator boxes around the microscope (*see* **Section 3.4**).
- Option for fast hardware (laser-based) autofocus (*see* **Section 3.5.4**), including piezoelectric focus or stage drive.
- Sturdy and fast shutter (widefield systems) that tolerates both high temperatures from constant illumination (mercury or xenon arc lamps), and high shutter rates over several days.
- Robust and fast filter wheels for multicolor imaging.
- Stable light source for quantitative analysis of fluorescent protein intensities.
- Cooled CCD camera with high quantum efficiency and low read-out noise (widefield systems).
- Software support for implementation of imaging macros and loading of custom-designed stage position lists (e.g., for screening applications).
- Support for direct image storage to hard drive (as opposed to RAM memory) is essential, as data volumes can easily exceed 100 GByte/day.
- Open image formats for automated data analysis. Ability to directly acquire or convert images into 8-bit format to reduce hard disk space.

The choice between laser scanning confocal or widefield microscopes depends on the assays to be performed. Laser scanning microscopes provide higher resolution, particularly along the optical axis, without the need of image postprocessing. They offer higher flexibility, for example, for photobleaching or photoactivation assays (reviewed in *(22)*). On the other hand, widefield systems are significantly faster and cheaper, and high image resolution can be achieved by deconvolution procedures (reviewed in *(23)*). **Table 7.2** summarizes the advantages and limitations of these two imaging systems.

3.5.2. Phototoxicity

Exposure to light can severely perturb cell viability and progression through cell division. Particularly short wavelength illumination, such as needed for excitation of cyan fluorophores can cause high levels of phototoxicity. For this reason, we prefer fluorophore combinations that can be excited with longer wavelengths, for example, EGFP/mRFP instead of CFP/YFP. Photosensitivity frequently is dependent on the cell cycle stage; for example, high light exposure typically causes prometaphase/metaphase arrests, or prevents entry into mitosis.

Before addressing potential phototoxic effects, it is important to verify that cell viability is not impaired by the imaging medium or incorrect incubation (*see* **Section 3.4**).

Table 7.2
Comparison of confocal laser scanning and widefield epifluorescence microscopes for automated live cell imaging

Confocal Laser Scanning Microscope	Widefield Fluorescence Microscope
+ Flexibility to study protein dynamics by bleaching/activation assays (e.g., FRAP, FLIP, FRET, PA assays).	+ Lower costs than confocal systems.
+ Reduced out-of-focus fluorescence.	+ Fast acquisition when filter wheels and shutters are optimized.
+ Reflection-based autofocus can be implemented without specialized hardware.	+ Hardware-independent controller software available (e.g., MetaMorph) for universal macro programming running on different microscopes.
– Higher costs.	+ Higher sensitivity, specifically when working with lower NA objectives (e.g., 10X–20X high NA air objectives).
– Slower acquisition rate than widefield.	– Higher out-of-focus fluorescence, higher autofluorescence from imaging media.
– No hardware-independent software platform available.	– Fast laser-based autofocus not available for all widefield- systems.
	– Software-based AF usually too slow/phototoxic when repetitive AF is needed.

3.5.2.1. Illumination

Finding the balance between optimal image quality and cell viability is an empirical process. Even though optimizing imaging conditions for long-term experiments is often very tedious, it is absolutely crucial for reproducible live imaging experiments. As a general rule, cells should always be exposed to a minimal amount of light required for a specific assay. Furthermore, constant imaging conditions must be maintained where different perturbations are compared. If the number of dead cells rises during time-lapse imaging, or if the mitotic index changes, it is important to compare this with adjacent regions of the coverslip that were not repetitively exposed for time-lapse imaging. This allows us to test if cell viability might have been compromised by parameters not associated with light exposure (e.g., incubation or culture medium problems; *see* **Section 3.4**). We recommend first implementing a live imaging assay with very low light exposure to determine the baseline of cell death and cell division errors (**Fig. 7.5**). Next, the image quality should be increased, for example, by higher light exposure, the use of higher magnifying objectives, and so on, to increase the sensitivity of the assay readout.

A frequently underestimated photodamage can be induced when fields of interest are interactively assigned before starting

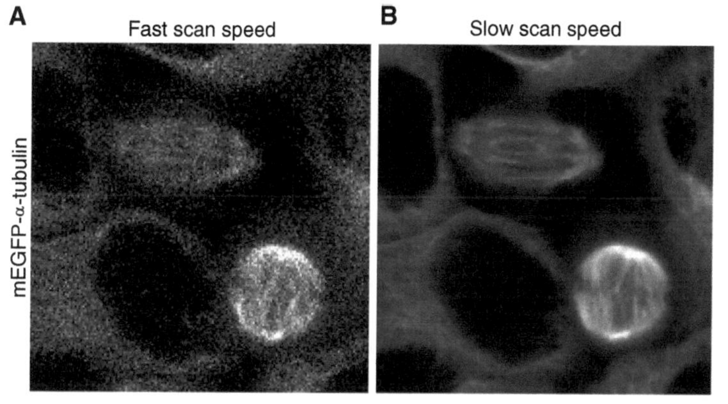

Fig. 7.5 Typical noise levels for live cell imaging on laser scanning confocal microscope with 40 × oil immersion 1.3 NA. (**A**) Typical noise levels from low-light level imaging, necessary for cell viability in long-term experiments (high scan speed). (**B**) High-resolution image with identical settings, but slower scan speed. These settings would lead to cell death in a long-term experiment with the given cell line.

a time-lapse movie. For example, an initial exposure to high-level light for several seconds in the epifluorescence mode when searching cells through the ocular of a confocal system can cause significantly more phototoxicity than subsequent long-term recording in the laser scanning mode. We strongly advise using low-transmitting neutral density filters to attenuate the excitation light (e.g., 6–25% transmission with 50 W mercury arc lamp). In addition, the field diaphragm should be closed so that only the imaging field of the ocular or camera is illuminated. Another important parameter to minimize phototoxicity on widefield systems is a fast responding excitation shutter. This is important to avoid specimen illumination during times where the camera is not recording. One strategy to minimize phototoxic effects with slow shutters is to attenuate the illumination intensity by neutral density filters and to compensate by longer exposure times. To optimize imaging settings on laser scanning confocals, it is important to consider that much higher noise levels must be taken into account than typical fixed cell imaging settings would generate. Because of the high degeneration of images by noise, it is often advisable to use significantly larger pinholes than those which would yield an optimal resolution along the optical axis (we typically record with optical slice thickness of 1.5–3 μM). Again, we suggest starting with very poor imaging quality (fast scanning, low laser output, low zoom factor, small image sampling size) to determine a baseline of cell viability (**Fig. 7.5**). The image quality should then be improved in subsequent experiments.

3.5.3. Objective Choice

Choosing the right objective for a particular assay depends on many parameters (also reviewed in *(21)*). Here, we aim to provide some considerations that are particularly important for automated live microscopy:

- The highest possible numerical aperture (NA) should be used. Even a small increase in NA significantly improves brightness *(24, 25)*.
- The use of immersion media is not always compatible with automated multilocation time-lapse imaging. For example, water evaporation needs to be compensated by constant liquid supply, which is available only by few commercial systems. Oil cannot automatically be replenished, but if appropriately applied can be used for multilocation imaging of confined regions (maximum about 2*4 cm). For many applications the optical quality of air objectives is sufficient. When using air objectives, we recommend high NA short-distance objectives (e.g., 0.8 NA, 20X) instead of low NA long-distance objectives.
- Objectives should be plan-corrected for a flat imaging plane.
- For multicolor imaging and co-localization studies, achromatic, fluorite (Fluar), or apochromatic (Apo) lenses should be used as they are corrected for spherical and chromatic aberrations. However, highly corrected apochromatic objectives reduce the amount of transmitted light due to additional optical elements in the objective *(25)*. It can thus sometimes be beneficial to use less corrected objectives to increase sensitivity.
- Objectives designed for phase contrast should not be used, as the additional elements in the objective significantly reduce light transmission. DIC objectives are a good alternative, but the analyzer and Wollaston prism should be removed from the light path when they are not needed *(26)*.

It is important to know that brightness decreases with increasing magnification. For example, a 40X objective will result in an approximately 2.5-fold increase in brightness compared to a 63X objective when all other parameters are identical. An additional method to reduce noise and increase brightness in widefield epifluorescence microscopy is binning, whereby a square group of pixels is averaged to one larger pixel.

Table 7.3 summarizes the strategy to optimize imaging conditions for long-term experiments.

3.5.4. Autofocus

Maintaining a stable focus during the time course of a long-term imaging experiment is often a severe challenge for the microscope setup. It is not advisable to compensate for potential focus drifts by increasing the number of z-sections, because it would unnecessarily increase photodamage and would require tedious postacquisition analysis. Temperature fluctuations are the major cause of focus drifts during a time-lapse experiment (*see* **Section 3.4.1**). Other factors contributing to focal drift are mechanical

Table 7.3
Optimizing imaging settings for live cell microscopy

1. Validate cell viability on microscope incubator (see **Section 3.4**). – Check with independent external sensors if environmental control works correctly (stable temperature, constant CO_2, correct humidity). – Verify that imaging medium is well tolerated by cells (test cell viability/proliferation without imaging on microscope)	
2. Run test experiment with minimal light exposure. – Start test experiment with a low magnification (e.g., 10X or 20X high NA air objective). – Excitation and emission filters should be optimized to the respective fluorophores. – Avoid long (>2 s) illumination of regions that will be imaged by long-term time-lapse. – Adjust field diaphragm to the camera/detector field of view. – Search stage locations for imaging in transmitted light mode (in fluorescence mode use neutral density (ND) filters). – Make final imaging adjustments directly with camera/CCD and the acquisition software. – Acquire images with a low time-resolution (e.g., 15 min time lapse). – Acquire only a single focal plane.	
Parameters for Low Light Exposure	
Laser Scanning Confocal	**Widefield**
– Work with minimal laser output power (0.1–2%).	– Use CCD camera with high quantum efficiency and low read- out noise.
– Fully open pinhole(s).	– Use high-speed shutters.
– Adjust image intensity with gain (don't over saturate image).	– Attenuate fluorescent light as much as possible with ND filters.
– Start with highest scan-speed.	– Minimize exposure times (exposure time should be less than 100 ms for each wavelength/fluorophore)
– Switch off line averaging.	Use binning if applicable.
– Use binning if applicable.	
3. Adjust settings for improved image quality and compare cell viability with minimal exposure. If significant perturbations occur, reduce light exposure (Fig. **7.5**).	

instabilities that control the z-position of the objective revolver, tension on the glass of the imaging chamber, or a loosely fitted imaging chamber on the stage. To minimize focus drifts, we recommend maintaining a constant microscope incubator temperature and room air conditioning even when the system is not in use. In addition, the reproducibility of long-term experiments can be significantly improved by the use of autofocus procedures. Two autofocus approaches are commercially available.

3.5.4.1. Image-Based Autofocus (Software Autofocus)

The principle of an image-based autofocus is to acquire multiple z-sections of an object and to analyze these images with a software algorithm that determines the optimal focal plane based on the lowest signal-to-noise ratio. This is offered by several commercial software packages (e.g., MetaMorph), but in general is rather slow and requires extensive specimen illumination, which limits its applicability for high-throughput and long-term imaging approaches, especially when a repetitive autofocus is required for each timepoint. In addition, image-based autofocus is sensitive to the morphological changes of the probe, for example, cellular debris, or apoptotic or mitotically arrested cells.

3.5.4.2. Reflection-Based Autofocus (Hardware Autofocus)

Reflection-based autofocus devices are much faster, more reliable, and induce minimal phototoxicity. The principle of most commercial hardware autofocus implementations is to detect the glass/plastic bottom surface of the imaging chamber with a reflected infrared laser beam and to image at a user-defined offset to acquire the optimal focal plane. Reflection-based autofocus is independent of cell shape and fluorescence intensity of the marker. Reflection-based autofocus solutions are commercially available for widefield microscopes, and can be implemented by macro programming on laser scanning systems without additional hardware *(27)*. In any case, a fast focus device (e.g., piezoelectric objective or stage stepper) is indispensable to achieve fast and accurate measurements.

3.5.5. Increasing Acquisition Speed

To optimize performance in screening applications, acquisition throughput can be enhanced by optimizing software and hardware configurations as outlined below.

3.5.5.1. Software Optimization

1. Minimize stage movements. Can be calculated by standard algorithms (traveling salesman problem).
2. Minimize filter changes (widefield): reverse filter settings per location: GFP-RFP; RFP-GFP; GFP-RFP; RFP-GFP; and so on.
3. Write images to a local hard drive if fileserver performance is low.
4. Record directly in 8-bit, or convert 12- or 16-bit images into 8-bit. This also reduces file size.
5. Turn off direct display of acquired images on the computer monitor.
6. Disable/uninstall any unnecessary software running in the background of the computer controlling the microscope. If applicable, disable anti-virus scanning for file types that are saved by the image acquisition software (e.g., tiff files).

3.5.5.2. Hardware Optimization

1. High-speed filter wheels instead of filter cubes minimize time intervals between changing of two different wavelengths (widefield).
2. Place excitation/emission filters that are often used in combination next to each other: for example, GFP next to RFP, CFP next to YFP.

3. Fast motorized stage with linear encoders for high precision. However, moving the stage too fast can cause detaching of rounded-up mitotic cells. Try to avoid long-distance stage movements between subsequent locations for the same reason.
4. Piezoelectric stage to acquire fast z-sequences, especially if autofocus routines are used.
5. High-speed shutters (widefield) to minimize intervals between two exposures. However, the shutter must resist heating by continuous illumination with excitation light over several days.

4. Notes

1. *If too many colonies are observed*: Add selective medium after 24 h of incubation. Make more dilutions when plating on 15 cm dish (e.g., 1, 5, and 94%). Use less DNA for transfection (e.g., 0.5 or 1.0 μg). Check if antibiotic concentration is appropriate (*see* **Note 2**). Make a limiting dilution if everything stated above fails. *If no colonies are observed:* Check if DNA concentration is correct by an independent UV photospectrometer. Test if the transfection reagent is still working by transiently transfecting an empty GFP-vector. Incubate DNA/FuGENE6 transfection mix for 1 h. Increase amount of transfection reagent. Try another transfection reagent (e.g., JetPI [Polyplus] and follow the manufacturer's protocol). Transfect directly into a 10 or 15 cm dish (cells should be 20% confluent). Increase the amount of DNA and transfection reagent accordingly. Check if cells can be transiently transfected on (chambered) coverslips (+/−DNA) and image cells overnight to see if (a) the transfection itself causes cell death (−DNA), or (b) the reporter gene is deleterious to the cell (+DNA). As an additional control, image nontransfected cells in a separate well to exclude that cell death is caused by phototoxicity/imaging problems. If the reporter gene is deleterious, exchange the CMV promoter of the pIRES vector with a weaker (e.g., SV40), the gene's endogenous, or a tetracycline inducible promoter (Clontech). Although one of the above solutions can work, different combinations might have to be tried.
2. To determine the lowest drug concentration for a new antibiotic or a different cell line, generate a death curve. For this, plate 3×10^4 cells/well into two 6-well dishes. After 24 h, exchange the medium of 11 wells with selective media, containing different antibiotic concentrations

(choose a broad range, at least fivefold difference between each drug concentration). Use stock solutions (e.g., prepare 15 mL Falcon tubes); do not pipette the antibiotic directly to the well. One well should contain normal D-MEM and serve as a negative control. Replenish the selective media every 2–3 days. After 12 days, select the lowest antibiotic concentration based on the well in which all cells have died. The procedure can be repeated to determine the drug concentration more precisely by using a narrower range of drug concentrations based on the first determined value.

3. When colonies are picked too small, they might not survive the transfer to the multiwell dish. When colonies tend to be small, transfer them into a 96-well dish instead of the 48-well dish. In this case use 50 μL of trypsin, add 100 μL of cells/PBS, incubate for 5 min, and add 150 μL of D-MEM (containing 20% FBS). Resuspend well. Again, it is critical to replace the medium the following day with fresh selective D-MEM. Be careful not to aspirate the colony when replacing the medium (e.g., place a 0.1–10 μL nonbarrier tip on top of the Pasteur pipette while aspirating the medium). For RPE cells (and other cells with high cell motility), isolate and transfer clonal colonies to a 48-well dish. Aspirate 200 μL of cells/PBS with a 200 μL Gilson pipette (do not use a 1 mL pipette). Try to scratch a larger region of the clone with the tip while aspirating. When many clones have to be screened, plate directly into 96-well glass bottom dishes. However, be aware that many cell types adhere much more weakly to glass, especially when seeded at low concentrations. The colony isolating step can also be performed on an inverted microscope at low magnification to facilitate visual control of scratching and aspirating the cells. However, be aware that this can lead to bacterial or fungal contamination if performed outside the sterile hood. We recommend always testing new stable cell lines for mycoplasm contamination. Some markers are toxic when overexpressed and therefore very difficult to stably express. If no positive clones were obtained, we recommend making a stable cell line in parallel with a marker that is easy to express and to detect (e.g., H2B-mRFP) as a positive control to test if the experimental procedures were appropriate.
4. When using a motorized stage, be aware that fast stage movements may cause spilling of the imaging media, especially when moving over large distances and using single-chambered coverslips. Under such conditions reduce either stage speed or use less imaging media and refill evaporated media regularly.
5. Silicon grease can easily be applied by filling it into a 15 mL plastic syringe (without tip). Smearing the silicon grease on

the outside rim of the chambered coverslip prevents contact with the imaging media. Alternatively, silicon grease can also be applied to the inside rim of the lid, but this can be toxic to cells.

Acknowledgments

We thank P. Steigemann for helpful discussions and critical reading of this manuscript. We thank M. Held for software programming and S. Maar for technical help. We thank G. Csucs and J. Kusch from the light microscopy center for technical support; Toni Lehmann for making LabTek stage holders, and technical assistance; R. Y. Tsien for mRFP, mCherry, and MyrPalm-EYFP; J. Lippincott-Schwartz for mEGFP; J. Ellenberg for H2B-mRFP; and S. Narumiya for HeLa "Kyoto" cells. This work was supported by a European Young Investigator (EURYI) award of the European Science Foundation to D.G., and a Roche Ph.D. fellowship to M.S. M. S. is a fellow of the Life Science Zurich Graduate School, Zurich, Switzerland.

References

1. Gerlich, D., Beaudouin, J., Gebhard, M., Ellenberg, J., and Eils, R. (2001) Four-dimensional imaging and quantitative reconstruction to analyse complex spatiotemporal processes in live cells. *Nat Cell Biol* **3**, 852–5.
2. Gerlich, D. and Ellenberg, J. (2003) 4D imaging to assay complex dynamics in live specimens. *Nat Cell Biol* **Suppl**, S14–9.
3. Dailey, M.E., Manders, E., Soll, D.R., and Terasaki, M. (2006) Confocal Microscopy of Living Cells, in *Handbook of Biological Confocal Microscopy* (Pawley, J.B., ed.), Springer, New York, NY, pp. 381–403.
4. Bodnar, A.G., Ouellette, M., Frolkis, M., Holt, S.E., Chiu, C.P., Morin, G.B., Harley, C.B., Shay, J.W., Lichtsteiner, S., and Wright, W.E. (1998) Extension of life-span by introduction of telomerase into normal human cells. *Science* **279**, 349–52.
5. Neumann, B., Held, M., Liebel, U., Erfle, H., Rogers, P., Pepperkok, R., and Ellenberg, J. (2006) High-throughput RNAi screening by time-lapse imaging of live human cells. *Nat Methods* **3**, 385–90.
6. Kanda, T., Sullivan, K.F., and Wahl, G.M. (1998) Histone-GFP fusion protein enables sensitive analysis of chromosome dynamics in living mammalian cells. *Curr Biol* **8**, 377–85.
7. Snapp, E.L., Hegde, R.S., Francolini, M., Lombardo, F., Colombo, S., Pedrazzini, E., Borgese, N., and Lippincott-Schwartz, J. (2003) Formation of stacked ER cisternae by low affinity protein interactions. *J Cell Biol* **163**, 257–69.
8. Shu, X., Shaner, N.C., Yarbrough, C.A., Tsien, R.Y., and Remington, S.J. (2006) Novel chromophores and buried charges control color in mFruits. *Biochemistry* **45**, 9639–47.
9. Campbell, R.E., Tour, O., Palmer, A.E., Steinbach, P.A., Baird, G.S., Zacharias, D.A., and Tsien, R.Y. (2002) A monomeric red fluorescent protein. *Proc Natl Acad Sci USA* **99**, 7877–82.
10. Stepanova, T., Slemmer, J., Hoogenraad, C.C., Lansbergen, G., Dortland, B., De Zeeuw, C.I., Grosveld, F., van Cappellen, G., Akhmanova, A., and Galjart, N. (2003) Visualization of microtubule growth in cultured neurons via the use of EB3-GFP (end-binding protein 3-green fluorescent protein). *J Neurosci* **23**, 2655–64.
11. Piel, M., Meyer, P., Khodjakov, A., Rieder, C.L., and Bornens, M. (2000) The respective contributions of the mother and daughter centrioles to centrosome activity and behavior in vertebrate cells. *J Cell Biol* **149**, 317–30.

12. Sugimoto, K., Fukuda, R., and Himeno, M. (2000) Centromere/kinetochore localization of human centromere protein A (CENP-A) exogenously expressed as a fusion to green fluorescent protein. *Cell Struct Funct* **25**, 253–61.
13. Zacharias, D.A., Violin, J.D., Newton, A.C., and Tsien, R.Y. (2002) Partitioning of lipid-modified monomeric GFPs into membrane microdomains of live cells. *Science* **296**, 913–6.
14. Leonhardt, H., Rahn, H.P., Weinzierl, P., Sporbert, A., Cremer, T., Zink, D., and Cardoso, M.C. (2000) Dynamics of DNA replication factories in living cells. *J Cell Biol* **149**, 271–80.
15. Zaal, K.J., Smith, C.L., Polishchuk, R.S., Altan, N., Cole, N.B., Ellenberg, J., Hirschberg, K., Presley, J.F., Roberts, T.H., Siggia, E., Phair, R.D., and Lippincott-Schwartz, J. (1999) Golgi membranes are absorbed into and reemerge from the ER during mitosis. *Cell* **99**, 589–601.
16. Terasaki, M., Jaffe, L.A., Hunnicutt, G.R., and Hammer, J.A., 3rd (1996) Structural change of the endoplasmic reticulum during fertilization: evidence for loss of membrane continuity using the green fluorescent protein. *Dev Biol* **179**, 320–8.
17. Rizzuto, R., Brini, M., De Giorgi, F., Rossi, R., Heim, R., Tsien, R.Y., and Pozzan, T., et al. (1996) Double labelling of subcellular structures with organelle-targeted GFP mutants in vivo. *Curr Biol* **6**, 183–8.
18. Muhlhausser, P. and Kutay, U. (2007) An in vitro nuclear disassembly system reveals a role for the RanGTPase system and microtubule-dependent steps in nuclear envelope breakdown. *J Cell Biol* **178**, 595–610.
19. Rao, P.N. and Engelberg, J. (1965) Hela cells: Effects of temperature on the life cycle. *Science* **148**, 1092–4.
20. Rieder, C.L. and Cole, R.W. (2002) Cold-shock and the Mammalian cell cycle. *Cell Cycle* **1**, 169–75.
21. Waters, J.C. (2007) Live-cell fluorescence imaging. *Methods Cell Biol* **81**, 115–40.
22. Lippincott-Schwartz, J., Altan-Bonnet, N., and Patterson, G.H. (2003) Photobleaching and photoactivation: following protein dynamics in living cells. *Nat Cell Biol* **Suppl**, S7–14.
23. Swedlow, J.R. (2007) Quantitative fluorescence microscopy and image deconvolution. *Methods Cell Biol* **81**, 447–65.
24. Abramowitz, M., Spring, K.R., Keller, H.E., and Davidson, M.W. (2002) Basic principles of microscope objectives. *Biotechniques* **33**, 772–4, 776–8, 780–1.
25. Keller, H.E. (2006) Objective lenses for confocal microscopy, in *Handbook of Biological Confocal Microscopy* (Pawley, J.B., ed.), Springer, New York, pp. 145–161.
26. Inoué, S. and Spring, K. (1986) *Video Microscopy*. Plenum Press, New York.
27. Rabut, G. and Ellenberg, J. (2004) Automatic real-time three-dimensional cell tracking by fluorescence microscopy. *J Microsc* **216**, 131–7.

Chapter 8

Electron Tomography of Microtubule End-Morphologies in *C. elegans* Embryos

Eileen O'Toole and Thomas Müller-Reichert

Abstract

In this chapter we describe the preparation of early mitotic *C. elegans* embryos for the tomographic reconstruction of end-morphologies of spindle microtubules. Early embryos are prepared by high-pressure freezing and freeze-substitution for thin-layer embedding in Epon/Araldite. We further describe data acquisition, tomographic reconstruction, and 3-D modeling of microtubules in serially sectioned mitotic spindles. The presented techniques are applicable to other model systems.

Key words: Mitosis, *C. elegans*, early embryo, high-pressure freezing, freeze-substitution, cryofixation, EM specimen preparation, electron tomography, three-dimensional reconstruction, 3-D modeling.

1. Introduction

Microtubules are polar polymers composed of head-to-tail aligned α- and β-tubulin subunits *(1)*. At the microtubule ends, tubulin subunits are either added during polymerization (growth) or removed during depolymerization (shrinkage). In vitro studies have shown that the dynamic state of an individual microtubule is correlated with the morphology of its ends. Growing microtubules exhibit a flared sheetlike structure *(2)*, whereas shrinking microtubules have curled protofilaments with a distinct radius of curvature at their ends *(3, 4)*.

In animal cells, microtubules commonly emanate from centrosomes with a typical orientation: microtubule minus ends are located in the centrosome and microtubule plus ends grow away

Andrew D. McAinsh (ed.), *Mitosis: Methods and Protocols*, Methods in Molecular Biology *545*,
DOI 10.1007/978-1-60327-993-2_8, © Humana Press, a part of Springer Science+Business Media, LLC 2009

from the centrosome *(5, 6)*. The centrosome provides nucleation sites and a kinetically favorable environment for microtubule polymerization. Minus ends found in yeast *(7, 8)* and *Drosophila (9)* show a distinct, capped-shaped morphology, presumably due to γ-tubulin in a complex with other proteins. Similarly, the closed minus ends of microtubules nucleated from γ-tubulin in vitro are cone-shaped *(10–12)*. Recently we have reported the occurrence of closed and open microtubule minus ends within the mitotic centrosome of *C. elegans*, with the open minus-end morphology preferentially associated with kinetochore microtubules *(13)*.

Here we describe the preparation of mitotic *C. elegans* embryos for tomographic reconstruction of end-morphologies of spindle microtubules. Currently, we are using the EM PACT2 + RTS high-pressure freezer (Leica Microsystems, Vienna, Austria) for correlative light and electron microscopy of isolated early embryos *(14–16)*. Because the RTS allows quick loading of embryos, contained in cellulose capillary tubes, the time window between light microscopic observation and freezing can be considerably reduced, thus allowing precise staging of embryos prior to cryo-immobilization. Electron tomography can then be applied to study the three-dimensional architecture of spindles and cellular organelles with a resolution of ~6 nm in 3-D *(17–19)*. Conceptually similar to CT scans in medical imaging, this method is based on the use of serial tilted views of a semi-thick section to create a computer-generated volume that can be sliced and imaged in any direction. Using electron tomography, the mitotic spindle organization in *S. cerevisiae (12, 20)* and *C. elegans (14)*, and the architecture of interphase microtubules in *S. pombe (21)*, have been analyzed. Spindle pole bodies in *S. cerevisiae (12, 22)*, basal body structure in *Chlamydomonas (23)*, centrosomes in *C. elegans (14, 24)*, *D. melanogaster (25)*, and *Spisula (26)*, and the mammalian kinetochore *(27, 28)* have also been reconstructed in 3-D. Recently, we have applied electron tomography to analyze the structural pathway of centriole assembly and the role of katanin in female meiosis in *C. elegans (24, 29)*.

2. Materials

2.1. High-Pressure Freezing and Freeze-Substitution

1. EM PACT2 + RTS (Leica Microsystems, Vienna, Austria) (*see* **Note 1**).
2. Stereomicroscope with light source.
3. Light microscope (phase contrast, DIC, epifluorescence).
4. Cellulose capillary tubes with an inner diameter of 200 μm. This tubing is available from Spectrum, 23022 La Cadena

Dr., Suite 100, Laguna Hills, CA 92653, USA and Leica Microsystems, Vienna, Austria (*see* **Note 2**).

5. Micropipettor (0.5–10 μl size) and gel loader tips.
6. Nail polish.
7. *C. elegans* strain expressing either GFP::histone, GFP:: γ-tubulin or both (optional, *30*).
8. 20% BSA (w/v) (Sigma) made with M-9 buffer. This buffer has the following composition: 22 mM potassium phosphate monobasic (KH_2PO_4), 19 mM NH_4Cl; 48 mM sodium phosphate dibasic (Na_2HPO_4), 9 mM NaCl (*see* **Note 3**).
9. Tools, for example, two syringe needles for cutting open the worms to release the embryos.
10. Small (2.5–6 cm) plastic Petri dishes.
11. Microscope slides.
12. Fine forceps.
13. Specimen holders for high-pressure freezer: 100-μm deep membrane carriers (*15*).
14. Scalpels for cutting dialysis tubing.
15. Filter paper wedges for removing fluids.
16. EM grade acetone.
17. Freeze-substitution "cocktail" composed of 1% OsO_4 + 0.1% uranyl acetate in acetone (*31*).
18. Automated freeze-substitution device (Leica EM AFS2; *see* **Note 4**).
19. Epon/Araldite resin. Our routine Epon/Araldite formula is: 6.2 g Epon 812 substitute, 4.4 g Araldite, 12.2 g DDSA, and 0.55 ml DMP-30.
20. Teflon®-coated glass slides for thin-layer embedding (*see* **Note 5**). Slides are coated with Teflon using either a spray (MS-122DF) or a solution (MS-143 V, Miller-Stephenson Chemical Co., Inc., Danbury, CT, USA).
21. "Dummy" blocks for remounting and fast glue.

2.2. Electron Tomography

1. For electron tomography of semi-thick (200–400 nm) sections we use a microscope operating at intermediate voltage (300 kV; TECNAI F30; FEI Company, Eindhoven, The Netherlands). Automated image acquisition programs are available from commercial vendors (FEI Company, Eindhoven, The Netherlands; Tietz Video and Image Processing Systems (TVIPS), Gauting, Germany). Several software packages are freely available (TOM, UCSF Tomography, and SerialEM, Boulder Laboratory for 3-D electron microscopy of cells).
2. Copper slot grids.
3. 0.7% (w/v) Formvar in ethylene dichloride.
4. 2% uranyl acetate in 70% methanol; rinse solutions of 70%, 50%, 20% methanol.

5. Reynold's lead citrate.
6. 10- or 15-nm colloidal gold.

2.3. Three-Dimensional Reconstruction and Modeling

1. We routinely use the IMOD software package (http://bio3d.colorado.edu/imod), which contains all of the programs needed for calculating tomograms and for the display and modeling of mitotic features within the reconstructed volume *(32)*. The IMOD software package runs on multiple platforms, including Linux, Mac OSX, and Windows. The programs used for tomographic reconstruction are managed by a graphical user interface, eTomo. The eTomo interface facilitates the ease with which users step through the various steps of the process, much as with a flow chart. Image display and modeling are carried out with the "3dmod" viewing program from the IMOD software package. This program can be run by command line; it contains windows for image display and the slicer tool for rotating slices of image data and for modeling features of interest in the reconstruction.

3. Methods

3.1. High-Pressure Freezing and Freeze-Substitution

The combination of high-pressure freezing with freeze-substitution has evolved into a routine technique to prepare *C. elegans* for electron tomography. In principle, either whole worms or isolated embryos can be cryo-immobilized within milliseconds *(33, 34)*. To avoid serial sectioning through whole worms, and the subsequent searching for early embryos at the right mitotic stage, however, it is advantageous to freeze isolated staged embryos for cellular electron tomography. We give details of this method below.

3.1.1. Cryo-Immobilization of Isolated Embryos

1. Prepare a "loading device" for collecting isolated early embryos into cellulose capillary tubes by mounting a piece of tubing about 2 cm long into a pipette tip and using nail polish to seal. Dry before using *(14)*.
2. Cut worms open in small Petri dishes using M-9 buffer containing 20% BSA. Select an early embryo under a dissecting scope and suck the selected embryo into the capillary tubing.
3. Submerge tubing into BSA-containing M-9 buffer in a small plastic Petri dish. Using the crimping tool, cut the region of the cellulose capillary tubing containing the early embryo into a length that will fit into the specimen carrier. The ends should be crimped so that the early embryo will not leak out.

4. Transfer capillary tubing from the Petri dish to a droplet of BSA-containing buffer on a glass slide and observe the early development of the embryo using either DIC or fluorescence microscopy. In the latter case, a line expressing like GFP::histone;GFP::γ-tubulin can be used for live cell imaging *(30)*.
5. At an appropriate mitotic stage, transfer the embryo-containing tubing from the microscope slide to a specimen carrier, prefilled with BSA-containing buffer, freeze, and store in LN_2.

3.1.2. Freeze-Substitution, Embedding, Ultramicrotomy, and Screening of Serial Sections

1. Transfer samples to precooled (–90°C) cryovials containing the freeze-substitution cocktail, composed of 1% OsO_4 plus 0.1% uranyl acetate in acetone. Maintain temperature for 8–24 h at –90°C.
2. Warm samples to room temperature at a rate of 5°C/h.
3. Rinse samples in pure acetone and infiltrate with Epon/Araldite.
4. Process samples for thin-layer embedding and remount selected specimens on dummy blocks for ultramicrotomy.
5. Cut ribbons of serial semi-thick sections (300–400 nm) through *C. elegans* embryos and collect ribbons of sections on Formvar-coated copper slot grids.
6. Stain sections with uranyl acetate, followed by Reynold's lead citrate.
7. Examine thick sections in a standard TEM operating at 100 or 120 kV to identify the sections containing spindle components, such as kinetochore regions, centrosomes, or fractions of the mitotic spindle. Map the features of interest through the serial sections by imaging at low magnification (*see* **Note 6**).
8. In preparation for electron tomography, apply 10- or 15-nm gold fiducials to samples by placing the selected slot grids on top of a drop of gold solution for 5 min, blotting excess fluid, turning the grid over, and repeating for the other side.

3.2. Acquisition of Tomographic Data

1. Image a tilt series of semi-thick sections in an intermediate voltage (200–300 kV) EM equipped with a eucentric tilting stage. Collect serial tilted views of the section every degree over a ±60° or 70° range.
2. After the first tilt series has been acquired, rotate the grid 90° to image a second tilt series over a ±60° or 70° range.
3. Calculate a double-tilt tomogram *(35)*.

3.3. Modeling of Spindle Microtubules and Microtubule End-Morphologies

1. Open the reconstructed volume using "3dmod" and go into model mode *(36)*.
2. Create a model object and edit the object type as an open contour.

3. Open the slicer window; choose a microtubule by clicking on the left mouse button. Rotate the x, y, and z sliders at the top of the slicer window to orient the microtubule along its long axis (*see* **Note** 7). Deposit model points along the microtubule using the middle mouse button. Each new microtubule modeled is a new contour in the object. In addition, create new objects for either open or closed microtubule plus and minus ends. Deposit model points at each microtubule end. A projection of the 3-D model can then be opened (**Fig. 8.1**).
4. Different classes of microtubules (i.e., kinetochore microtubules, astral microtubules) are organized as separate objects and can be distinguished using different colors (**Fig. 8.2**).

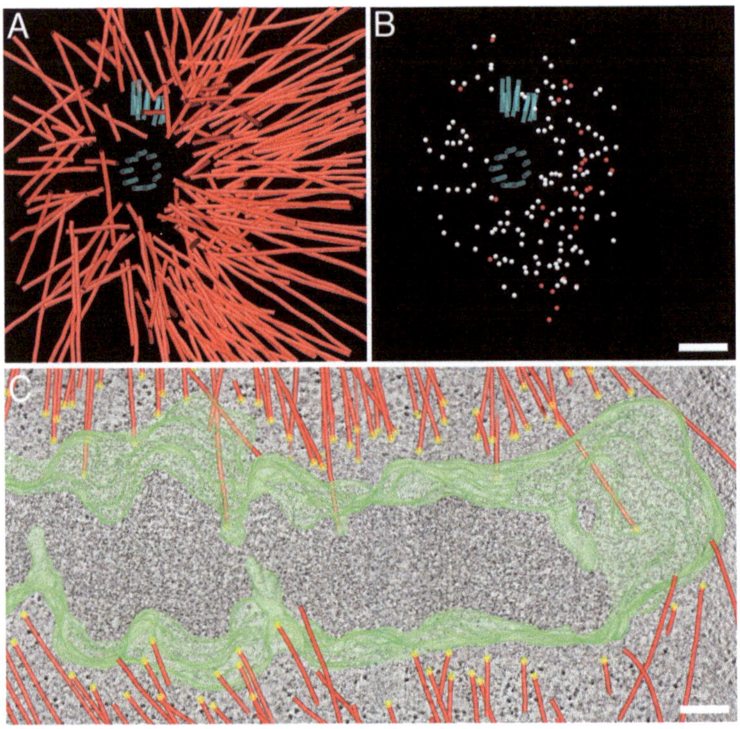

Fig. 8.1. Three-dimensional reconstruction and modeling of spindle components. (**A**) Partial reconstruction of a centrosome, showing a pair of centrioles (*blue cylinders*) and microtubules (*red*). (**B**) Model showing the pair of centrioles and the distribution of closed (*white spheres*) and open microtubule minus ends (*red spheres*) in the mitotic centrosome. (**C**) Partial reconstruction of the holocentric kinetochore in *C. elegans*. The surface of the DNA is outlined in green. The kinetochore microtubules are outlined in red, their plus ends indicated by yellow spheres. Modified from O'Toole et al. *(13)*. Scale bars, 250 nm.

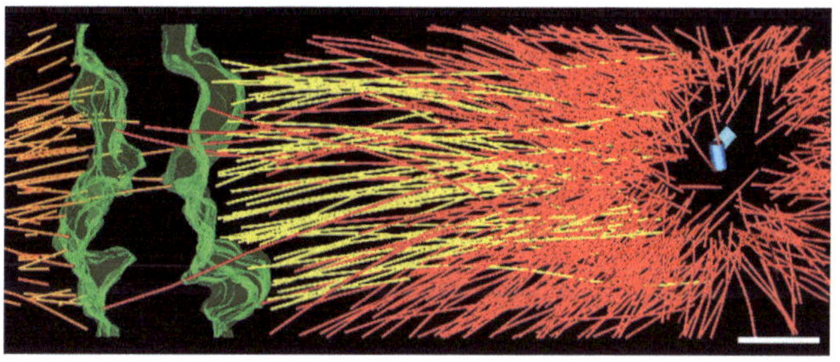

Fig. 8.2. Partial reconstruction of a metaphase spindle. The surface of the DNA is outlined in green, and kinetochore microtubules are outlined in white. Other spindle microtubules are shown in either red or orange. The centriole pair is shown as blue cylinders. 3-D reconstruction allows identification of kinetochore microtubules within the spindle and analysis of their plus and minus ends. Modified from O'Toole et al. *(13)*. Scale bar, 1 μm.

5. Once the microtubules in the volume have been modeled, a program can be run, mtrotlong, that will extract a series of subvolumes that contain the microtubules in longitudinal orientation. The operator can then step through successive tomographic slices of the subvolumes to analyze the microtubule end-morphology in detail (**Fig. 8.3**).

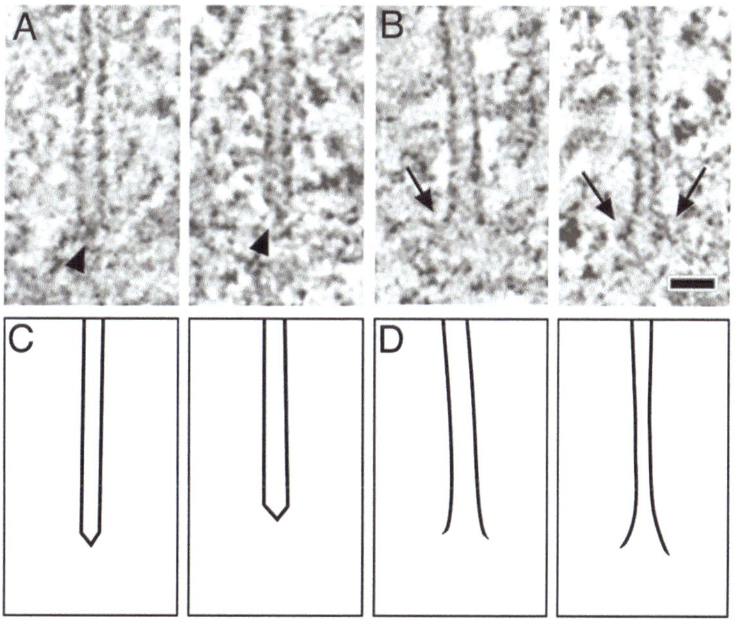

Fig. 8.3. Illustration of microtubule end-morphologies in *C. elegans* embryos as observed by electron tomography. (**A**) Closed microtubule minus ends visualized in the mitotic centrosome with a capped, pointed morphology (*arrowheads*). (**B**) Open microtubule minus ends with a flared morphology (*arrows*). (**C**), (**D**) Corresponding schematic representation of the end-morphologies. Bar, 50 nm.

6. Model objects can be displayed together, or subsets of objects can be turned off or on, to highlight the 3-D relationships of particular features within the cell.

3.4. Summary and Outlook

In this chapter we have described the tomographic analysis of microtubule minus and plus ends in mitotic *C. elegans* wild-type embryos. Such a structural analysis can be expanded to RNAi-treated embryos, where a specific gene product has been depleted prior to high-pressure freezing *(24, 37)*. Note that the described methodology is applicable to other model systems, such as yeast *(8, 20)*, *Chlamydomonas (36)*, the *Drosophila* embryo *(9)*, and mammalian tissue culture cells *(27, 28)*, and we are certainly only at the beginning of exploiting these model systems for tomographic structure-function studies.

4. Notes

1. The EM PACT2 + RTS (Leica) high-pressure freezer is a portable machine that can be easily moved to the site where staging of the isolated embryo is performed. The rapid transfer system (RTS) allows fast loading of the specimen into a preloaded high-pressure freezer under standardized conditions *(15, 16)*.
2. Cellulose capillary tubing has been used previously to freeze either cell suspensions or *C. elegans* hermaphrodites *(38)*.
3. The use of 20% BSA gives reproducibly good freezing results, also for other samples such as *Drosphila* embryos and tissue culture cells *(15)*.
4. Homemade devices can also be used for freeze-substitution *(15)*.
5. Samples are embedded in thin, optically clean layers of Epon/Araldite on microscope slides. Thin layer-embedding allows relocation of samples by light microscopy and remounting of selected specimens on dummy blocks *(14, 33)*.
6. Trapezoid-shaped sections allow searching for the first and last section of each ribbon. This way sections can be easily identified when using the intermediate voltage electron microscope for data acquisition.
7. The "image slicer" window in "3dmod" is used routinely to extract a slice of image data one voxel thick and to adjust the 3-D volume in any position or direction to contain the axis of a microtubule in a single view *(8)*.

Acknowledgments

The authors wish to thank Jana Mäntler of the MPI-CBG EM Facility for excellent help with electron microscopy. Work by E. O'T. was supported in part by grant RR-00592 from the National Center for Research Resources of the National Institutes of Health to A. Hoenger and work by T.M.-R. was supported by grant MU 1423/2-1 from the Deutsche Forschungsgemeinschaft (DFG) to T. Müller-Reichert.

References

1. Hyams, J.S., and Lloyd, C.W. (1994) "Microtubules", Wiley-Liss, New York.
2. Chrétien, D., Fuller, S.D., and Karsenti, E. (1995) Structure of growing microtubule ends: two-dimensional sheets close into tubes at variable rates. *J. Cell Biol.* **129**, 1311–1328.
3. Mandelkow, E.M., Mandelkow, E., and Milligan, R.A. (1991) Microtubule dynamics and microtubule caps: a time-resolved cryo-electron microscopy study. *J. Cell Biol.* **114**, 977–991.
4. Müller-Reichert, T., Chrétien, D., Severin, F., and Hyman, A.A. (1998) Structural changes at microtubule ends accompanying GTP hydrolysis: information from a slowly hydrolyzable analogue of GTP, guanylyl (α,β) methylene-diphosphonate. *Proc. Natl. Acad. Sci. USA* **95**, 3661–3666.
5. Doxsey, S., McCollum, D., and Theurkauf, W. (2005) Centrosomes in cellular regulation. *Annu. Rev. Cell Dev. Biol.* **21**, 411–434.
6. Wittmann, T., Hyman, A., and Desai, A. (2001) The spindle: a dynamic assembly of microtubules and motors. *Nat. Cell Biol.* **1**, E28–34.
7. Byers, B., Shriver, K., and Goetsch, L. (1978) The role of spindle pole bodies and modified microtubule ends in the initiation of microtubule assembly in *Sacharomyces cerevisiae*. *J. Cell Sci.* **30**, 331–352.
8. O'Toole, E.T., Winey, M., and McIntosh, J.R. (1999) High-voltage electron tomography of spindle pole bodies and early mitotic spindles in the yeast *Saccharomyces cerevisiae*. *Mol. Biol. Cell.* **10**, 2017–2031.
9. Moritz, M., Braunfeld, M.B., Fung, J.C., Sedat, J.W., Alberts, B.M., and Agard, D.A. (1995) Three-dimensional structural characterization of centrosomes from early *Drosophila* embryos. *J. Cell Biol.* **130**, 1149–1159.
10. Keating, T.J., and Borisy, G.G. (2000) Immunostructural evidence for the template mechanism of microtubule nucleation. *Nat. Cell Biol.* **2**, 352–357.
11. Moritz, M., Braunfeld, M.B., Guenebaut, V., Heuser, J., and Agard, D.A. (2000) Structure of the γ-tubulin ring complex: a template for microtubule nucleation. *Nat. Cell Biol.* **2**, 365–370.
12. Wiese, C., and Zheng, Y. (2000) A new function for the for the g-tubulin ring complex as a microtubule minus-end cap. *Nat. Cell Biol.* **2**, 358–364.
13. O'Toole, E.T., McDonald, K.L., Mäntler, J., McIntosh, J.R., Hyman, A.A., and Müller-Reichert, T. (2003) Morphologically distinct microtubule ends in the mitotic centrosome of *Caenorhabditis elegans*, *J. Cell Biol.* **163**, 451–456.
14. Müller-Reichert, T., Srayko, M., Hyman, A., O'Toole, E., and McDonald, K. (2007) Correlative light and electron microscopy of early *Caenorhabditis elegans* embryos in mitosis. *Meth. Cell Biol.* **79**, 99–117.
15. McDonald, K.L., Morphew, M., Verkade, P., and Müller-Reichert, T. (2007) Recent advances in high-pressure freezing. *Meth. Mol. Biol.* **369**, 143–174.
16. Verkade, P. (2008) Moving EM: the rapid transfer system as a new tool for correlative light and electron microscopy and high throughput for high pressure freezing. *J. Microsc.* **230**, 317–328.
17. McEwen, B.F., and Marko, M. (2001) The emergence of electron tomography as an important tool for investigating cellular ultrastructure. *J. Histochem. Cytochem.* **49**, 553–563.
18. Frank, J. (1992) "Electron tomography" Plenum Publishing Corporation, New York.
19. McIntosh, J.R., Nicastro, D., and Mastronarde, D. (2005) New views of cells in 3-

D: an introduction to electron tomography. *Trends Cell Biol.* **15**, 43–51.
20. Yoder, T.J., McElwain, M.A., Francis, S.E., Bagley, J., Muller, E.G., Pak, B., O'Toole, E.T., Winey, M., and Davis, T.N. (2005) Analysis of a spindle pole body mutant reveals a defect in biorientation and illuminates spindle forces. *Mol. Biol. Cell* **16**, 141–152.
21. Höög, J.L., Schwartz, C., Noon, A.T., O'Toole, E.T., Mastronarde, D.N., McIntosh, J.R., and Antony, C. (2007) Organization of interphase microtubules in fission yeast analyzed by electron tomography. *Dev Cell.* **12**, 349–361.
22. Bullitt, E., Rout, M.P., Kilmartin, J.V., and Akey, C.W. (1997) The yeast spindle pole body is assembled around a central crystal of Spc42p. *Cell* **89**, 1077–1086.
23. O'Toole, E.T., Giddings, T.H., McIntosh, J.R., and Dutcher, S.K. (2003) Three-dimensional organization of basal bodies from wild-type and δ-tubulin deletion strains of *Chlamydomonas reinhardtii. Mol. Biol. Cell.* **14**, 2999–3012.
24. Pelletier, L., O'Toole, E., Schwager, A., Hyman, A.A., and Müller-Reichert, T. (2006) Centriole assembly in *Caenorhabditis elegans. Nature* **444**, 619–623.
25. Moritz, M., Braunfeld, M.B., Fung, J.C., Sedat, J.W., Alberts, B.M., and Agard, D.A. (1995) Three-dimensional structural characterization of centrosomes from early *Drosophila* embryos. *J. Cell Biol.* **130**, 1149–1159.
26. Schnackenberg, B.J., Khodjakov, A., Rieder, C., and Palazzo, R.E. (1998) The disassembly and reassembly of functional centrosomes *in vitro. Proc Natl Acad Sci USA* **95**, 9295–9300.
27. VandenBeldt, K.J., Barnard, R.M., Hergert, P.J., Meng, X., Maiato, H., and McEwen, B.F. (2006) Kinetochores use a novel mechanism for coordinating the dynamics of individual microtubules. *Curr Biol.* **16**, 1217–1223.
28. Dong, Y., Vanden Beldt, K.J., Meng, X., Khodjakov, A., and McEwen, B.F. (2007) The outer plate in vertebrate kinetochores is a flexible network with multiple microtubule interactions. *Nat Cell Biol.* **9**, 516–522.
29. Srayko, M., O'Toole, E.T., Hyman, A.A., and Müller-Reichert, T. (2006) Katanin disrupts the microtubule lattice and increases polymer number in *C. elegans* meiosis. *Curr. Biol.* **16**, 1944–1949.
30. Oegema, K., Desai, A., Rybina, S., Kirkham, M., and Hyman, A.A. (2001) Functional analysis of kinetochore assembly in *Caenorhabditis elegans. J. Cell Biol.* **153**, 1209–1226.
31. McDonald, K., and Müller-Reichert, T. (2002) Cryomethods for thin section electron microscopy. *Meth. Enzymol.* **351**, 96–123.
32. Kremer, J.R., Mastronarde, D.N., and McIntosh, J.R. (1996) Computer visualization of three-dimensional image data using IMOD. *J. Struct. Biol.* **116**, 71–76.
33. Müller-Reichert, T., O'Toole, E.T., Hohenberg, H., and McDonald, K.L. (2003) Cryoimmobiliztion and three-dimensional visualization of *C. elegans* ultrastructure, *J. Microsc. (Oxford)*. **212**, 71–80.
34. Müller-Reichert, T., Mäntler, J., Srayko, T., M O'Toole, E. (2008) Electron microscopy of the early *Caenorhabditis elegans* embryo. *J. Microsc.* **230**, 297–307.
35. Mastronarde, D.N. (1997) Dual-axis tomography: An approach with alignment methods that preserve resolution. *J. Microsc. (Oxford)* **120**, 343–352.
36. O'Toole, E.T., Giddings, T., and Dutcher, S.K. (2007) Understanding microtubule organizing centers by comparing mutant and wild-type structures with electron tomography. *Meth. Cell Biol.* **79**, 126–143.
37. Özlü, N., Srayko, M., Kinoshita, K., Habermann, B., O'Toole, E.T., Müller-Reichert, T., Schmalz, N., Desai, A., and Hyman, A.A. (2005) An essential function of the *C. elegans* orthologue of TPX2 is to localize activated Aurora A kinase to mitotic spindles. *Dev. Cell.* **9**, 237–248.
38. Hohenberg, H., Mannweiler, K., and Müller, M. (1994) High-pressure freezing of cell suspensions in cellulose capillary tubes. *J. Microsc. (Oxford)*. **175**, 34–43.

Chapter 9

Dissecting Mitosis with Laser Microsurgery and RNAi in *Drosophila* Cells

António J. Pereira, Irina Matos, Mariana Lince-Faria and Helder Maiato

Abstract

Progress from our present understanding of the mechanisms behind mitosis has been compromised by the fact that model systems that were ideal for molecular and genetic studies (such as yeasts, *C. elegans*, or *Drosophila*) were not suitable for intracellular micromanipulation. Unfortunately, those systems that were appropriate for micromanipulation (such as newt lung cells, PtK1 cells, or insect spermatocytes) are not amenable for molecular studies. We believe that we can significantly broaden this scenario by developing high-resolution live cell microscopy tools in a system where micromanipulation studies could be combined with modern gene-interference techniques. Here we describe a series of methodologies for the functional dissection of mitosis by the use of simultaneous live cell microscopy and state-of-the-art laser microsurgery, combined with RNA interference (RNAi) in *Drosophila* cell lines stably expressing fluorescent markers. This technological synergism allows the specific targeting and manipulation of several structural components of the mitotic apparatus in different genetic backgrounds, at the highest spatial and temporal resolution. Finally, we demonstrate the successful adaptation of agar overlay flattening techniques to human HeLa cells and discuss the advantages of its use for laser micromanipulation and molecular studies of mitosis in mammals.

Key words: Mitosis, *Drosophila*, S2 cells, HeLa cells, live cell microscopy, RNAi, agar overlay, laser microsurgery.

1. Introduction

During every cell cycle the genetic information must be replicated in order to be equally distributed in the form of chromosomes to the progeny cells at the time of mitosis. In the last two decades mitosis research has attracted unprecedented attention

Andrew D. McAinsh (ed.), *Mitosis: Methods and Protocols*, Methods in Molecular Biology *545*,
DOI 10.1007/978-1-60327-993-2_9,

because of its fundamental relationship to the etiology of cancer. Nevertheless, despite enormous recent progress, a solid view of how chromosomes are segregated during cell division remains elusive. It is, however, well established that the remarkable movements of chromosomes during division are mediated and monitored by the kinetochore, a minute structure that forms the interface between chromosomes and a microtubule-based apparatus known as the mitotic spindle *(1)*.

Over the last years laser microsurgery has been established as one of the most important tools to investigate how mitosis works. The mitotic apparatus is an appealing context for the use of such spatial domain techniques because it contains discrete structures that can be manipulated in order to address their respective roles in the production of pulling or pushing forces to move chromosomes during mitosis. Accordingly, laser microsurgery has been seminal to elucidate the mechanistic basis of the spindle-assembly checkpoint *(2)*, the role of the centrosome in spindle assembly *(3)*, and the role of kinetochores in chromosome movement *(4–6)*. We believe that we can significantly broaden this scenario by combining powerful live cell imaging and state-of-the-art laser technology with molecular tools. Recent advances in the study of how genes function in living cells, when combined with increasingly sophisticated lasers and microscopes with higher spatial (nanometers) and temporal (milliseconds) resolutions, represents the next new challenge for cell surgeons and will provide a powerful approach to unravel the molecular mechanisms behind many concurrent processes that drive mitosis.

Here we propose the systematic use of *Drosophila melanogaster* somatic cells in culture to perform intracellular micromanipulation towards a functional dissection of mitosis. We envision that questions regarding acentrosomal spindle formation, the spindle checkpoint signalling mechanism, the role of kinetochore-microtubule dynamics for chromosome movement, as well as the fate of cells with a compromised mitosis could be addressed by the synergistic combination of laser microsurgery and RNAi tools. Indeed, this approach has already been successfully used in *Drosophila* S2 cells to investigate the role of kinetochores in spindle morphogenesis *(7)*, as well as to elucidate the molecular mechanism of spindle microtubule flux at kinetochores *(8)*.

A first step towards the application of laser microsurgery to *Drosophila* is the establishment of cell lines stably expressing different Green-Fluorescent Protein (GFP) and/or red variant fluorescent-tagged components of the mitotic apparatus (e.g., microtubules, centrosomes, kinetochores, and chromosomes *(9, 10)*). By combining Differential Interference Contrast

(DIC) and epi-fluorescence microscopy one will be able to image at least two of those constituents simultaneously in living cells.

Drosophila provides several key advantages for cell division studies. First, the genome is fully sequenced and more than 60% of *Drosophila* proteins are conserved with humans *(11)*. Furthermore, 75% of genes associated with human diseases have orthologues in *Drosophila (12)*. Second, *Drosophila* has only 4 pairs of chromosomes (in comparison with 23 pairs in the case of human cells), and individual kinetochore-microtubule fibers (K-fibers) can be easily observed at the light microscopy (LM) level. Third, *Drosophila* is one of the most powerful genetic systems known, with an accumulated knowledge encompassing more than 100 years of research, and extensive collections of mutants for genes involved in mitosis are freely available. Fourth, protein redundancy at the cellular level is minimal in *Drosophila*, whereas humans, for example, usually express several isoforms or closely related proteins. Fifth, specific gene silencing can be easily achieved by RNA interference (RNAi), by simply adding dsRNA to the culture medium *(13)*. This is particularly useful for high-throughput genomewide screenings, which are now well established *(14–16)*. One major disadvantage, however, has been related with the fact that most *Drosophila* cell lines are semi-adherent and thus cells are round, making high-resolution microscopy difficult. To overcome this problem we have modified the agar-overlay technique recently described for *Drosophila* culture cells by Fleming and Rieder (2003) *(17)* so that we can control the degree of cell flattening. Unlike cells flattened by growing on a concanavalin A substrate *(18)*, which is particularly useful for routine microscopy studies, the agar overlay approach allows one to select cells that are sufficiently flattened to be imaged without compromising the ability of cells to progress through mitosis and undergo normal cytokinesis. By taking advantage of stable cell lines expressing fluorescent components of the mitotic apparatus, and by focusing high-energy pulses of laser light through a high resolution lens, it would be possible to selectively destroy, cut, and otherwise manipulate K-fibers, chromosomes, kinetochores, and centrosomes in living cells and in different genetic backgrounds.

To conclude, we discuss the possibility of extending some of the above-mentioned methodologies to human cells in culture, particularly the adaptation of the agar overlay technique to HeLa cells. This will improve the resistance and viability of mitotic cells during 3D time-lapse recordings by reducing the phototoxic effects of light and also stabilize the focal plane throughout mitosis. This will facilitate the use of RNAi studies combined with laser micromanipulation during mitosis in a mammalian model system.

2. Materials

2.1. Cell Culture

2.1.1. *Drosophila* Schneider 2 Cells (see Note 1)

1. Disposable 25 cm^2 tissue culture flasks and/or 6-well plates
2. Heat-inactivated Fetal Bovine Serum (FBS; GIBCO #10500-064)
3. Schneider's *Drosophila* medium (Sigma #S0146)
4. Disposable sterile pipettes (1, 5, 10, and 25 ml)

2.1.2. Human HeLa Cells (see Note 1)

1. Dulbecco's Modified Eagle's Medium (DMEM; GIBCO #11880) supplemented with 10% of heat-inactivated FBS
2. Sterile Phosphate Buffered Solution (PBS) pH 7,4
3. 0.05% Trypsin, 0.53 mM EDTA (Trypsin-EDTA, GIBCO #25300)
4. Disposable 25 cm^2 tissue culture flasks (T-flasks) and/or 6-well plates
5. Disposable sterile pipettes (1, 5, 10, and 25 ml)

2.2. Stable Transfections of S2 Cells

1. S2 cells exponentially growing in 6-well plates (1×10^6 cells/ml; *see* **Note 2**)
2. Serum-Free Schneider's *Drosophila* medium (SFM)
3. Complete Schneider's *Drosophila* medium (Schneider's medium supplemented with 10% of heat-inactivated FBS)
4. Insect DNA expression vector (1 μg; *see* **Note 3**)
5. Sterile 1.5 ml microcentrifuge tubes
6. Cellfectin insect transfection reagent (Invitrogen #10362-010)
7. Selection antibiotics (Hygromycin-B, Sigma #H3274, or Blasticidin S HCl – Fluka BioChemika #15205)

2.3. RNAi in S2 Cells

1. S2 cells exponentially growing in 6-well plates (10^6 cells/ml).
2. Schneider's *Drosophila* SFM.
3. Complete Schneider's *Drosophila* medium.
4. Target DNA specific PCR primers containing the T7 RNA polymerase-binding site (**TAATACGACTCAC-TATAGGG**).
5. Linear DNA fragments approximately 700 bp in length (obtained by PCR from the target cDNA or genomic DNA).
6. PCR Clean-Up kit (MoBio Laboratories, Inc. #12500-100; *see* **Note 4**).
7. MEGAscript T7 Kit (Ambion #AM1334).

2.4. Live Cell Microscopy

1. 22 mm nr. 1½ coverslips (Corning #2870-22; *see* **Note 5**).
2. 25 mm nr. 1½ coverslips (Corning #2870-25).

3. Modified Rose chambers *(19)* (*see* **Note 6**).
4. 5 ml syringes.
5. 26G needles.
6. Low melting agarose (SIGMA #A-9414).
7. Complete Schneider's *Drosophila* medium (for S2 cells).
8. Leibovitz's L-15 medium without phenol red (GIBCO #31415) supplemented with 10% FBS (for HeLa cells; *see* **Note 7**).
9. Concanavalin A coated coverslips (for S2 cells; *see* **Note 8**).
10. Poly-L-lysine coated coverlips (for HeLa cells; *see* **Note 9**).

2.5. Laser Microsurgery

1. Nd:YAG laser 2nd harmonic (532 nm) single-mode, 8-ns pulses (ULTRA-CFR TEM00 Nd:YAG from Big Sky Laser, Quantel). Laser pulse transverse profile approximates the fundamental (Gaussian) mode (M-squared factor, which is a measure of beam quality ~1.5).
2. Attenuating stage comprising a half-wave plate and a polarizing cube (CVI Corporation, QWPM-532-05-2-532 and PBSO-532-050) and a beam dump.
3. Pulse energy fine-tuning stage comprising a half-wave plate (CVI Corporation, QWPM-532-05-2-532) mounted on a rotating stage and a $\lambda/8$ plane glass window at Brewster's angle (approximately 56°) and a beam dump.
4. Beam expander (LINOS Photonics) with variable magnification (2–8X) tuned to match the laser beam diameter to the objective pupil diameter.
5. Half-wave plate (Thorlabs, WPMH05M-532) used to align the polarization axis of the beam with one of the principal axes of the DIC Wollaston prism.
6. Periscope (2 mirrors).
7. Dichroic mirror (Semrock, FF493/574-Di01): reflective at 532 nm and transmissive at the GFP emission spectral window.
8. Nikon TE2000U microscope with 'stage-up kit'.

3. Methods

3.1. Cell Culture

Drosophila S2 cells grow at 25°C in Schneider's complete medium.

Human HeLa cells grow at 37°C with 5% CO_2, in DMEM medium supplemented with 10% of heat-inactivated FBS.

3.2. Establishment of Stable S2 Cell Lines Expressing Fluorescently Tagged Proteins

Day 1: Preparation of cells

1. Prepare exponentially growing S2 cells for transfection by seeding 10^6 cells/ml in a 6-well plate in 2 ml complete medium.

Day 2: Transient transfection(*see* **Note 10**)

2. Prepare the following transfection mix (per well) just before transfection:
   Solution **A**: 1 μg of the plasmid DNA
   25 μl of SFM
   Solution **B**: 5 μl of Cellfectin
   25 μl of SFM
3. Slowly add solution **A** dropwise to solution **B** with continuous mixing and leave at room temperature for 45–60 min.
4. Centrifuge the cells at 1000 rpm for 5 min and replace the cell culture medium with 450 μl of fresh SFM.
5. Mix the solution prepared in step 2 and add dropwise to each well. Swirl to mix in each drop.
6. Incubate 3–4 h at 25°C.
7. Replace the SFM with complete medium after washing once.
8. Incubate for 1–2 days.

Day 4: Selection (stable transfection; *see* **Note 10**)

9. Add 1 ml of fresh complete medium and start the selection using the appropriate selection antibiotic, starting with the above-mentioned concentrations. Replace the medium every 4–5 days until resistant cells start expanding. Note that it is normal that those cells which do not incorporate the plasmid will die (*see* **Note 11**).

+2–3 Weeks: Expansion

10. Centrifuge cells and resuspend in complete medium containing the appropriate selection agent. Pass cells at a 1:2 dilution when they reach a density of 6–20 × 10^7 cells/ml. When cells become more concentrated they may be passed at higher dilutions, typically 1:5 or 1:10.
11. Expand resistant cells into 25 cm^2 flasks and test for expression of the protein of interest by Western blot and/or fluorescence microscopy.

3.3. RNAi

3.3.1. Preparation of dsRNA

1. Design 18-mer sequence-specific oligonucleotides to make a PCR product of ~700 bp from the cDNA or genomic DNA of interest (*see* **Note 12**). Don't forget to incorporate the 5' T7 RNA polymerase binding site into your primers.
2. Prepare 10–12 individual PCR reactions on ice containing the following reagents:

- 0.5–1 ng of template cDNA from a plasmid or 0.5–1 μg of genomic DNA
- 1 μM of each primer
- 0.2 mM of dNTP mix
- Taq polymerase enzyme buffer
- Enzyme buffer
- 2 mM $MgCl_2$ (if not included in the enzyme buffer)
- Water to a final volume of 100 μl (take in consideration the volume of Taq polymerase that will be added afterwards)

3. Set up the following PCR program (this may have to be adjusted to your specific conditions):
   - 94°C – 2 min
   - Add 2.5 U of Taq polymerase to the PCR reaction
   - 94°C – 30 s
   - 55°C – 60 s × 30 cycles
   - 72°C – 60 s
   - 72°C – 10 min
4. Purify the PCR products by pooling 3 reactions, and running them through each column from a PCR Clean-Up Kit according to the manufacturer's instructions.
5. Test 1 μl of the clean product by electrophoresis on a 1% agarose gel and quantify the DNA by measuring Abs_{260}. The DNA yield is calculated as follows:
   Abs_{260} × dilution factor × 50 = DNA conc. in μg/ml
6. Use this DNA as template for at least 10 in vitro RNA synthesis reactions with the MEGAscript T7 Kit (Ambion) according to the manufacturer's instructions, except that the incubation time should be increased to at least 6 h (*see* **Note 13**).
7. Pool the reactions into a single tube and precipitate the RNA with LiCl and RNAse-free water according to the instructions included with the kit.
8. Carefully wash the pellet once with 70% ethanol and let it air-dry.
9. Resuspend the pellet in 100 μl of nuclease-free water and check the RNA concentration as before but using the following algorithm:
   Abs_{260} × dilution factor × 40 = RNA conc. in μg/ml
10. Denature RNA secondary structures by heating at 65°C for 30 min in a beaker containing 200 ml of previously warmed water and then let it cool down slowly to room temperature to make dsRNA duplexes.
11. Test 1 μl of the dsRNA by electrophoresis in 1% agarose gel. It should run like DNA as a clean band of ~700 bp. Sometimes a lower band is also visible. This corresponds to ssRNA that did not anneal into duplexes.
12. Store the dsRNA at –20°C.

3.3.2. Live Cell Imaging of Depleted Cells

1. Grow *Drosophila* S2 cells in T-flasks at 25°C in Schneider's complete medium for 4 days.
2. Check how many time points (how many wells) will be needed for the experiment and whether an efficient depletion (*see* **Note 14**) requires a second pulse of dsRNA.
3. Distribute 1 ml of serum free media containing 10^6 cells per each well.
5. Add 30 μg of specific dsRNA to half of the wells, and an equivalent amount of control dsRNA to the other half. Mix well by swirling and leave the cells at 25°C for 1 h to allow incorporation of the dsRNA.
6. Add 2 ml of complete Schneider's *Drosophila* medium to each well and put the cells back in the incubator at 25°C.
7. At the specific timepoint use a cell scraper to collect an S2 cell suspension for microscopy analysis and also take a sample for Western blot (typically 10^6 cells per time point) to monitor knockdown efficiency.

3.4. Flattening Cells for Live Cell Microscopy

3.4.1. S2 Cells (Concanavalin A; see ***Note 15****)*

1. Place 22 × 22 mm coverslips previously coated with concanavalin A in the bottom of a 6-well plate. Put 300 μl of an S2 cell suspension on top of the coated coverslip and allow cells to adhere to the glass at 25°C for 2 h.
2. Mount modified Rose chambers and observe under the microscope.

3.4.2. S2 Cells (Agar Overlay)

1. Prepare a 170 μm thick layer of agarose as follows (**Fig. 9.1A**):
   i. Place two rectangular coverslip fragments (obtained from a 25 × 25 mm coverslip) on opposite ends of a slide, to act as spacers. Put a drop of PBS or water to stick the spacers to the slide (prepare another spare slide that will be required on step 3, but using ¼ of a 25 × 25 mm coverslip as spacers and without sticking them to the slide).
   ii. Using a heater, melt 0.1 g of low melting agarose in 5 ml of Schneider's SFM.
   iii. After heating, supplement the mixture with 10% FBS and pipette the liquid agarose into the space between the coverslip fragments.
   iv. Place another slide on top of the agarose to form a sandwich and wait until it solidifies. This can be kept at 4°C for a week in a humid chamber.
   v. Carefully separate the two slides with a razor blade.

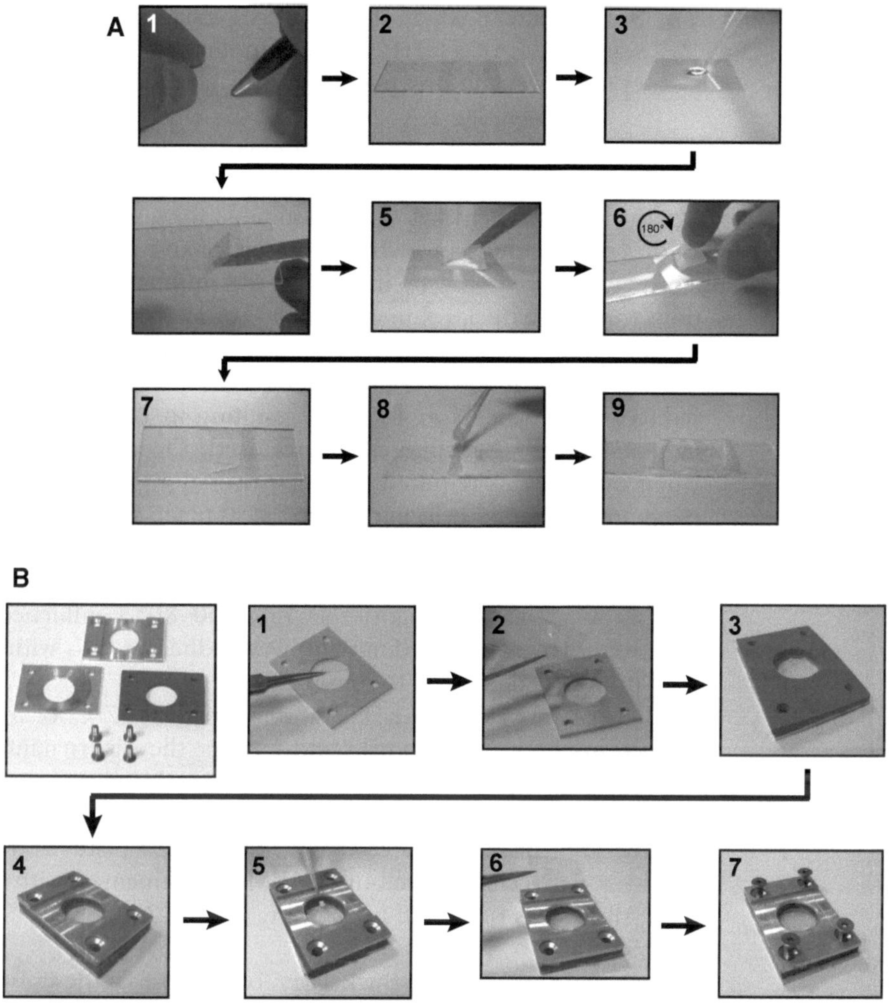

Fig. 9.1. Illustration of the agar overlay flattening procedure. (**A**) For slide preparations two ¼ fragments of a 25 × 25 mm coverslip are cut with a diamond pen (1). These fragments will work as spacers (2). A drop of cell suspension is on a 25 × 25 mm coverslip (3). A small agarose square piece (4) is placed on top of the cells sitting on the coverslip (5), which is then carefully flipped 180° and positioned in contact with one of the spacers on the slide (6). The medium should only make contact with one of the spacers to generate a gradient of flatness (7). With the help of a cotton bud the coverslip edges are sealed with warmed VALAP (8). When it solidifies the chamber is ready for observation (9). (**B**) For Rose chambers, HeLa cells are grown on a 22 × 22 mm poly-L-lysine coverslip and placed on the flat surface from the chamber part that will be proximal to the objective (1). A square piece of the agarose layer is positioned on top of the cells (2), followed by the silicon part of the chamber (3). The top part of the chamber that will be closer to the condenser is put with the flat surface in contact with the silicon (4). A small piece of filter paper is then soaked with medium to prevent fast evaporation (5) and the chamber is closed by placing another 22 × 22 mm coverslip in the top part of the chamber (6). Finally, four stainless steel bolts are used to seal the chamber (7).

2. Cut a 1 × 1 cm piece from the agarose layer and gently place it on top of a drop of an S2 cell suspension that was previously placed on a 25 × 25 mm coverslip.
3. Carefully invert the coverslip containing the cells on agarose onto the new slide (**step i**; *see* **Note 16**).

4. Seal the lateral edges of the coverslip/slide preparation with warm VALAP (1:1:1 vaseline:lanolin:paraffin) to prevent evaporation and observe under the microscope (**Fig. 9.2**).

3.4.3. HeLa Cells (Agar Overlay)

HeLa cells are one of the most popular model systems for the study of mitosis in humans. However, their round morphology during mitosis makes them unsuitable for several microscopy studies such as those involving fluorescence and DIC. This problem may be overcome by increasing the number of optical sections through a thick volume (typically 10–12 μm), which implies higher exposure to light. In order to avoid this we have adapted our agar overlay protocol to HeLa cells (**Fig. 9.1B**), which reduces the thickness to 4–5 μm, resulting in 2–3× decrease in light exposure when fluorescence is used. On the other hand, when only DIC is required, the agar overlay reduces cell movement and maintains cells in the same focal plane over time as they progress through mitosis (**Fig. 9.3**).

1. Grow HeLa cells in DMEM supplemented with 10% FBS at 37°C with 5% CO_2 until they reach 60–80% confluence.
2. Remove growth medium and wash adherent cells with PBS prewarmed at 37°C to remove FBS.
3. Add 1 ml of trypsin-EDTA and incubate at 37°C for 5 min.
4. Resuspend detached cells and transfer the supernatant to a tube with 2 ml of media with FBS (to inactivate trypsin). Place a poly-L-lysine treated 22 × 22 mm coverslip to each 35 mm plate (or to each well of a 6-well plate) and seed 4×10^5 cells per well in DMEM supplemented with FBS. Allow cells to adhere by growing for 18–24 h at 37°C with 5% CO_2.
5. Prepare a 170 μm thick layer of agarose as described in **Section 3.4.2, step ii** of the agar overlay protocol for S2 cells but replacing Schneider's SFM by Leibovitz's L-15 SFM.
6. Mount the coverslip with adherent cells in an open Rose chamber and add a small drop of L-15 supplemented with 10% FBS. Cut a 1 × 1 cm piece from the agarose layer and gently place it on top of the cells.
7. Carefully add enough L-15 supplemented with 10% FBS to the cells just to prevent dehydration but without detaching the agar.
8. Close the chamber with a clean 22 × 22 mm coverslip and observe the cells by DIC + fluorescence at 37°C.

3.5. Laser Microsurgery

1. The setup used for laser microsurgery is designed to ablate intracellular structures at the submicron level. The schematic representation of the components of the workstation is shown in detail in **Fig. 9.4**. We use an infinity-conjugated Nikon Eclipse TE2000-U microscope with a

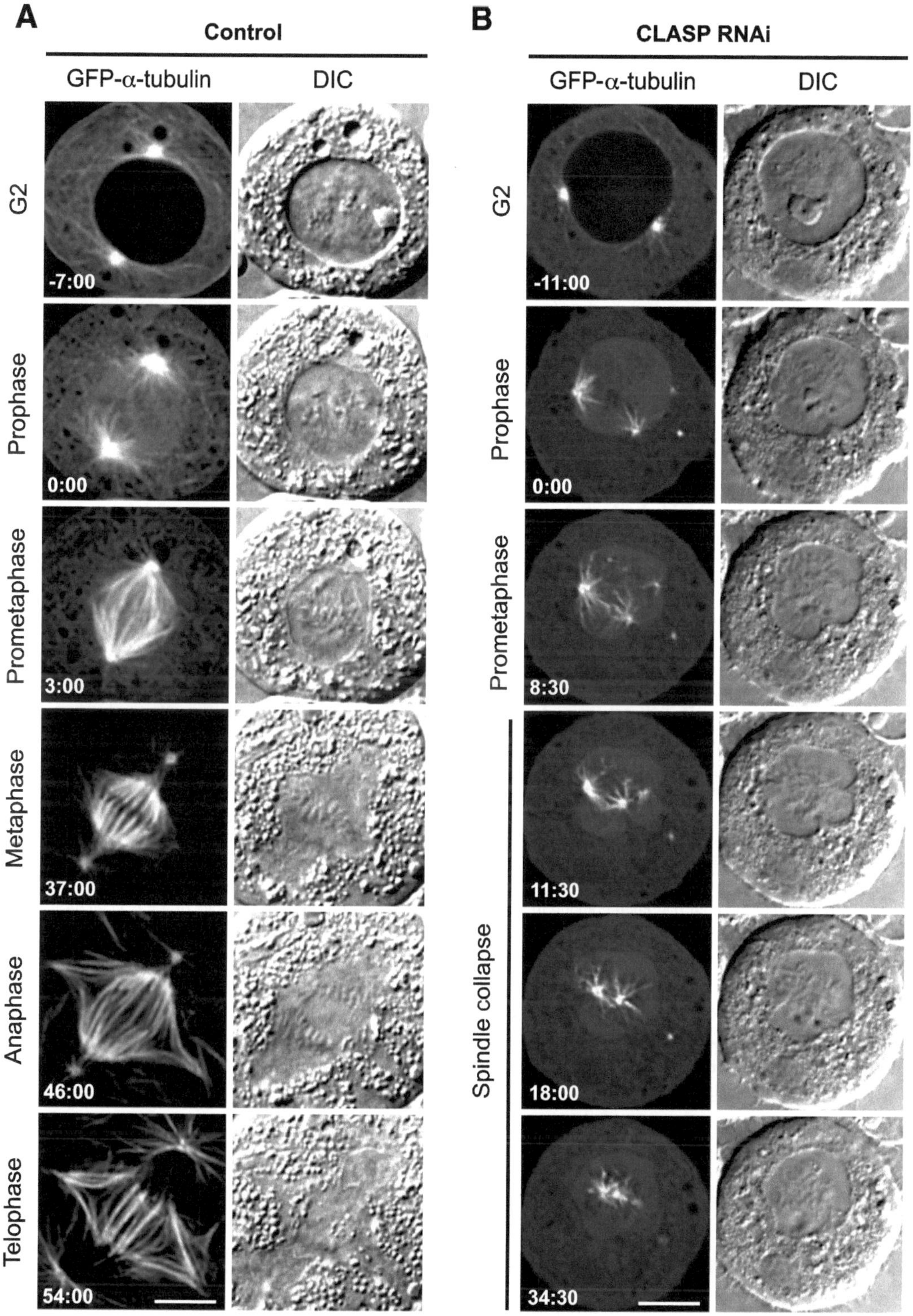

Fig. 9.2. The agar overlay technique in *Drosophila* S2 cells. (**A**) Time lapse sequence of control and (**B**) CLASP-depleted S2 cells expressing GFP-α-tubulin and DIC. Time in min:sec. Scale bar is 5 μm.

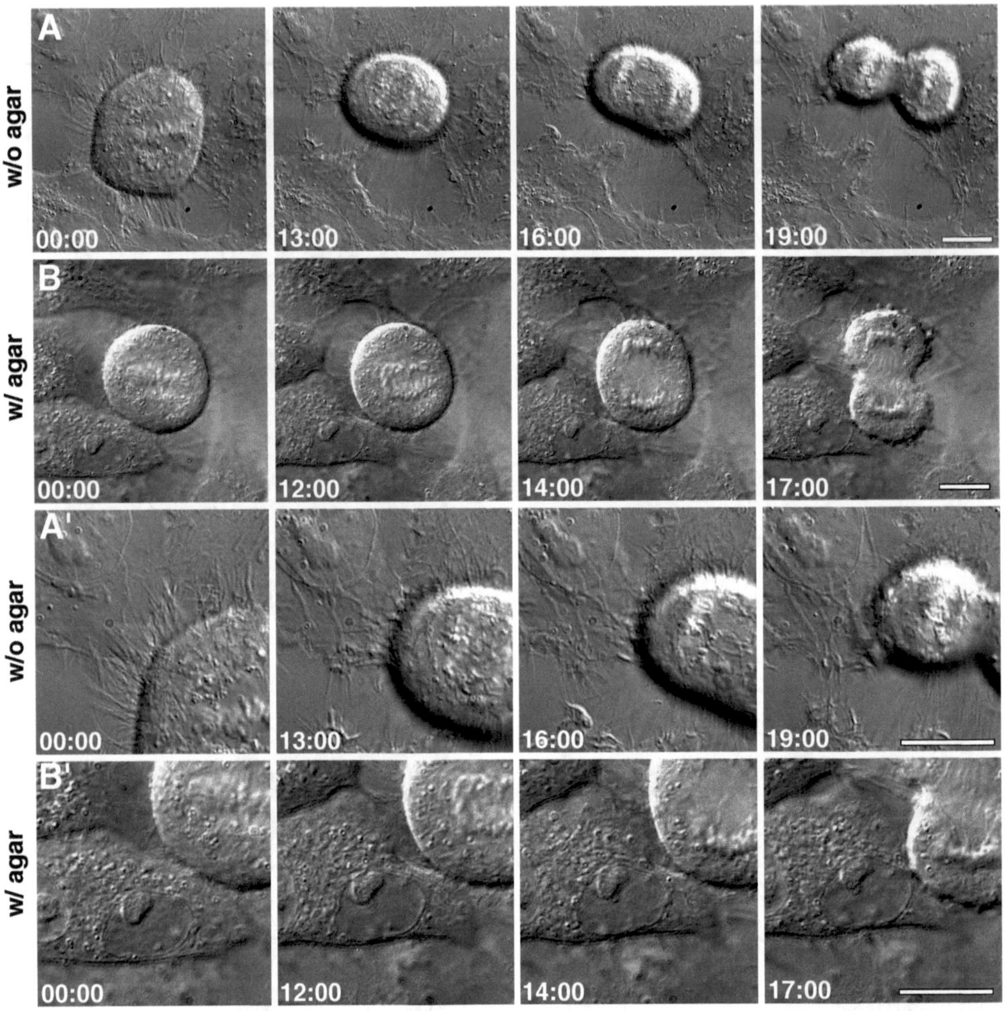

Fig. 9.3. Adaptation of the agar overlay technique to human HeLa cells. (**A**) Time-lapse sequence of a HeLa cell growing on poly-L-lysine coated coverslips, without agar, at the metaphase–anaphase transition. (**B**) Time-lapse sequence of a HeLa cell growing on poly-L-lysine coated coverslips, with agar overlay, at the metaphase–anaphase transition. (A', B') Higher magnification views from the same cells shown in A and B. The focal plane was kept at a fixed position in each case throughout the experiment. Note that mitotic cells filmed under the agar-overlay remained focused and fairly immobile, whereas mitotic cells without agar were highly mobile and chromosomes quickly went out of focus after anaphase onset. As a reference, a nucleus/nucleolus from a neighboring interphase cell is shown for each situation. Time is shown in min:sec. Scale bar is 10 μm.

'stage-up kit' and an extra filter turret (**Fig. 9.5**) added in the plane-wave region of the microscope (between the objective and the tube lens), which allows the injection of external, collimated, light beams into the microscope's light path.

2. We use 532 nm nanosecond pulses to perform microsurgery. This option is mainly driven by the fact that similar pulse parameters have been used successfully in the past

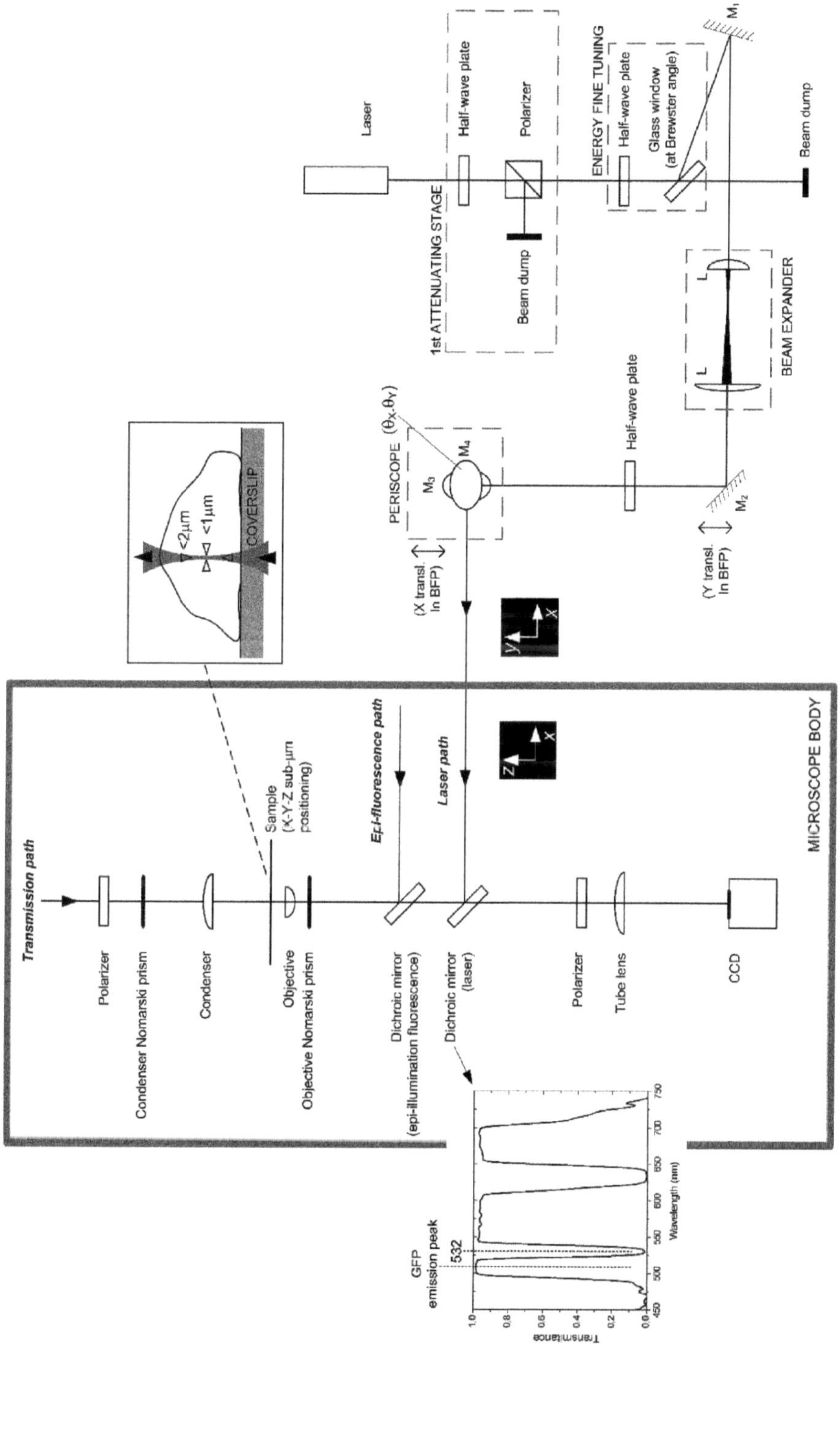

Fig. 9.4. Optical setup for laser microsurgery and imaging.

Fig. 9.5. Laser microsurgery workstation. (**A**) General picture of the 1 × 1 m workstation with the environmental control chamber. (**B**) and (**C**) General view of the external optics that control and steer the laser beam into the microscope. (**D**) Laser source.

(9, 10). It should be stressed, however, that, given the nonresonant character of the light–matter interaction processes relevant to microsurgery, other wavelengths, as well as other pulsewidths, may be used *(20)*.

3. A number of laser beam properties have to be modulated before steering into the microscope body, namely the beam diameter, energy, polarization, and pointing direction. Our laser is much more energetic (about 1000-fold) than needed (1–2 μJ), so we used a first strong attenuating stage composed of a half-wave plate and a polarizing cube. We have set this attenuation to approximately its maximum value (but not quite, as the maximal extinction regime may normally introduce aberrant phase profiles) and this is used as a constant attenuator in our setup. These two components are subjected to the highest irradiance in the setup: it should be guaranteed that the damage threshold of the optics coatings is not exceeded.
4. The second stage is used for fine-tuning the pulse energy. The polarizing cube is here substituted by a plane, uncoated, glass plate with surface quality $\lambda/8$ at the Brewster's angle, where polarization selection is highest. This is much cheaper than the polarizing cube and serves perfectly the purpose of varying beam attenuation upon half-wave plate rotation.

5. Beam expansion is performed using a variable expander although a custom 2-lens system may be used if the needed magnification is constant. Matching the beam diameter to the objective pupil diameter allows usage of the full numerical aperture of the objective, hence minimizing laser spot size: the expected diameter, *d*, for a (nontruncated) Gaussian beam and high-quality focusing optics is approximated by $d \approx \lambda/NA$, where λ is the wavelength inside the medium and NA is the numerical aperture of the objective. On the other hand, excessive expansion leads to strong truncation of the laser beam (at the objective pupil) which may introduce diffraction rings at possibly significant levels. However, if the diffraction rings are kept weak, it may be preferable to introduce them rather than losing focusing capacity of the central diffraction disk. The focal plane of the laser should coincide (within a fraction of the wavelength) with the focal plane of the imaging system; this can be guaranteed either by axial translation of the CCD or of the second lens of the beam expander.
6. Beam diameter, along with beam quality, is the determinant to approach diffraction-limited ablation. A typical bulk measure of beam quality is the M-squared parameter, which for our laser is approximately 1.5. This roughly means that the minimal spot diameter is increased by 50%. For lower-quality beams, spatial filtering should be considered to block propagation of high-order (non-Gaussian) beam modes.
7. Before steering the beam into the microscope body, a half-wave plate is used to align the polarization axis of the beam with one of the principal axes of the objective DIC crystal. This prevents the beam from being split into two 'DIC-like' beams.
8. Adjustment screws are used for optical alignment. Mirror M_2 and the periscope (M_3 and M_4) are translated to align the beam in the *x*- and *y*-axes of the objective's Back-Focal Plane (BFP), respectively. Mirror M_4 tilt adjustment allows the alignment of the beam in the *x*- and *y*-axis of the focal (sample) plane. Operationally, apart from casual alignments, we maintain mirror M_4 steering the beam roughly to the center of the field of view, and use the *xy*-stage to move the sample and choose the ablation target. In our particular case, we define a 512×512 pixel region-of-interest in the CCD centered in the actual beam spot. Fine-tuning the CCD region-of-interest center instead of carefully aligning the beam with the microscope optical axis is perfectly acceptable for small beam deviations.
9. The laser beam is steered into the microscope optical axis through a dichroic mirror with the convenient spectral

features: reflective at the laser line and transmissive at the GFP spectral window. Our particular dichroic mirror also has the following positive and negative features: (i) it is reflective at the 480 nm region, which allows the use of this spectral region for GFP point-bleaching laser injection; and (ii) it is not highly transmissive in the emission spectral region of TexasRed and similar fluorescent markers (*see* **Note 17**). In summary, if only the microsurgery 532 nm

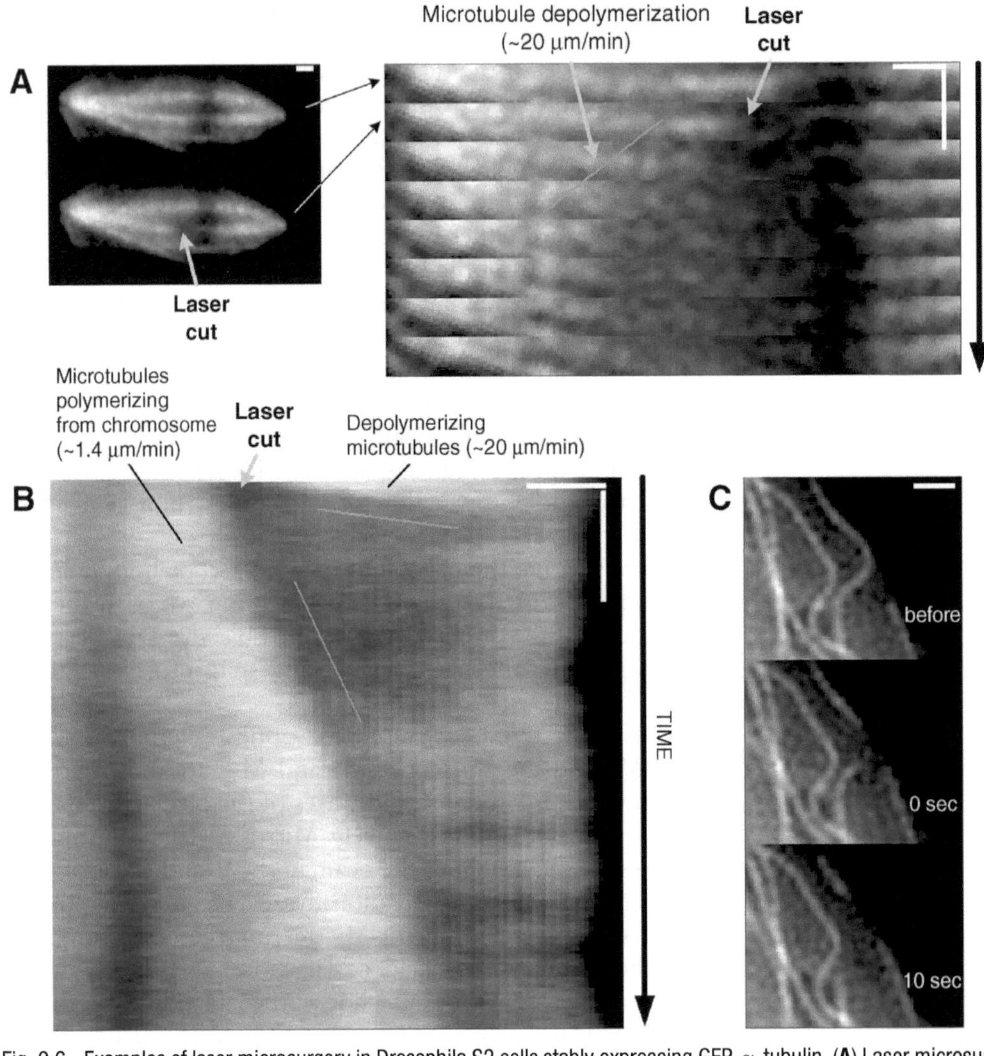

Fig. 9.6. Examples of laser microsurgery in Drosophila S2 cells stably expressing GFP-α-tubulin. (**A**) Laser microsurgery on a kinetochore-fiber in a metaphase spindle from S2 cells. The unstable, newly exposed, microtubule plus-ends prompt full depolymerization of the kinetochore-fiber at a very high rate (~20 μm/min). Horizontal bars: 1 μm, vertical bar: 10 s. (**B**) Similar experiment in another cell. The target is in this case more distant from the chromosome, leaving a small portion of the microtubule fiber attached to the kinetochore. In addition to the rapid catastrophe on the pole-directed fiber portion, polymerization is observed from the kinetochore at approximately 1.4 μm/min. Horizontal bar: 1 μm, vertical bar: 30 s. (**C**) Laser microsurgery on an interphase microtubule. Horizontal bar: 1 μm.

laser is to be coupled to the microscope, we advise the use of a 'laser-line' dichroic mirror, that is, one that reflects in the 532 region and is transmissive otherwise (*see* **Fig. 9.6** for examples of laser microsurgery).

10. Operationally, we use Nikon's NIS-Elements to control the whole system with the exception of the laser itself, which is controlled through a custom routine written in Labview where the laser pulse train properties are set, namely the number of pulses (typ. 1–5 pulses), the pulse train frequency (typ. 20 Hz), and the pulse energy (typ. 1–2 μJ).

4. Notes

1. Ideally tissue culture should be performed without the use of antibiotics, which may select particular resistant subclones. Nevertheless, if the conditions for cell culture are not ideal, especially regarding *Drosophila* S2 cells that do not require sophisticated environmental control, the use of an antibiotic/antimycotic cocktail is recommended (penicillin G + streptomycin sulphate + amphotericin B; Sigma #A5955).
2. Exponential growth of *Drosophila* S2 cells is usually achieved 2–3 days after each passage.
3. We have been successfully using pAc5.1/V5-His A, B, C (Invitrogen # V411020) and pMT/V5-His A, B, C (Invitrogen # V412020) for constitutive or inducible expression of fluorescent fusion proteins, respectively.
4. Alternatively, it is possible to gel-purify the PCR products using a gel extraction kit (Qiagen #28704).
5. Most objective lenses of the main microscope manufacturers are infinity-corrected assuming a glass coverslip with 0.17 mm thickness. This thickness corresponds to coverslip nr. $1\frac{1}{2}$ (range from 0.16 to 0.19 mm).
6. Rose chambers are a proven, cost-effective alternative to reusable chambers for live-cell imaging. We designed our own modified version of Rose chambers taking into account our specific objective features and to allow different assembly options (e.g., open or closed chambers). Nevertheless, there are other possible commercial alternatives. We recommend the MatTek glass bottom culture disks (MatTek #P35G-1.5-20-C). The same principles discussed previously regarding coverslip thickness also apply in this case.
7. Phenol red is a common pH indicator used in cell culture media that may interfere with live cell microscopy fluores-

cent assays. Therefore, we highly recommended the use of media without this dye that are usually available from most manufacturers.

8. For coating with concanavalin A, immerse acid-treated coverslips into a concanavalin A (Calbiochem #2345467) solution (0.5 mg/ml). Carefully place the coverslips against the sides of a Petri dish forming ~45° angles, with filter paper at the bottom and let them air-dry. Before using the coated coverslips, put them under UV light for 20 min.
9. For coating with poly-L-lysine, place a 1:10 dilution of a poly-L-lysine solution 0.1% w/v (Sigma #P8920) in PBS on top of coverslips and leave it at room temperature for 1 h. Wash extensively with PBS. Carefully place the coverslips against the sides of a Petri dish forming ~45° angles, with filter paper at the bottom and let them air-dry. Before using the coated coverslips, put them under UV light for 20 min.
10. In order to generate a stable cell line, we highly recommend transfecting exponentially growing S2 cells with a single-expression vector that includes both the antibiotic resistance gene for selection of transfected cells and the gene of interest. In the lab we have been using either Hygromycin (300 μg/ml) or Blasticidin (25 μg/ml) with 100% success in several different cell lines. It is also possible to obtain stable transfections by using two separate vectors, one expressing the antibiotic resistance gene and the other expressing the gene of interest. However, the efficiency drops considerably, given that not all the antibiotic resistant cells will be expressing the gene of interest.
11. *Drosophila* S2 cells grow better if kept concentrated in culture. However, after antibiotic selection of transfected cells, the vast majority of cells die and survivor cells take a long time to recover. In order to overcome this problem, it is recommended to change the selection medium with antibiotic for fresh complete medium after the first day of selection. Allow cells to recover for 2–3 days and restart the selection protocol.
12. In some cases, RNAi efficiency increased when this PCR product covered the codon of the first methionine but we and others have successfully knockdown genes using a PCR product exclusively covering the 5' untranslated region. It is important to note it is highly recommended to perform BLAST searches with the corresponding target sequence to rule out the possibility of off-target effects.
13. Work in an RNAse-free environment. The major source of RNAse is usually the operator, hence we highly recommend the use of gloves.

14. The depletion efficiency should be monitored by Western blot, which will give a rough estimate of the total protein present in a cell population, and also by immunofluorescence, which will give a more accurate idea of the protein present in a cell-by-cell basis.
15. Note that this type of coating is known to interfere with cytokinetic furrow ingression.
16. It is important to note that less fluid produces greater flattening, but also reduces viability over time. The ideal situation is when one leaves only enough media to make contact with only one of the spacers. As a result cells closer to the dry spacer remain relatively rounded, whereas cells closer to the wet spacer became extremely flat. This approach allows the formation of a gradient of flatness to select those cells that were ideal for high-resolution light microscopy analyses, but were not inhibited in their mitotic progression or cytokinesis.
17. It should be noted that although the 532 nm laser only ablates material within a well-defined region (once the pulse energy is properly tuned), the beam profile at the focal plane necessarily spreads out to distant regions and may be enough to bleach fluorescent markers that absorb (even if slightly) in the green region, as with mRFP, mCherry, and many spectrally similar markers *(21)*. In those cases where dual-color imaging is needed, other laser wavelengths should be considered to perform the microsurgery.

Acknowledgments

We are greatly indebted to Alexey Khodjakov and Valentin Magidson for their invaluable advice regarding the development of our laser microsurgery system. We would also like to thank all the colleagues working with *Drosophila* S2 cells, especially Gohta Goshima, Monica Bettencourt-Dias, and Mar Carmena, for sharing materials and for their intellectual contribution related with the development and optimization of the techniques described in this chapter. I.M. and M.L-F., respectively, hold a Ph.D. studentship (SFRD/BD/22020/2005) and a postdoctoral fellowship (SFRH/BPD/26780/2006) from Fundação para a Ciência e a Tecnologia of Portugal. The setup of the laser microsurgery system was supported by grants from the Luso-American Foundation (L-V-675/2005), Crioestaminal/Viver a Ciência and the Gulbenkian Programmes for Research Stimulation and Frontiers in the Life Sciences. Work in the lab of

H.M. is supported by grants POCI/SAU-MMO/58353/2004 (POCI2010 and FEDER) and PTDC/BIA-BCM/66106/2006 from Fundação para a Ciência e a Tecnologia of Portugal.

References

1. H. Maiato, J. DeLuca, E. D. Salmon, W. C. Earnshaw, The dynamic kinetochore-microtubule interface. *J Cell Sci* **117**, 5461 (Nov 1, 2004).
2. C. L. Rieder, R. W. Cole, A. Khodjakov, G. Sluder, The checkpoint delaying anaphase in response to chromosome monoorientation is mediated by an inhibitory signal produced by unattached kinetochores. *J Cell Biol* **130**, 941 (Aug, 1995).
3. A. Khodjakov, R. W. Cole, B. R. Oakley, C. L. Rieder, Centrosome-independent mitotic spindle formation in vertebrates *Curr Biol* **10**, 59 (Jan 27, 2000).
4. S. Brenner, D. Pepper, M. W. Berns, E. Tan, B. R. Brinkley, Kinetochore structure, duplication, and distribution in mammalian cells: analysis by human autoantibodies from scleroderma patients. *J Cell Biol* **91**, 95 (Oct, 1981).
5. A. Khodjakov, C. L. Rieder, Kinetochores moving away from their associated pole do not exert a significant pushing force on the chromosome. *J Cell Biol* **135**, 315 (Oct, 1996).
6. A. Khodjakov, R. W. Cole, A. S. Bajer, C. L. Rieder, The force for poleward chromosome motion in Haemanthus cells acts along the length of the chromosome during metaphase but only at the kinetochore during anaphase. *J Cell Biol* **132**, 1093 (Mar, 1996).
7. H. Maiato, C. L. Rieder, A. Khodjakov, Kinetochore-driven formation of kinetochore fibers contributes to spindle assembly during animal mitosis. *J Cell Biol* **167**, 831 (Dec 6, 2004).
8. H. Maiato, A. Khodjakov, C. L. Rieder, Drosophila CLASP is required for the incorporation of microtubule subunits into fluxing kinetochore fibres. *Nat Cell Biol* 7, 42 (Jan, 2005).
9. A. Khodjakov, R. W. Cole, C. L. Rieder, A synergy of technologies: combining laser microsurgery with green fluorescent protein tagging. *Cell Motil Cytoskeleton* **38**, 311 (1997).
10. V. Magidson, J. Loncarek, P. Hergert, C. L. Rieder, A. Khodjakov, Laser microsurgery in the GFP era: a cell biologist's perspective. *Methods Cell Biol* **82**, 239 (2007).
11. M. D. Adams *et al.*, The genome sequence of Drosophila melanogaster. *Science* **287**, 2185 (Mar 24, 2000).
12. M. E. Fortini, M. P. Skupski, M. S. Boguski, I. K. Hariharan, A survey of human disease gene counterparts in the Drosophila genome. *J Cell Biol* **150**, F23 (Jul 24, 2000).
13. H. Maiato, C. E. Sunkel, W. C. Earnshaw, Dissecting mitosis by RNAi in Drosophila tissue culture cells. *Biol Proced Online* **5**, 153 (2003).
14. M. Boutros *et al.*, Genome-wide RNAi analysis of growth and viability in Drosophila cells. *Science* **303**, 832 (Feb 6, 2004).
15. A. Echard, G. R. Hickson, E. Foley, P. H. O'Farrell, Terminal cytokinesis events uncovered after an RNAi screen. *Curr Biol* **14**, 1685 (Sep 21, 2004).
16. G. Goshima *et al.*, Genes required for mitotic spindle assembly in Drosophila S2 cells. *Science* **316**, 417 (Apr 20, 2007).
17. S. L. Fleming, C. L. Rieder, Flattening Drosophila cells for high-resolution light microscopic studies of mitosis in vitro. *Cell Motil Cytoskeleton* **56**, 141 (Nov, 2003).
18. S. L. Rogers, G. C. Rogers, D. J. Sharp, R. D. Vale, Drosophila EB1 is important for proper assembly, dynamics, and positioning of the mitotic spindle. *J Cell Biol* **158**, 873 (Sep 2, 2002).
19. G. Rose, A separate and multipurpose tissue culture chamber. *Texas Reports on Biol. and Med.* **12**, 1074 (1954).
20. M. W. Berns, A history of laser scissors (microbeams). *Methods Cell Biol* **82**, 1 (2007).
21. N. C. Shaner *et al.*, Improved monomeric red, orange and yellow fluorescent proteins derived from Discosoma sp. red fluorescent protein. *Nat Biotechnol* **22**, 1567 (Dec, 2004).

Chapter 10

Fluorescence Imaging of the Centrosome Cycle in Mammalian Cells

Suzanna L. Prosser and Andrew M. Fry

Abstract

The formation of a bipolar spindle is essential for the equal segregation of duplicated DNA into two daughter cells during mitosis. Spindle bipolarity is largely dependent on the mitotic cell possessing two centrosomes that can each establish one spindle pole. The centrosome is also now known to regulate many other aspects of cell cycle progression, including G1/S progression, spindle orientation and symmetry, cytokinesis, and checkpoint signalling. As a result, defects in centrosome arrangement or number can lead to loss of cell polarity, defective cell division, and abnormal chromosome segregation, all events that are typical of cancer cells. Indeed, cancer cells often exhibit overduplicated centrosomes and multipolar spindles. Here, we outline a number of fluorescence imaging methodologies that can be used to study events of the centrosome duplication cycle, as well as the dynamics of individual centrosome proteins. Specifically, we discuss the generation and imaging of cell lines with fluorescently labelled centrosomes, the use of photobleaching methods to measure the dynamics of centrosome proteins, and assays for observing centrosome overduplication and centrosome separation in fixed and live cells. These experimental approaches can provide important information on the regulation of centrosomes, their role in normal cell cycle progression and how their deregulation might contribute to the deleterious phenotypes of malignant cancer cells.

Key words: Centrosome, mitosis, cell cycle, centriole, fluorescence microscopy, time-lapse imaging, mitotic spindle, FRAP, FLIP.

1. Introduction

The centrosome is the primary site of microtubule nucleation in animal cells and therefore plays a major role in organizing the cytoskeleton during both interphase and mitosis *(1, 2)*. It is a non-membranous organelle, approximately one micron in diameter, that sits in the cytoplasm but is usually located close to and

Andrew D. McAinsh (ed.), *Mitosis: Methods and Protocols*, Methods in Molecular Biology *545*,
DOI 10.1007/978-1-60327-993-2_10, © Humana Press, a part of Springer Science+Business Media, LLC 2009

connected with the nucleus *(3, 4)*. At the beginning of the cell cycle, it is present as a single copy organelle within the cell. However, upon cell cycle progression it undergoes a single duplication event to produce two centrosomes prior to mitotic onset *(5, 6)*. As each centrosome can assemble one spindle pole, the duplication of a centrosome once and once only per cell cycle is critical to ensure that the mitotic spindle has a bipolar structure.

Morphologically, a centrosome consists of two centrioles surrounded by pericentriolar material (PCM). Each centriole is formed from a ninefold symmetrical array of triplet microtubules *(2, 7)*, whilst the PCM is a dynamic, but well-organised, lattice that anchors the γ-tubulin ring complexes responsible for microtubule nucleation *(8, 9)*. As a result of the duplication process in the previous cell cycle, there is an older centriole, known as the mother, and a younger centriole, called the daughter. These can be distinguished by electron microscopy as only the mother centriole possesses appendages towards its distal end. Throughout interphase, the mother and daughter centrioles remain joined at their proximal ends by a fibrous linkage that consists primarily of the proteins C-Nap1, rootletin, and Cep68 *(10–14)*.

Centrosome duplication involves the generation of two additional centrioles, initially called procentrioles, that lie perpendicular to and in very close proximity with the proximal ends of the mother and daughter centrioles. As such, centrosome duplication can be considered to be a semi-conservative event as each new centrosome contains one old and one new centriole. Procentriole formation is initiated at the G1/S transition, thereby linking centrosome duplication with DNA replication *(5, 15, 16)*. Procentriole extension occurs in a proximal to distal direction throughout the S and G2 phase, whereas, in late G2, centrosomes also undergo a maturation event that is characterised by an increase in size due to the recruitment of additional PCM components *(17)*. This includes recruitment of extra γ-tubulin ring complexes in preparation for increased microtubule-nucleating activity at mitosis. The duplicated centrosome continues to act as a single microtubule organising centre until mitotic prophase. At this time, the protein kinase Nek2 phosphorylates components of the intercentriolar linker inducing its disassembly in a process referred to as centrosome disjunction *(10, 18, 19)*. Motor proteins then exert forces on microtubules emanating from opposing spindle poles to push the centrosomes apart leading to the formation of the mitotic spindle *(20)*. Finally, the centrosome duplication cycle is completed in late mitosis when the two centrioles within each spindle pole lose their perpendicular orientation. It is this disorientation of the older centriole, that will become the new 'mother', from the younger centriole, that will become the new 'daughter', that licenses centrioles for the next round of duplication in the following interphase *(16, 21, 22)*.

Although the centrosome is a nonmembranous organelle, it maintains a highly complex structure with a composition that changes during cell cycle progression. In addition to recruiting core components during the processes of centrosome duplication and maturation, and losing linker proteins during centrosome disjunction and possibly centriole disorientation, centrosomes also act as scaffolds for the transient association of regulatory molecules that control diverse cellular activities *(23, 24)*. In this chapter, we describe the application of fluorescence imaging approaches to the study of events associated with the centrosome duplication cycle, as well as the measurement of the dynamics of individual centrosome proteins. In particular, we describe the generation of cell lines with fluorescent centrosomes, the use of fluorescence recovery after photobleaching (FRAP), and fluorescence loss in photobleaching (FLIP) methods to study centrosome protein dynamics, and assays for observing centrosome overduplication and centrosome disjunction. The use of all these methods has significantly contributed to our knowledge of centrosome regulation and function and will continue to be vital for the characterisation of novel centrosome proteins.

2. Materials

2.1. Cell Culture

1. Dulbecco's Modified Eagle's Media with GlutaMAX™-I (DMEM; Gibco) supplemented with 10% heat-inactivated foetal bovine serum (FBS; Gibco) and penicillin/streptomycin (Gibco) at 100 units/ml and 100 μg/ml, respectively.
2. Ham's F12 with GlutaMAX™-I (Gibco) with 10% FBS and penicillin/streptomycin at 100 units/ml and 100 μg/ml, respectively.
3. Phosphate buffered saline (PBS): Prepare 10 × stock with 1.37 M NaCl, 26.8 mM KCl, 2.7 mM Na_2HPO_4, 1.4 mM KH_2PO_4, and adjust pH to 6.8 with HCl. Prepare working solution by dilution of one part stock with nine parts water, autoclave, and store at room temperature.
4. PBS-EDTA: 0.5 mM EDTA in 1× PBS, autoclave, and store at room temperature.

2.2. Transient Transfections and Generation of Stable Cell Lines

1. Lipofectamine™ 2000 Reagent (Invitrogen).
2. Opti-MEM with GlutaMAX™-I (Gibco).
3. G 418 (Sigma-Aldrich), prepare 50 mg/ml solution in ddH_2O. Filter-sterilise through 0.22 μm or 0.45 μm pore size filter. Aliquot and store at –20°C, stable for 12 months.
4. Sterile cloning disks, size 3 mm (Sigma-Aldrich).

2.3. Live Cell Imaging and Photobleaching

1. Glass-bottomed culture dishes: 35 mm dishes with No. 1.5 coverslips either 14 mm or 20 mm in diameter (MatTek Corporation, MA); or 12 mm or 27 mm in diameter (Iwaki, Japan; *see* **Note 1**).

2.4. Centrosome Overduplication Assay

1. Microscope coverslips (20 mm in diameter; VWR), No. 1.5. Acid-etch in 1 M HCl for 30 min, rinse in 100% ethanol, air-dry, and bake at 250°C for 4 h.
2. Hydroxyurea (HU): prepare at 100 mM in ddH_2O and filter-sterilise. Aliquot and store at –20°C. Stable for 1 month.

2.5. Indirect Immunofluorescence Microscopy

1. 100% ice-cold (–20°C) methanol.
2. Paraformaldehyde (PFA): prepare a 4% (w/v) solution in 1 × PBS. Add 100 ml 1 × PBS to 4 g PFA (Sigma-Aldrich). Gently warm to 65°C in a fume hood. Add 1 drop of concentrated NaOH to aid PFA solubility. Allow to cool; adjust pH to 6.8. Aliquot and store at –20°C for up to 3 months.
3. Permeabilisation buffer: 0.25% (v/v) Triton X-100 in 1 × PBS.
4. Blocking solution: 1% (w/v) BSA (Fluka Biochemika, UK) in 1 × PBS.
5. Antibody dilution buffer: 3% (w/v) BSA in PBS.
6. Primary antibodies: Mouse anti-γ-tubulin (0.15 μg/ml; Sigma-Aldrich; T6557); rabbit anti-centrin (2 μg/ml; Abcam; ab11257).
7. Secondary antibodies: AlexaFluor-488 and -594 goat anti-rabbit and goat anti-mouse IgGs (1 μg/ml; Invitrogen).
8. Nuclear stain: Hoeschst 33258 (0.2 μg/ml; Calbiochem).
9. Anti-fade mounting solution: 80% (v/v) glycerol, 3% (w/v) N-propyl-gallate (Sigma-Aldrich) in 1 × PBS. Store at 4°C; protect from light.

2.6. Centrosome Disjunction Assay

1. Doxycycline (1 μg/ml; Sigma-Aldrich), prepare at 1 mg/ml in water (warming may be required to fully dissolve). Protect from light; store at 4°C for 3 days or –20°C for 2 months.

3. Methods

3.1. Generation and Imaging of Cell Lines with Fluorescent Centrosomes

Proteomic approaches have shown that the mammalian centrosome consists of over 200 proteins *(25)*. Some of these are structural, contributing to the basic architecture of the centrosome, whilst others are regulatory and associate only transiently and often in a cell cycle-dependent manner *(23)*. To observe centrosomes throughout the cell cycle by time-lapse microscopy, we generally recommend the generation of cell lines that express fluorescently tagged versions of stable, intrinsic centrosome proteins, such as the centriolar component centrin. Indeed, a beautiful study with cells stably expressing centrin1-EGFP first revealed the different dynamics of mother and daughter centrioles during interphase *(26)*. From a practical perspective, generation of such cell lines involves first creating a eukaryotic expression plasmid encoding an N- or C-terminal fusion of a cDNA encoding a centrosome protein to the cDNA of a fluorescent protein, such as green fluorescent protein (GFP) *(27, 28)*. However, the specific choice of fluorescent protein should be considered carefully in terms of the proposed experiments as the range of soluble, monomeric fluorescent proteins now available is expanding rapidly, providing the user with an ever-increasing choice of wavelengths, as well as proteins specifically designed for photobleaching, photoactivation, photoconversion, or FRET (fluorescence resonance energy transfer)-based interaction assays *(29)*. Moreover, combined with state-of-the-art confocal imaging systems, these provide the potential for simultaneous spectral unmixing of multiple fluorescent proteins within individual live cells. Finally, it is worth noting that, once an expression construct has been generated, it is important to demonstrate that the fluorescently tagged protein exhibits similar behaviour to the endogenous protein by, for example, determining its subcellular localization, activity, and stability.

3.1.1. Generation of Stable Cell Lines

1. We routinely use Chinese Hamster Ovary (CHO) and human U2OS osteosarcoma cells to study the centrosome cycle due to their simple culture conditions, good imaging properties (e.g., high cytoplasmic:nuclear ratio and distinct microtubule network), and high transfection efficiency. CHO cells are grown in Ham's F12 media, whilst U2OS cells are grown in DMEM. Cell lines are maintained in a humidified 5% CO_2 atmosphere at 37°C and passaged before reaching confluence with PBS-EDTA. The day before transfection, cells are seeded at 1×10^5 cells/cm^2 to give a confluency of 80% the next day. For microscopy, cells are plated in 6-well dishes with acid-etched glass coverslips, whilst to generate stable cell lines, cells are seeded in 100 mm or 60 mm plates.
2. On the day of transfection, prepare the following for each 100 mm dish: in tube 1, add 4 μg of plasmid DNA to 1 ml Opti-MEM (serum-free). In tube 2, add 16 μl Lipofectamine to 1 ml Opti-MEM (serum-free; *see* **Note 2**). For 60 mm dishes halve all volumes; for one well of a 6-well plate use a quarter of all volumes. Incubate both tubes at

room temperature for 5 min. Add contents of tube 1 to contents of tube 2, mix gently and incubate for 20–30 min at room temperature (*see* **Note 3**). Replace the media on the cells with 8 ml of serum-free standard growth media. Add the lipofectamine-DNA mix dropwise to the cells and return to the incubator for 4–6 h (*see* **Note 4**). Remove media and replace with fresh growth media (with serum). Incubate for 24–48 h at 37°C.

3. To generate stable cell lines: on the day following transfection, passage cells to give a very low density, such that individual transfected cells can form individual colonies (*see* **Note 5**). The following day, the appropriate drug to kill untransfected cells, for example, G 418 at 600 μg/ml if the plasmid carries a neomycin resistance gene, is added and maintained in the culture media. For cells expressing GFP-fusion proteins, a simple fluorescence microscope with a 488 nm excitation filter can be used to monitor the progress of colony formation. Once distinct isolated colonies, each containing around 30–50 cells, are formed they can be removed from the dish using cloning discs soaked in PBS-EDTA. These are then placed into separate wells of a 24-well plate with fresh growth medium containing G 418. The colonies are allowed to grow to near confluence before cells are lifted and transferred to a 12-well plate. As cells next approach confluency, they are serially plated in a 6-well plate, 60 mm dish, and finally a 100 mm dish, at which point stocks should be frozen in liquid nitrogen.

3.1.2. Live Cell Imaging

1. For optimum visualisation of centrosomes and centrosome protein dynamics in live cells, conditions should be as close to ideal for cell growth as possible *(30)*. This includes the use of a humidified chamber on the microscope that can be maintained at 37°C (*see* **Note 6**) with 5% CO_2. The chamber should be at a stable temperature at least 3 h before imaging starts and this may require the chamber to be switched on the day before imaging (*see* **Note 7**).
2. On the day prior to imaging, the desired cells are seeded onto glass-bottomed culture dishes (*see* **Note 8**), as described in **Section 3.1.1**, and incubated overnight. The required density of cells depends on the duration of imaging; cells 50–70% confluent can be followed for up to two mitotic events depending on the stage of the cell cycle when imaging is started.
3. If cells are cultured in a medium that normally contains phenol-red (e.g., standard DMEM contains 15 mg/l phenol-red), it may be advisable to change the growth media to an equivalent reduced or phenol-red-free alternative as phenol-red can quench GFP fluorescence. This is

not necessary with Ham's F12 media, which contain only 1.2 mg/l phenol-red. If the media are changed before imaging, this should be performed at least 2 h before the cells are transferred to the stage.

4. On the day of imaging, cells are transferred to the microscope stage. Cells are viewed under brightfield (to locate cells) and widefield or confocal fluorescence microscopy. On a widefield microscope, an appropriate filter set is selected to visualise the GFP fluorescence (excitation 450–490 nm; emission 500–550 nm). Typically, a Plan Apo 60 × (NA 1.4) DIC oil objective is used coupled to a high-speed Piezo focus drive (Orbit II) to capture z-sections consisting of 5 optical sections at intervals of 0.5–1 μm at each timepoint using a CCD camera. Neutral density shutters are employed to reduce exposure of cells to excessive levels of illumination. On a laser scanning confocal microscope (LSCM), excitation of GFP fluorescence is normally achieved with an argon 488 nm laser line, with emitted light collected between 500 and 550 nm (peak emission at 507 nm). To lessen photodamage to cells, low laser intensity is selected (typically between 5 and 20%). Level of zoom and number of z-steps will also affect the survival of cells, with each condition needing to be optimised on the specific microscope system being used. Generally, a scan zoom of 2 with a 63 × oil-immersion objective (NA 1.4), coupled with a z-section consisting of 30–40 steps, each 0.3–1 μm in size, captured every 5–10 min is tolerated well by CHO cells for up to 50 h (*see* **Note 9,** and **Fig. 10.1** for detailed imaging conditions with a Leica TCS SP5 LSCM).
5. The use of a motorized stage to capture multiple positions during an imaging session can greatly increase the amount of data collected. This is useful, firstly, to image a sufficient number of cells as to generate statistically significant data and, secondly, if reagents are particularly precious. However, the limiting factor is often the capability of the system to handle large datasets.

3.2. Photobleaching Assays for Measuring Centrosome Protein Dynamics

A variety of structural and regulatory proteins associates, either stably or transiently, with the centrosome at specific points in the cell cycle *(23)*. Many of these proteins are in a dynamic exchange between cytoplasmic and centrosomal pools, and the rate of dynamics can depend on a variety of factors, such as whether their transport is microtubule-dependent. In general, the dynamics of proteins that adopt specific localisations can be studied using FRAP and/or FLIP in cells expressing GFP-tagged proteins of interest. Using these photobleaching methods, it can be determined which proteins are stably associated with the centrosome, and therefore which are likely to have a structural role, as compared to those that undergo rapid turnover and are more likely to be regulatory. It is also possible to investigate how these proteins are recruited to the centrosome, for example, by performing photobleaching experiments with and without nocodazole to determine whether depolymerization of microtubules alters the protein dynamics *(31)*. For detailed descriptions of the principles and calculations involved in photobleaching methods, the reader is referred to Lippincott-Schwartz et al. and Presley et al. *(32, 33)*.

1. The turnover of a protein at the centrosome can be measured using FRAP (**Fig. 10.2**). This procedure can be

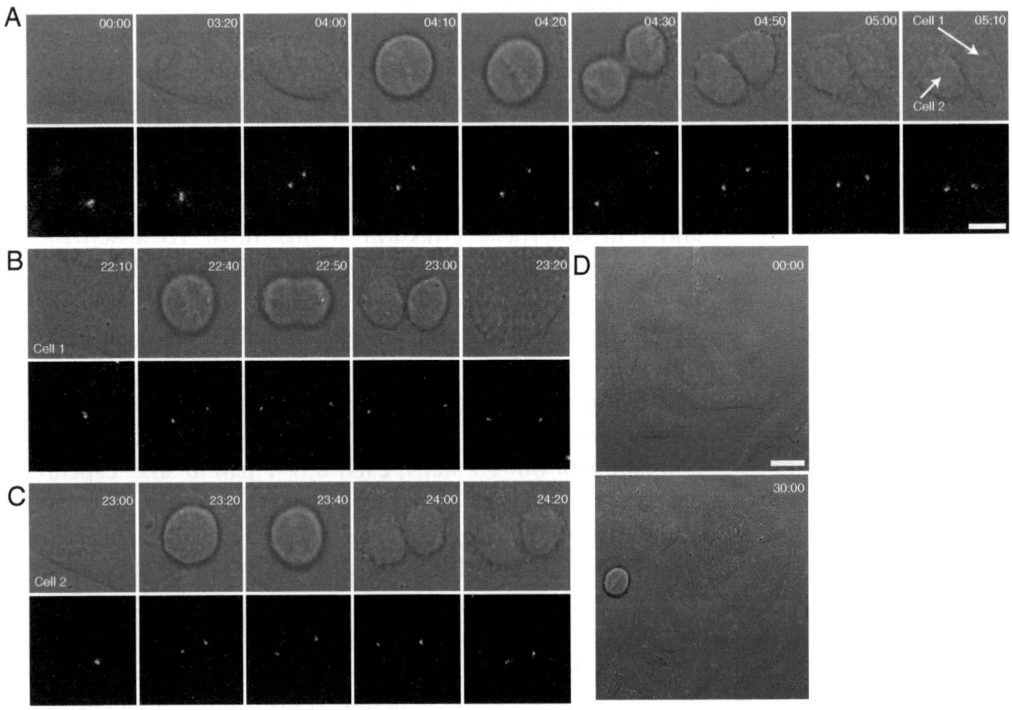

Fig. 10.1. Time-lapse imaging of cell lines with fluorescent centrosomes using an LSCM. On the day prior to imaging, CHO cells stably expressing centrin1-GFP were plated onto glass-bottomed culture dishes. Images were captured on a Leica TCS SP5 LSCM equipped with a Leica DMI 6000B inverted microscope and SuperZ Galvo stage using an HCX Plan Apo 63 × oil objective (NA 1.4). The SP5 uses an acousto-optic tunable filter (AOTF) to select the excitation wavelength and an acousto-optical beamsplitter (AOBS) instead of filter blocks to send the excitation light to the sample and reflect the emitted light from the sample into the scan head. This is a more light-efficient way than using glass di- or multichroic mirrors and filters. Cells were cultured on the stage at 5% CO_2 and 37°C using a microscope temperature control system (The Cube and The Box, Life Imaging Services). (**A**) Cells were located under brightfield (*upper images*) and the corresponding GFP fluorescence monitored with 10% power of a 488 nm argon laser (lower images). An area of cells was selected for imaging with a scan zoom of 2. Typically, z-section parameters of 30–40 steps of 0.3–1 μm were set so that rounded mitotic cells would be visible. A z-stack was collected every 10 min for the duration of the experiment. Images are shown as the maximum intensity projections of the z-stack collapsed into a single image per timepoint. A cell is shown going through mitosis within the first 5 h of imaging. (**B**) & (**C**) The daughter cells (Cell 1 and Cell 2) from the division shown in panel (**A**) are shown going through a subsequent mitosis between 22 and 24 h of imaging giving a cell cycle duration of 18–19 h. (**D**) The whole field of view in brightfield is shown at the beginning of imaging (*upper panel*) and after 30 h (*lower panel*). In total, these cells were imaged for 50 h under these conditions without significant toxicity. Time is shown in hours:minutes. Scale bar in A, 10 μm; scale bar in D, 20 μm.

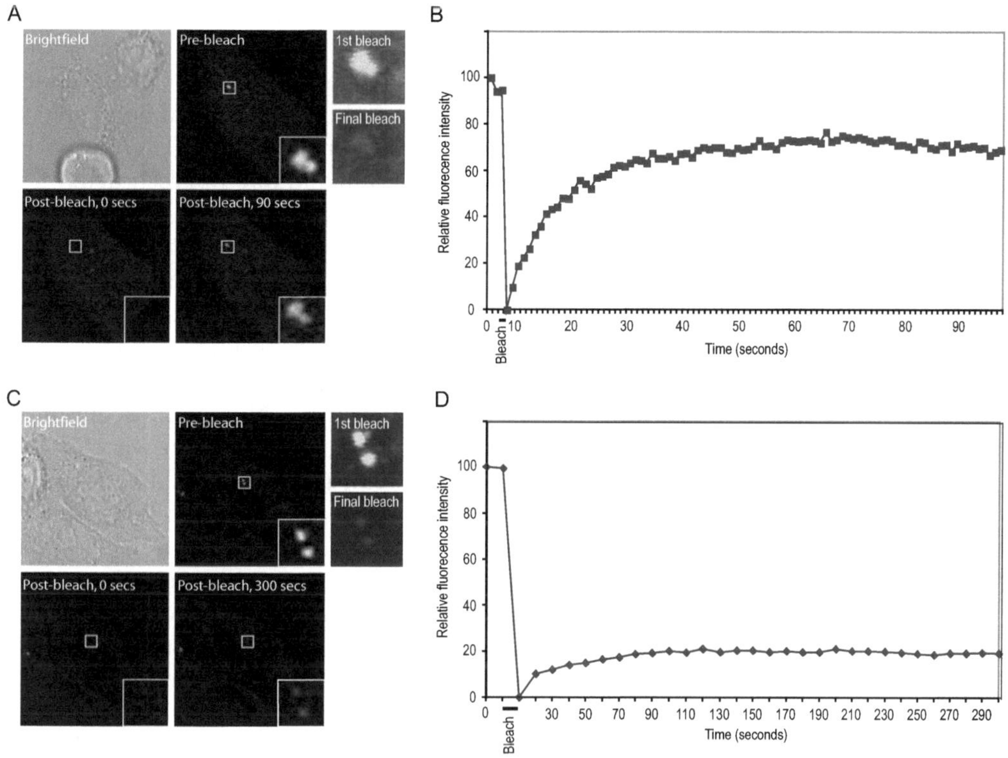

Fig. 10.2. FRAP analysis of centrosome protein dynamics. Photobleaching experiments were performed on a Leica TCS SP5 confocal microscope equipped with a Leica DMI 6000B inverted microscope using a 63 × oil objective (NA 1.4) and a scan zoom of 3. Cells were cultured in glass-bottomed dishes and maintained on the stage at 5% CO_2 and 37°C. Two images were captured prior to a region of interest (ROI) 3 μm × 3 μm, encompassing the centrosome, being bleached with 4 iterations and 100% laser power (488 nm argon laser). Bleaching was performed by zooming in to the ROI so that fewer iterations were required. (**A**) CHO cells transiently expressing GFP-Nek2A-WT were bleached as described and an image captured every second for 90 s postbleaching. A brightfield image and three fluorescence images (prebleach, 0 s postbleach, and 90 s postbleach) are shown. Magnified images of the centrosomes after the first and final bleach iteration are also shown to demonstrate the extent of fluorescence loss before recovery. (**B**) For each timepoint, the relative fluorescent intensity was calculated and plotted against time. The result is characteristic of a highly dynamic protein with a short half-time ($t_{1/2} = 5$ s). (**C**) CHO cells stably expressing centrin1-GFP were bleached as described and an image captured every 10 s for 5 min postbleaching. A brightfield image and three fluorescence images (prebleach, 0 s postbleach, and 300 s postbleach) are shown. Again, magnified images of the centrosomes after the first and final bleach iteration are shown. (**D**) The relative fluorescent intensity was calculated and plotted against time. The result is characteristic of a relatively stable centrosomal component showing little recovery over the course of the experiment. Results are shown as the average of at least 10 cells per bleaching experiment.

carried out equally well with cells transiently transfected with or stably expressing the GFP fusion construct. However, stable cell lines may be more desirable as the efficiency of transfection is not an issue and the level of expression is usually more constant. If transfection is required, the standard protocol can be used with cells plated on glass-bottomed dishes. In this case the volumes are as for one well of a 6-well plate.

2. To maintain ideal conditions, the microscope and cells are prepared for live cell imaging as described above before transfer to the microscope stage. Cells are viewed under brightfield (to locate cells) and under confocal fluorescence microscopy. Excitation at 488 nm induces GFP fluorescence, with emitted light between 500 and 550 nm recorded.
3. A region of interest (ROI) is selected that is sufficient to contain the centrosome, and this size is maintained between different cells (*see* **Note 10**). At first, three images are acquired before bleaching using standard imaging parameters. Bleaching of the ROI is then performed using 100% laser power. The number of bleaches will have to be optimised for the system being used (*see* **Note 11**). Acquisition of images after bleaching depends on the recovery of the protein. For a dynamic protein, an image should be acquired every second, whilst for a more stable protein longer intervals can be used (*see* **Note 12**).
4. For each timepoint, the fluorescence intensity (P) of the photobleached ROI is determined. The fluorescence intensity of the background (B), at a point between cells using a ROI of the same dimensions, is also measured and the corrected fluorescence intensity (P–B) calculated. The amount of fluorescence recovery is calculated as the corrected fluorescence intensity of a given frame divided by the corrected fluorescence intensity of the first frame before photobleaching. This can be expressed as a percentage.
6. In a similar manner, FLIP can be performed, however, in this case the whole of the cell except a region encompassing the centrosome is bleached. As a much larger area is being bleached, a series of bleach cycles is undertaken, the time of which will depend upon the scan rate. Generally, five images are acquired, each five seconds apart, prior to bleaching and between each bleach cycle. A bleach cycle consists of 10 bleaches at 100% laser power. At least 10 cycles are performed for each cell. Fluorescence loss can be calculated in the same way as for FRAP. For details of the FLIP methodology as applied to centrosome proteins, see **Fig. 10.3**.

3.3. Centrosome Overduplication Assay

Centrosome duplication is a carefully controlled process that is normally tightly coupled with the DNA replication cycle. However, Balczon et al. first described in 1995 a centrosome overduplication assay in which they observed that CHO cells, treated with hydroxyurea (HU), accumulate multiple centrosomes during prolonged S phase arrest *(34)*. Hence, in these cells, centrosome duplication can be uncoupled from the DNA replication cycle. Later studies revealed that it is the loss of p53 that enables cells to undergo centrosome overduplication during an S-phase arrest *(35, 36)*. Using cells with compromised

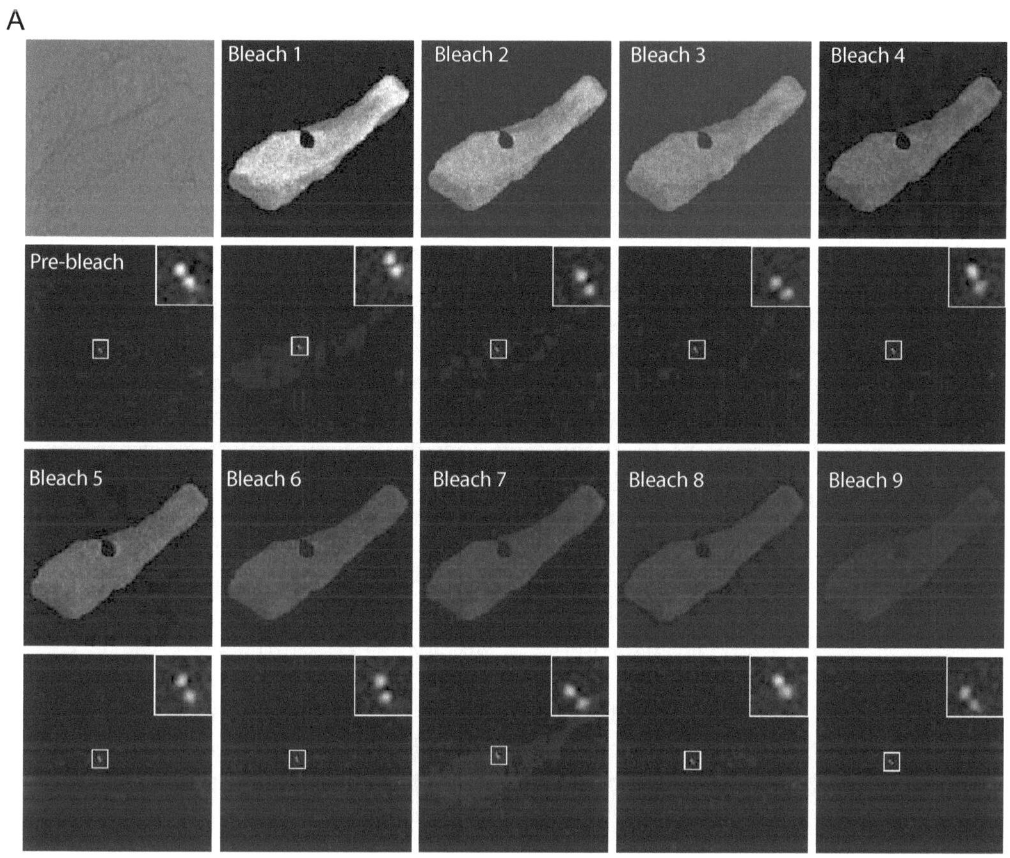

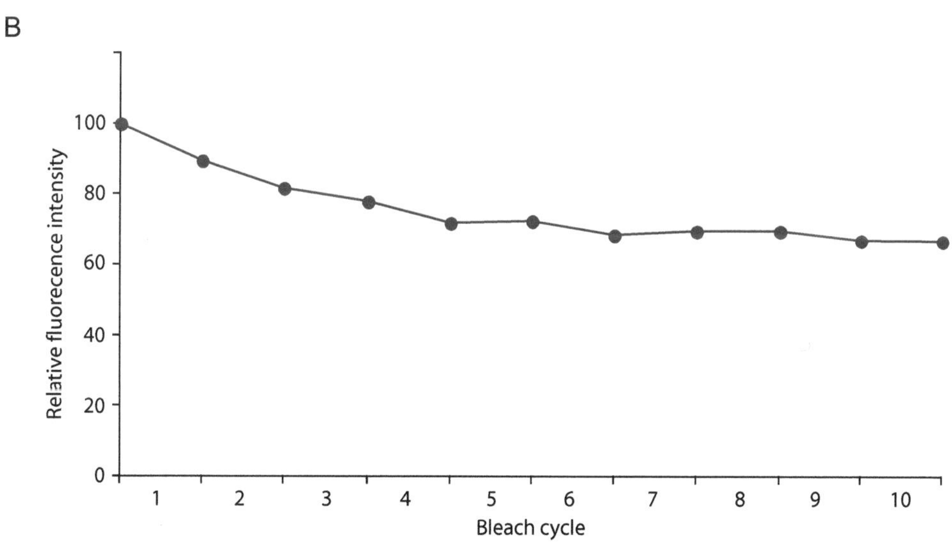

Fig. 10.3. FLIP analysis of centrosome protein dynamics. FLIP experiments were performed on a Leica TCS SP5 confocal microscope equipped with a Leica DMI 6000B inverted microscope using a 63 × oil objective (NA 1.4) and a scan zoom of 3. Cells were cultured in glass-bottomed dishes and maintained on the stage at 5% CO_2 and 37°C. (**A**) FLIP was performed on CHO:centrin1-GFP cells. An area was selected for bleaching that encompassed the entire cell except for

p53 function (e.g., CHO or U2OS), it is therefore possible to elucidate events involved in the overduplication of centrosomes. Specifically, it is possible to use pharmacological inhibitors, dominant-negative constructs or RNAi, along with a range of centrosomal markers, to look for changes in the recruitment of specific proteins under defined conditions to overduplicating centrosomes *(37, 38)*. The generation of multiple centrosomes is clearly of major clinical relevance as many cancer cells exhibit supernumerary centrosomes and these are highly likely to contribute to the tumour phenotypes of aneuploidy and chromosome instability *(39–41)*. However, as a cautionary note, it is important to appreciate that the mechanisms and pathways involved in centrosome overduplication may well differ in specific respects from the semi-conservative duplication of centrosomes that occurs during unperturbed cell cycle progression.

1. Multiple centrosomes can be observed to accumulate in CHO and U2OS cells that are blocked in S-phase with HU for up to 72 h. However, overduplication is much more rapid in CHO than U2OS cells and therefore easier to observe. Cells are plated onto glass coverslips at a density of 1×10^5 cells/cm^2. The following day, the cells are washed once with 1 × PBS and placed in fresh growth media containing 2 mM HU for CHO cells and 16 mM HU for U2OS. The cells are then fixed and processed for immunofluorescence microscopy 48 or 72 h post-HU treatment as described below (**Fig. 10.4**). The accumulation of multiple centrosomes can also be followed using live cell fluorescence imaging in cells expressing a GFP-tagged centrosomal protein.
2. For fixation, cells are rinsed once with 1 × PBS before adding ice-cold methanol pre-cooled at –20°C and incubated at –20°C for 10 min (*see* **Note 13**). Alternatively, cells may be fixed with 4% PFA, washed three times for 5 min with 1 × PBS and permeabilised with 0.25% Triton-PBS (*see* **Note 14**).

Fig. 10.3. (continued) the centrosome.Ten bleach cycles were performed on this area with 100% laser power for 10 iterations (upper panels). Prior to bleaching and between each bleach cycle, 5 images were captured 5 s apart using standard imaging conditions (10% laser) on the whole cell (lower panels). Insets in lower panels show enlargements of centrosomes. (**B**) Fluorescent intensity measurements were taken for a ROI containing the centrosome for each image. The fluorescence of the centrosome prebleach and between each bleach cycle was averaged for each set of 5 images and the relative fluorescent intensity calculated. This is plotted as a datapoint between each bleach cycle. The results are typical of a stable centrosomal component with a low degree of exchange and little loss during the experiment. Results are shown as the average of 15 cells.

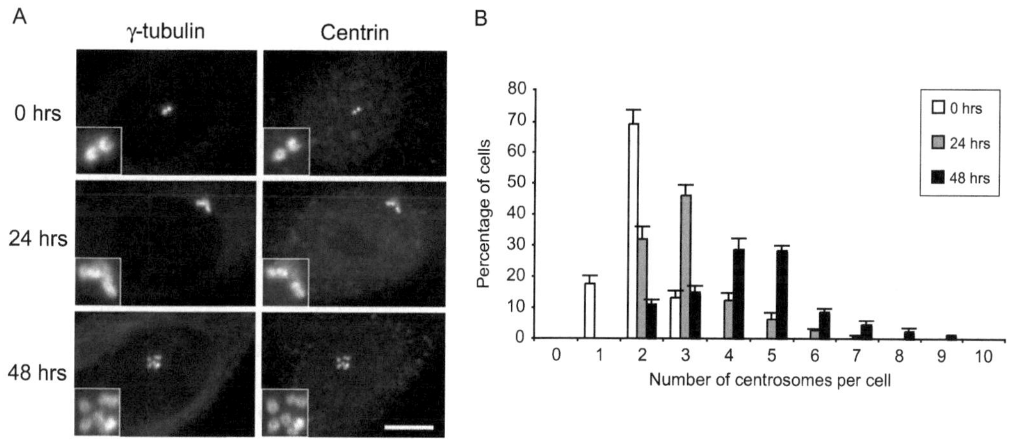

Fig. 10.4. Centrosome overduplication assay. (**A**) CHO cells were plated onto glass coverslips and the following day transferred to media containing 2 mM HU. The cells were fixed in ice-cold methanol at 0, 24, and 48 h post-HU treatment and processed for immunofluorescence microscopy with antibodies against the PCM component, γ-tubulin, and the centriole component, centrin. Fluorescence images were captured on a Nikon TE300 inverted microscope using a 100 × oil immersion objective, ORCA ER CCD camera (Hamamatsu, Japan) and Openlab 5.0 software (Improvision, UK). Images were processed using Adobe Photoshop 7. Scale bar, 10 μm. (**B**) The number of centrosomes per cell were counted for each timepoint based on γ-tubulin staining illustrating the overduplication of centrosomes with time. At least 200 cells were counted per experiment and results expressed as the average of 3 experiments. Error bars show standard deviation.

3. Cells are then rehydrated/washed four times for 5 min with 1 × PBS.
4. Coverslips are blocked by the addition of 1% BSA-PBS solution for 10 min at room temperature.
5. Blocking solution is replaced with primary antibody diluted in 3% BSA-PBS and incubated for 30–60 min in a humid chamber (*see* **Note 15**).
6. Coverslips are then washed three times for 5 min in 1 × PBS.
7. Coverslips are incubated with secondary antibody plus nuclear stain diluted in 3% BSA-PBS for 30–60 min in a humid chamber protected from light.
8. Coverslips are then washed again three times for 5 min with 1 × PBS.
9. Finally, dip the coverslips in water to remove excess PBS and gently touch the edge on a piece of tissue paper to remove excess water. To mount the coverslip, carefully invert onto a drop of mounting solution on a microscope slide. Blot excess mounting solution and seal the edges of the coverslip with clear nail varnish (*see* **Note 16**). Store slides in the dark at 4°C.

3.4. Centrosome Disjunction Assay

Formation of a bipolar spindle occurs following centrosome disjunction at the G2/M transition. This is in part dependent upon the kinase activity of Nek2A which triggers dissolution of the fibrous linkage between the mother and daughter centrioles allowing the two duplicated centrosomes to separate. A useful assay arises from the observation that overexpression of active Nek2A causes premature centrosome disjunction in interphase cells leading to a distinct and sustained separation between the two centrosomes *(42)*. In contrast, overexpression of kinase-inactive Nek2A (e.g., the K37R point mutant) hinders centrosome disjunction during mitotic onset resulting in monopolar spindle formation and defects in cell division. Hence, this assay can be used to score the state of Nek2A activation in cells under different conditions or treated with different inhibitors. However, the scoring of centrosome disjunction can be used more generally to investigate any condition that might regulate the cohesion of interphase centrosomes: for example, RNAi depletion of linker proteins, depolymerization of the microtubule network, or addition of certain growth factors *(14, 43–46)*.

1. Tetracycline-inducible U2OS cell lines that express GFP-Nek2A or GFP-Nek2A-K37R or indeed any protein of interest can be generated as detailed above. Alternatively, untransfected U2OS or HeLa cells can be used to investigate

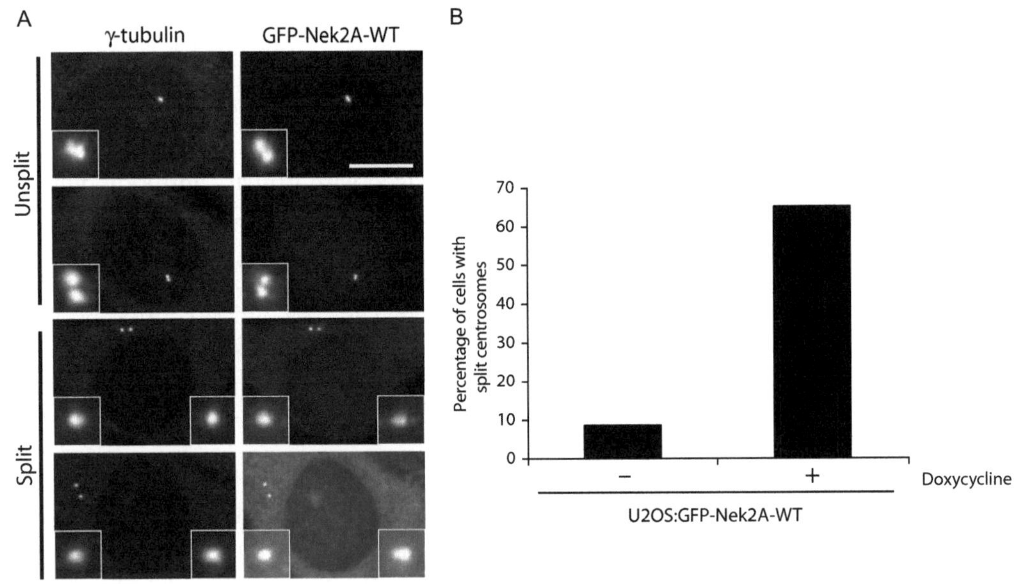

Fig. 10.5. Centrosome disjunction assay. (**A**) Inducible U2OS:GFP-Nek2A-WT cells were plated onto glass coverslips and, after 24 h, expression was induced by the addition of 1 μg/ml doxycycline. The next day cells were fixed and analysed by indirect immunofluorescence microscopy. The centrosome was detected with anti-γ-tubulin antibodies and GFP fluorescence. Fluorescence images were captured on a Nikon TE300 inverted microscope using a 100 × oil immersion objective, ORCA ER CCD camera (Hamamatsu, Japan), and Openlab 5.0 software (Improvision, UK). Images were processed using Adobe Photoshop 7. Two examples of cells with unsplit and two examples of cells with split centrosomes are shown. Scale bar, 5 μm. (**B**) The percentage of cells with split centrosomes (i.e., separated by >2 μm) is given for U2OS:GFP-Nek2A-WT cells following 24 h treatment with or without doxycycline.

other conditions or treatments that might stimulate centrosome disjunction in interphase.

2. If using inducible cell lines, cells are plated on coverslips or glass-bottomed culture dishes 24 h prior to induction of protein expression. Expression is induced by the addition of the stable tetracycline analogue, doxycycline, to the culture media at a final concentration of 1 μg/ml with further supplementation every 48 h if required.
3. Cells are either viewed by live cell fluorescence imaging to visualise centrosomes based on GFP-Nek2A localization, or fixed and processed for immunofluorescence imaging with antibodies against standard centrosome markers that are not part of the linker structure, for example, γ-tubulin. Centrosome disjunction is assessed in individual cells as the distance between the two centrosomes: for example, if the distance is >2 μm, the centrosomes are classed as split, and if the distance is <2 μm, the centrosomes are classed as unsplit (**Fig. 10.5**).

4. Notes

1. Choice of glass-bottomed dish may depend on cell type; for example, CHO cells attach much better to dishes manufactured by Iwaki than MatTek.
2. Transfection efficiency can be optimised by varying the confluency of cells at the time of transfection along with the ratio of DNA to transfection reagent.
3. This reflects the minimum time required for complexes to form. The complexes are stable at room temperature for up to 2 h.
4. The transfection mixture is toxic to cells and if left any longer will cause cell death.
5. Individual transfected cells can be obtained either with a low transfection rate or passage of cells following transfection. In the case of the latter, a range of dilutions is recommended from 1:10 to 1:40. This may also shorten the time required for transfected colonies to form as there will be fewer nontransfected cells to kill.
6. On our widefield microscopes we use incubators and temperature control systems from Solent Scientifc (UK), whilst on our confocal microscope we use The Cube and The Box from Life Imaging Services (Switzerland). It is recommended to check the temperature of the chamber at the exact place the cells will be located. The temperature can vary greatly throughout the chamber and may fluctuate by

up to 5°C from the set temperature. Such a change will be sufficient to compromise the growth of cells.

7. Ensuring the chamber is at a stable temperature avoids exposing cells to temperature fluctuations as the chamber heats up and importantly ensures there will be no distortions in imaging as metal components of the microscope expand.
8. Whilst dishes may be described as being of the same size, particularly in viewing surface, they do differ in their construction; for example, some have thicker or sloping sides. Therefore, it is important to check that the dish will actually fit securely on the microscope stage.
9. The number of z-steps, zoom size, and time interval between image acquisitions will all influence the amount of light that cells are exposed to and therefore cell survival. The higher the zoom, the more laser intensity in a small area, whereas the more z-steps, the more light that cells are exposed to for each timepoint taken. These parameters will need adjusting according to the cell type and system being used and the desired experimental outcome. Compromises may have to be made; for example, if faster acquisition is required then fewer or larger z-steps could be used.
10. A ROI should be large enough to encompass the centrosome; however, it should not be so large that, during bleaching, the total available pool of protein is also bleached. The size of the ROI should be maintained throughout the experiment for each cell and to allow comparison between experiments. During the recording of fluorescent intensities throughout an experiment, it may be observed that the centrosome has a tendency to move. If the movement is into a different focal plane, then the data for that cell may have to be discarded. If movement is only through x and y the ROI can be placed to include the centrosome for the course of the experiment, although it may have to be moved for each datapoint.
11. Bleach efficiency varies between different systems depending on how they perform the bleach. Some systems zoom to the ROI so there is more intensity and the bleach is performed faster. Others do not have this feature, in which case more iterations are required to achieve a satisfactory bleach. Optimisation will be required to determine the number of iterations required.
12. The dynamics of the protein will determine the best interval for image acquisition following bleaching. For those that recover rapidly, image acquisition should be fast so that the complete recovery profile can be followed. For less dynamic proteins the interval can be longer, especially as rapid imaging may cause additional bleaching.

13. Long-term storage of up to two months of coverslips fixed in methanol is possible without seriously affecting subsequent staining and imaging. In this instance, the coverslips are maintained in methanol at –20°C. This is best done in plastic culture dishes sealed with parafilm to ensure that the methanol does not evaporate and the coverslip dry out.
14. The best fixation method should be determined according to the primary antibody being used. Some give a strong staining pattern with one method, but no signal with the other method. Each method should be tested for each new primary antibody used. PFA tends to enhance the fluorescence seen from GFP-tagged proteins, whereas methanol can lead to loss of soluble GFP-tagged proteins.
15. A humid chamber is used for each antibody incubation step to ensure that coverslips do not dry out.
16. Let nail varnish dry completely before putting in cold-room, otherwise it may remain tacky increasing the chance that the coverslip will come away from the slide.

Acknowledgments

We particularly thank Dr. Kees Straatman (Leicester) for technical support and advice with fluorescence imaging applications. We are also grateful to all members of our laboratory for useful discussion and to The Wellcome Trust, the Biotechnology and Biological Sciences Research Council, Cancer Research UK, the Association for International Cancer Research, and AstraZeneca for supporting our research.

References

1. Bornens, M., Centrosome composition and microtubule anchoring mechanisms. *Curr Opin Cell Biol* 2002. **14**(1): 25–34.
2. Doxsey, S., Re-evaluating centrosome function. *Nat Rev Mol Cell Biol* 2001. **2**(9): 688–98.
3. Bornens, M., Is the centriole bound to the nuclear membrane? *Nature* 1977. **270**(5632): 80–2.
4. Manneville, J.B. and S. Etienne-Manneville, Positioning centrosomes and spindle poles: looking at the periphery to find the centre. *Biol Cell* 2006. **98**(9): 557–65.
5. Hinchcliffe, E.H. and G. Sluder, "It takes two to tango": understanding how centrosome duplication is regulated throughout the cell cycle. *Genes Dev* 2001. **15**(10): 1167–81.
6. Lange, B.M., et al., Centriole duplication and maturation in animal cells. *Curr Top Dev Biol* 2000. **49**: 235–49.
7. Marshall, W.F., Centrioles take center stage. *Curr Biol*, 2001. **11**(12): R487–96.
8. Ou, Y., M. Zhang, S. Chi, J.R. Matyas, and J.B. Rattner, Higher order structure of the PCM adjacent to the centrioles. *Cell Motil Cytoskeleton* 2003. **55**(2): 125–33.
9. Megraw, T.L., et al., The centrosome is a dynamic structure that ejects PCM flares. *J Cell Sci* 2002. **115**(Pt 23): 4707–18.
10. Bahe, S., et al., Rootletin forms centriole-associated filaments and functions in centrosome cohesion. *J Cell Biol* 2005. **171**(1): 27–33.

11. Mayor, T., et al., The mechanism regulating the dissociation of the centrosomal protein C-Nap1 from mitotic spindle poles. *J Cell Sci* 2002. **115**(Pt 16): 3275–84.
12. Mayor, T., et al., The centrosomal protein C-Nap1 is required for cell cycle-regulated centrosome cohesion. *J Cell Biol* 2000. **151**(4): 837–46.
13. Yang, J., M. Adamian, and T. Li, Rootletin interacts with C-Nap1 and may function as a physical linker between the pair of centrioles/basal bodies in cells. *Mol Biol Cell* 2006. **17**(2): 1033–40.
14. Graser, S., Y.D. Stierhof, and E.A. Nigg, Cep68 and Cep215 (Cdk5rap2) are required for centrosome cohesion. *J Cell Sci* 2007. **120**(Pt 24): 4321–31.
15. Sluder, G. and E.H. Hinchcliffe, The apparent linkage between centriole replication and the S phase of the cell cycle. *Cell Biol Int* 1998. **22**(1): 3–5.
16. Nigg, E.A., Centrosome duplication: of rules and licenses. *Trends Cell Biol*, 2007. **17**(5): 215–21.
17. Blagden, S.P. and D.M. Glover, Polar expeditions–provisioning the centrosome for mitosis. *Nat Cell Biol* 2003. **5**(6): 505–11.
18. Fry, A.M., et al., C-Nap1, a novel centrosomal coiled-coil protein and candidate substrate of the cell cycle-regulated protein kinase Nek2. *J Cell Biol* 1998. **141**(7): 1563–74.
19. Fry, A.M., P. Meraldi, and E.A. Nigg, A centrosomal function for the human Nek2 protein kinase, a member of the NIMA family of cell cycle regulators. *EMBO J* 1998. **17**(2): 470–81.
20. Sharp, D.J., G.C. Rogers, and J.M. Scholey, Microtubule motors in mitosis. *Nature* 2000. **407**(6800): 41–7.
21. Tsou, M.F. and T. Stearns, Mechanism limiting centrosome duplication to once per cell cycle. *Nature* 2006. **442**(7105): 947–51.
22. Tsou, M.F. and T. Stearns, Controlling centrosome number: licenses and blocks. *Curr Opin Cell Biol* 2006. **18**(1): 74–8.
23. Lange, B.M., Integration of the centrosome in cell cycle control, stress response and signal transduction pathways. *Curr Opin Cell Biol*, 2002. **14**(1): 35–43.
24. Doxsey, S., D. McCollum, and W. Theurkauf, Centrosomes in cellular regulation. *Annu Rev Cell Dev Biol*, 2005. **21**: 411–34.
25. Andersen, J.S., et al., Proteomic characterization of the human centrosome by protein correlation profiling. *Nature* 2003. **426**(6966): 570–4.
26. Piel, M., et al., The respective contributions of the mother and daughter centrioles to centrosome activity and behavior in vertebrate cells. *J Cell Biol* 2000. **149**(2): 317–30.
27. Straight, A.F., Fluorescent protein applications in microscopy. *Methods Cell Biol* 2007. **81**: 93–113.
28. Lippincott-Schwartz, J. and G.H. Patterson, Development and use of fluorescent protein markers in living cells. *Science* 2003. **300**(5616): 87–91.
29. Shaner, N.C., P.A. Steinbach, and R.Y. Tsien, A guide to choosing fluorescent proteins. *Nat Methods* 2005. **2**(12): 905–9.
30. Khodjakov, A. and C.L. Rieder, Imaging the division process in living tissue culture cells. *Methods* 2006. **38**(1): 2–16.
31. Hames, R.S., et al., Dynamic recruitment of Nek2 kinase to the centrosome involves microtubules, PCM-1, and localized proteasomal degradation. *Mol Biol Cell* 2005. **16**(4): 1711–24.
32. Lippincott-Schwartz, J., E. Snapp, and A. Kenworthy, Studying protein dynamics in living cells. *Nat Rev Mol Cell Biol* 2001. **2**(6): 444–56.
33. Presley, J.F., Measurement of protein motion by photobleaching, in D. Stephens, Ed. *Cell Imaging*, 2006. 119–142.
34. Balczon, R., et al., Dissociation of centrosome replication events from cycles of DNA synthesis and mitotic division in hydroxyurea-arrested Chinese hamster ovary cells. *J Cell Biol* 1995. **130**(1): 105–15.
35. Tarapore, P. and K. Fukasawa, Loss of p53 and centrosome hyperamplification. *Oncogene* 2002. **21**(40): 6234–40.
36. Bennett, R.A., H. Izumi, and K. Fukasawa, Induction of centrosome amplification and chromosome instability in p53-null cells by transient exposure to subtoxic levels of S-phase-targeting anticancer drugs. *Oncogene* 2004. **23**(40): 6823–9.
37. Warnke, S., et al., Polo-like kinase-2 is required for centriole duplication in mammalian cells. *Curr Biol* 2004. **14**(13): 1200–7.
38. Meraldi, P., et al., Centrosome duplication in mammalian somatic cells requires E2F and Cdk2-cyclin A. *Nat Cell Biol* 1999. **1**(2): 88–93.
39. Nigg, E.A., Centrosome aberrations: cause or consequence of cancer progression? *Nat Rev Cancer* 2002. **2**(11): 815–25.
40. D'Assoro AB, L.W., Salisbury JL., Centrosome amplification and the development of cancer. *Oncogene* 2002. **21**(40): 6146–53.

41. Doxsey, S., Duplicating dangerously: linking centrosome duplication and aneuploidy. *Mol Cell* 2002. **10**(3): 439–40.
42. Faragher, A.J. and A.M. Fry, Nek2A kinase stimulates centrosome disjunction and is required for formation of bipolar mitotic spindles. *Mol Biol Cell* 2003. **14**(7): 2876–89.
43. Schliwa, M., K.B. Pryzwansky, and G.G. Borisy, Tumor promoter-induced centrosome splitting in human polymorphonuclear leukocytes. *Eur J Cell Biol* 1983. **32**(1): 75–85.
44. Jean, C., et al., The mammalian interphase centrosome: two independent units maintained together by the dynamics of the microtubule cytoskeleton. *Eur J Cell Biol* 1999. **78**(8): 549–60.
45. Schliwa, M., K.B. Pryzwansky, and U. Euteneuer, Centrosome splitting in neutrophils: an unusual phenomenon related to cell activation and motility. *Cell* 1982. **31**(3 Pt 2): 705–17.
46. Sherline, P. and R.N. Mascardo, Epidermal growth factor induces rapid centrosomal separation in HeLa and 3T3 cells. *J Cell Biol* 1982. **93**(2): 507–12.

Chapter 11

Visualization of Fluorescence-Tagged Proteins in Fission Yeast: The Analysis of Mitotic Spindle Dynamics Using GFP-Tubulin Under the Native Promoter

Masamitsu Sato, Mika Toya and Takashi Toda

Abstract

Mitotic spindle microtubules pull chromosomes toward each pole to generate two daughter cells. Proper spindle formation and function are required to prevent tumorigenesis and cell death. The fission yeast *Schizosaccharomyces pombe* has been widely used as a model organism to understand the molecular mechanism of mitosis due to its convenience in genetics, molecular biology, and cell biology. The development of fluorescent protein systems and microscopy enables us to investigate the "true" behavior of proteins in living fission yeast cells using a strain with a fluorescence-tagged gene under its native promoter. In this way the level of expression of tagged protein is similar to the level of wild-type nontagged protein. In this chapter we illustrate standard methods to generate strains expressing fluorescently tagged proteins and to observe them under the microscope. Specifically, we introduce a GFP-tubulin strain to analyze the dynamic behavior of spindle microtubules. Observation of GFP-tubulin under its native promoter has illuminated the process of kinetochore–microtubule attachment process in fission yeast.

Key words: Fluorescent protein, gene targeting, fluorescence microscopy, spindle microtubule, tubulin, fission yeast.

1. Introduction

One of the most dynamic events in mitosis is chromosome segregation driven by spindle microtubules. This phenomenon is regulated by a number of proteins including kinetochore components, microtubule-associated proteins (MAPs), and cell-cycle regulators. Note that most of these fundamental factors which orchestrate mitosis in higher eukaryotes, including humans, are conserved in yeast. Therefore, it is convenient to use yeast as a model system to investigate the molecular mechanisms of mitosis

Andrew D. McAinsh (ed.), *Mitosis: Methods and Protocols*, Methods in Molecular Biology *545*,
DOI 10.1007/978-1-60327-993-2_11, © Humana Press, a part of Springer Science+Business Media, LLC 2009

because of the historical accumulation of genetic materials and methods, and the ease of observing living yeast cells. In fact, many of these mitotic regulators were originally identified in yeast, with the fission yeast *Schizosaccharomyces pombe* being one of the most widely used species. Because fission yeast contains only three chromosomes in contrast to 16 in the budding yeast *Saccharomyces cerevisiae*, it is easier to identify and capture the behavior of all the chromosomes during mitosis in living cells.

In addition, one of the practical advantages of yeast is the established methodology for gene targeting (*see* **Note 1**): it is possible to delete your gene of interest or fuse it to peptide tags such as GFP or HA by chromosomal insertion with various autotrophic or antibiotic-resistant marker genes via homologous recombination. In this chapter we illustrate how to generate such integrated strains.

Visualization of microtubules in living cells is also important for observation of mitosis. Fission yeast has two α-tubulin genes and a single β-tubulin gene. Because it is generally thought that fluorescent-tagged tubulin has compromised function, especially when tagged at the carboxyl-terminus, the nonessential a-tubulin gene *atb2*$^+$ is often chosen for GFP-tagging at its amino-terminus. There are several GFP-tubulin constructs available, some of which have expression plasmids containing *GFP-atb2*$^+$ and others are integrated *GFP-atb2*$^+$ strains. We constructed a GFP-Atb2 integrated strain whose expression is regulated under its own promoter (*see* **Note 2**). This strain has enabled us to uncover the detailed structure and dynamics of mitotic spindles in fission yeast, which had been difficult to visualize using other GFP-tubulin constructs.

2. Materials

2.1. Strains and Media

1. Fission yeast laboratories share a single original wild-type strain 972. This strain is *h*$^-$ for mating type and prototrophic with no auxotrophic requirements. A convenient strain for studies on interphase and mitosis is 513 (heterothallic) with the *leu1-32* and the *ura4-D18* mutations. In this chapter, we generally use strains in these backgrounds.
2. Homothallic *h*^{90} strains for study of meiosis.
3. YE5S medium *(1)* for observation of interphase and mitotic cells.
4. EMM-N (Minimal medium with glutamate as nitrogen source instead of ammonium sulfate) supplemented with appropriate amino acid supplements (adenine, leucine, uracial, histidine, lysine, and/or arginine, depending on the auxotrophy of the strain) for induction of sexual differentiation (mating, meiosis, and sporulation) *(1)*.

2.2. Construction of C-Terminal Tagged Strains

1. 100-base oligos (PAGE or HPLC purified; SIGMA)
2. LA Taq (TaKaRa)
3. PCR purification kit, for example, Wizard purification kit (Promega)
4. PCR machine such as PTC-200 (Bio-rad).
5. $Vent_R$® DNA polymerase (New England Biolabs)

2.3. Transformation of Fission Yeast

The following reagents are used as described previously in *(2)*.

1. LiOAc/TE: 0.1 M lithium acetate, 0.01 M Tris-HCl, 1 mM EDTA, pH 7.5
2. PEG/LiOAc/TE: 40% polyethylene glycol #4000 in TE/LiOAc, filter sterilized
3. Herring testis carrier DNA (10 mg/ml, Clontech)
4. YE5S plates without or with selection drugs
5. Concentration of selection drugs *(4)*:

Antibiotics	Final Concentration in YE5S(mg/L)	Stock Solution (mg/ml)
G418 (SIGMA)	100	100
clonNAT (Werner)	100*	100*
hygromycin B (Roche)	300	50

* clonNAT stock solution needs to be filter-sterilized, as this reagent is supplied as solid materials.

2.4. Functional Test of Patb2-Atb2

1. TBZ (15 μg/ml) (SIGMA)

2.5. Observation Under the Microscope

1. Lectin-coated glass-bottom dish (MatTek)
2. IX71 microscope (Olympus) with DeltaVision system (Applied Precision)
3. SoftWoRx software (Applied Precision)

2.6. FRAP Experiments

1. QLM-Laser bleaching system attached to Delta Vision microscope (Applied Precision)

3. Methods

3.1. Construction of C-Terminal Tagged Strains

3.1.1. Vectors for Chromosomal Integration

Here we illustrate the principle of the PCR-based gene targeting method as summarized in **Fig. 11.1**. For example, when the *GFP* gene is targeted at the C-terminus of a certain gene, homologous recombination operates and the expression of the resulting *GFP*-fusion gene will still be regulated under its own promoter, because the authentic flanking region including the 5'-UTR remains intact. Thus, we can accomplish a natural expression pattern of GFP-tagged protein that is controlled by its

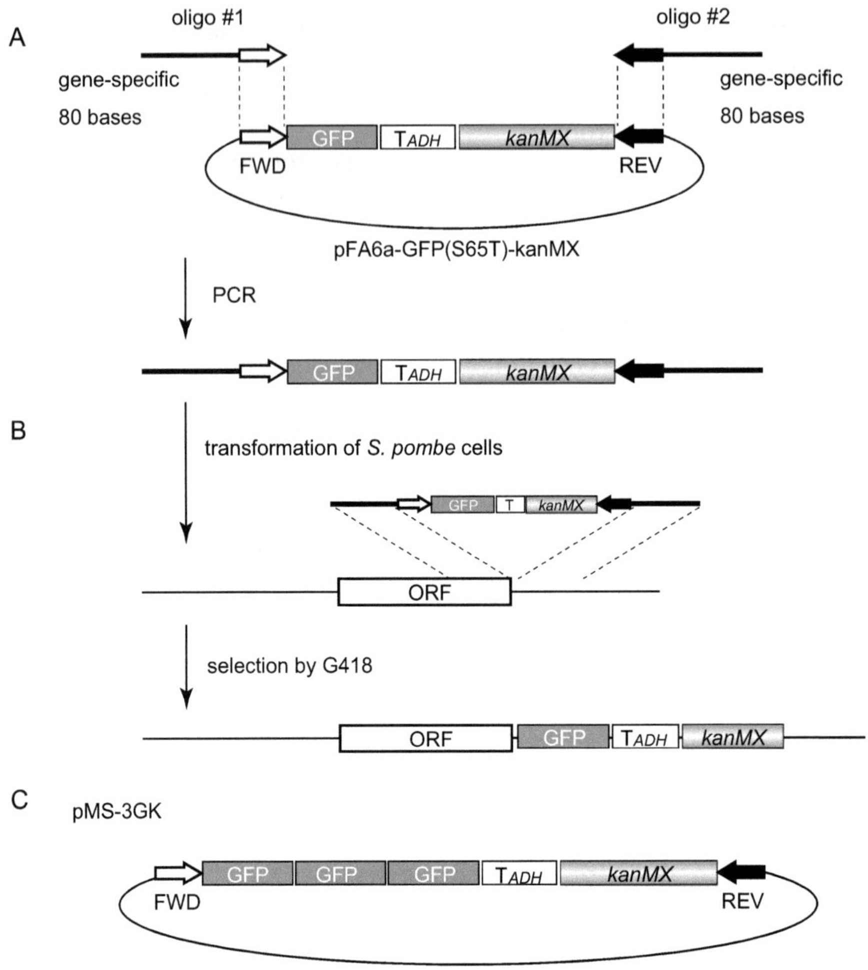

Fig. 11.1. Construction of GFP-tagged genes in fission yeast. **(A, B)** Bähler et al. *(2)* describe the standard method to tag GFP or other tags to the gene of interest. An example of GFP-tagging is shown. **(A)** A pair of 100-base oligos and the plasmid pFA6a-GFP(S65T)-kanMX need to be prepared. Each oligo contains the 80-base gene-specific sequence and the particular 20-base sequence (FWD and REV). PCR product contains *GFP*, *ADH1* terminator, and the *kan* (G418) resistance selection marker gene flanked by a pair of 80-base targeting sequences. **(B)** Fission yeast cells were transformed with the amplified PCR product. G418-resistant colonies can be selected as candidates for the integrant. **(C)** pMS-3GK, a vector for integration of tandem three copies of GFP. Similar vectors are summarized in **Table 11.1**.

own promoter. Vectors for PCR-based chromosomal integration method were first described in *(2)*. These vectors are based on pFA6a with molecular tags (GFP, HA, etc.) and a selection marker *kanMX* (the G418 resistant gene) flanked by a pair of common 20 base-sequences (white arrow: FWD and black: REV in **Fig. 11.1A**). For instance, the plasmid pFA6a-GFP(S65T)-kanMX can be used for GFP-tagging with the kanamycin selection marker gene (confers resistance to G418; **Fig. 11.1A**). A number of modified vectors with new selection markers have been published so far *(3, 4)*. Many similar vectors originally designed for budding yeast *(5)* can also be utilized for fission yeast. These plasmids, however, generally contain an S2 sequence which is different from the Forward 20 base-sequence described in *(2)*. Recently, vectors for tagging of tandem repeated GFPs (and other colors) began to be introduced in budding yeast and fission yeast *(6–9)*. A variety of colors of fluorescent proteins contributes to multicolor imaging in living cells. The use of multicolor fluorescent proteins in yeast has been widely propagated. We normally use GFP (or YFP), ECFP, and mCherry (or mRFP) as a standard three-color set in order to label three different proteins of interest simultaneously. All of those fluorescent colors can be tagged into any gene in a way similar to that described earlier. As the number of fluorescent colors and antibiotics increases, strain constructions through genetic crossing become complicated and confusing: for instance, two different genes are tagged with different colors, but with same drug-resistant markers. To avoid the color and/or marker conflicts, we have established our local rule in the lab to stick to GFP with *kan*, mRFP/mCherry with *hph*, and CFP with *nat*, and additional gene deletion can still be achieved by autotrophic markers such as *ura4*⁺ or *LEU2*. Multiple tandem copies of fluorescent proteins are often useful, especially when the fluorescent signal from a single copy is not bright enough. We also developed plasmids to integrate multi-GFP, multi-mRFP, or multi-mCherry fluorescent tags for fission yeast *(8)*. These plasmids are listed in **Table 11.1**. These plasmids basically contain a similar structure to the vectors above with the identical 20-base sequences for Forward and Reverse (see below and **Fig. 11.1A**). This enables us to

Table 11.1
Plasmids for multiple tandem fluorescent protein tagging

Number	Plasmid	Tag	Marker	Origin
1.	pMS-2GK	2GFP	kan	This study
2.	pMS-3GK	3GFP	kan	*(8)*
3.	pMS-4RK	4mRFP	kan	*(8)*
4.	pMS-4RN	4mRFP	nat	*(8)*
5.	pMS-4RH	4mRFP	hph	*(8)*
6.	pMS-2mChK	2mCherry	kan	This study
7.	pMS-3mChK	3mCherry	kan	This study
8.	pMS-4mChK	4mCherry	kan	This study
9.	pMS-2mChH	2mCherry	hph	This study
10.	pMS-3mChH	3mCherry	hph	This study
11.	pMS-4mChH	4mCherry	hph	This study

All of these plasmids contain the FWD and REV 20 bp sequences which are identical to those in pFA6a vectors described in Bähler et al. *(2)*.

Table 11.2
Forward and Reverse 20 bp sequences for primer design

Primer	Sequence
Forward	5'-(gene specific sequence)-CGG ATC CCC GGG TTA ATT AA-3'
Reverse	5'-(gene specific sequence)-GAATTCGAGCTCGTTTAAAC-3'

These 20 base sequences need to be added to the 80 base gene-specific sequences to design the forward and reverse primers for gene targeting *(2)*. For forward primer, the reading frame for the tag sequence is indicated.

utilize the same oligos already designed for a single GFP-tagging, and, for example, 3 mCherry-tagging.

3.1.2. Design of Oligos for PCR

1. A pair of synthesized 100-base oligos needs to be designed for PCR. Each oligo must contain a gene-specific 80-base sequence combined with a 20-base particular sequence (FWD and REV in **Fig. 11.1A**). The Forward oligo (oligo #1) contains the 5'-flanking 80-base sequence prior to the initiator ATG and the Reverse oligo (oligo #2) contains the 3'-terminal 80 bases from the 3'-UTR region of the gene.
2. The 20-base sequences for Forward and Reverse oligos serve as the annealing sites to the template plasmids. Nucleotide sequences are shown in **Table 11.2**. The sequences are common among all the plasmids and do not depend upon the gene of interest or tags.
3. 100 bp oligos should be purified by PAGE or HPLC to maximize the efficiency of PCR and transformation.

3.1.3. PCR

1. The template plasmid pFA6a-GFP(S65T)-kanMX contains the *GFP* gene, the terminator of the *ADH1* gene, and the kanamycin resistant marker gene, which are flanked by the Forward and the Reverse 20-base sequences to which the 3'-end of the 100-base oligos anneal.
2. A generic PCR protocol is:

Step	Temperature (°C)	Time (min:s)
1.	95	3:00
2.	95	1:00
3.	48	2:00
4.	74	4:00
5.	Go to step 2 and repeat steps 2–4 four times.	
6.	95	1:00
7.	52	2:00
8.	74	2:00
9.	Go to step 6 and repeat steps 6–8 thirty-four times.	
10.	4	Hold indefinitely

3. The amplified fragment can be purified by one of commercially available PCR purification kits for a better efficiency of transformation.

For multiple tandem copies of fluorescent proteins we normally use the following PCR program which was modified from Janke et al. *(5)*, sometimes with a mixture of LA Taq and $Vent_R$ DNA polymerase for amplification.

Step	Temperature (°C)	Time (min:s)
1.	95	3:00
2.	95	0:20
3.	53	0:30
4.	68	4:00
5.	Go to step 2 and repeat steps 2–4 nine times.	
6.	95	0:30
7.	53	0:30
8.	68	4:00*
9.	Go to step 6 and repeat steps 6–8 fourteen times.	
10.	4	Hold indefinitely

* With incremental 20 s per additional cycle.

3.1.4. Transformation of Fission Yeast with PCR Products

The following method is recommended based on a common method described in *(2)*.

1. Harvest cells from a YE5S culture and resuspend in LiOAc/TE solution.
2. Mix cells with the PCR products from **Section 3.2.2** and carrier DNA (derived from herring testis) as well as PEG/LiOAc/TE, and incubate for 1 h.
3. Shift temperature to 42°C for 15 min for heat shock, and plate onto YE5S nonselective media.
4. After 18 h at 30°C, replicate the cells onto YE5S plates containing the appropriate antibiotics or FOA (5-fluoroorotic acid).
5. Confirm using PCR that resistant colonies have correctly integrated the gene.

3.2. Construction of an N-Terminal GFP-Tagged α2-Tubulin (Atb2) Strain

As an example of the gene targeting method described in **Section 3.2** we describe below the creation of a GFP-atb2 strain:

3.2.1. Replacement of the ura4+ Cassette by GFP-atb2 Construct

1. As illustrated in **Fig. 11.2A**, the *GFP-Atb2+* fragment flanked by upstream and downstream regions of the *atb2+* gene was amplified by PCR.
2. This fragment was introduced into an *atb2::ura4+* strain in which the *atb2+* ORF had been deleted by the *ura4+* marker gene cassette *(10)*, to induce gene replacement from *ura4+* to *GFP-atb2+*.

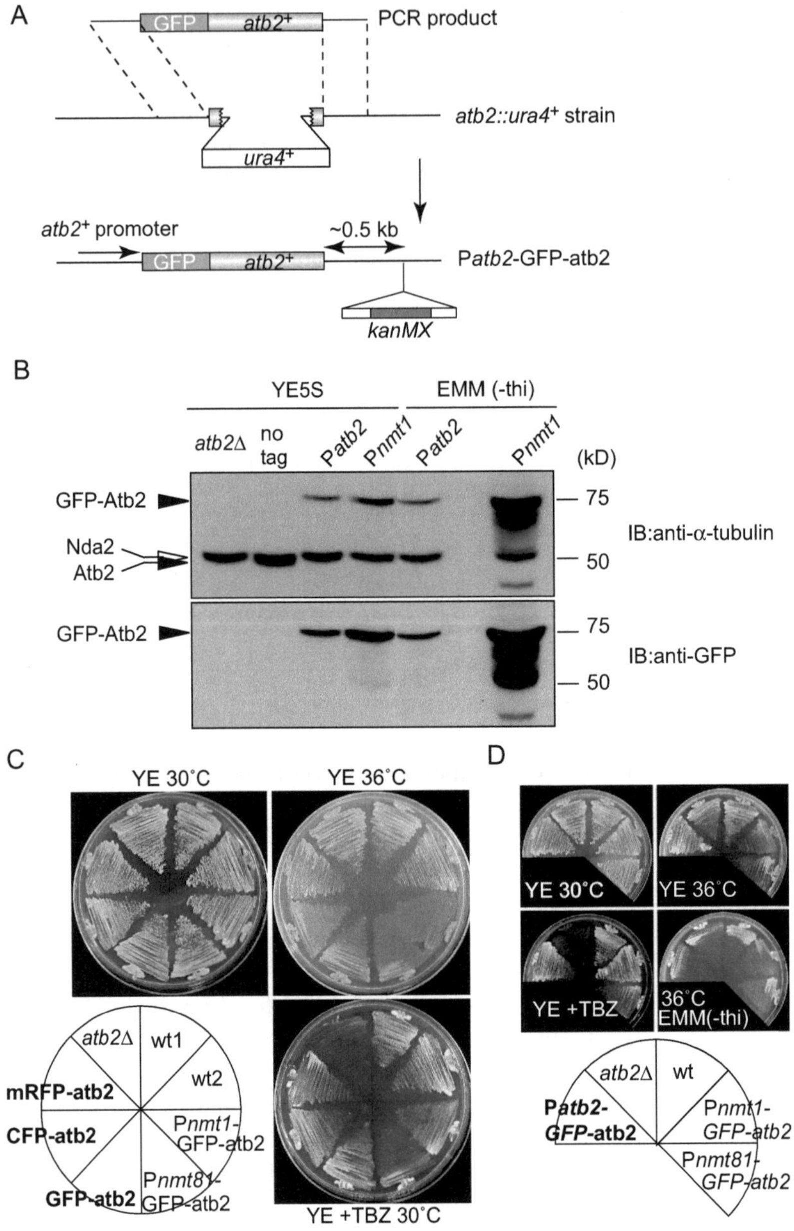

Fig. 11.2. Construction of the GFP-atb2 strains with own promoter and terminator. **(A)** The gene-targeting fragment was amplified by PCR. PCR product was targeted to the *atb2* locus which had been disrupted by the *ura4*⁺ autotrophic marker gene. A resulting strain, P*atb2*-GFP-atb2, consists of the *GFP-atb2* gene with its own promoter and terminator. Additional kanamycin (G418) resistant marker was inserted 500 bp-downstream of the *atb2*⁺ termination codon. (**B**) Expression level of GFP-Atb2 confirmed by Western blotting. Protein extract was prepared from the following strains: wild-type (no tag), *atb2* disruptant (*atb2*Δ), GFP-*atb2* strain regulated by the *nmt1* promoter (P*nmt1*), and P*atb2*-GFP-atb2 strain (P*atb2*). Cells were grown in YE5S (containing thiamine) or EMM (without thiamine) media. Immunoblotting was performed with anti-a-tubulin and anti-GFP antibodies. **(C, D)** Sensitivity to temperature and TBZ (15 μg/ml) of each strain was tested on plates in comparison with P*nmt1* and P*nmt81*-driven GFP-atb2 strains. The red dye Phloxin B was added into YE plates to stain dead cells dark red. **(C)** A functional test for P*atb2*-GFP-atb2, P*atb2*-CFP-atb2, and P*atb2*-mRFP-atb2 strains without selection markers. **(D)** A functional test for P*atb2*-GFP-atb2-kan strain.

3. After transformation procedures (*see* **Section 3.1.3.**), cells were spread onto YE5S plates followed by replica plating onto YE5S containing FOA, which is toxic to the cells expressing the *ura4*$^{+}$ gene product.
4. Colonies resistant to FOA were chosen to confirm correct integration by microscopy and PCR. This strain was named GFP-atb2.
5. Similarly, mRFP-atb2 and CFP-atb2 strains were also constructed.

3.2.2. Additional Insertion of kan Marker Gene

The strain produced in **Section 3.3.1** expresses the GFP-*atb2*$^{+}$ gene from the native *atb2*$^{+}$ promoter and terminator. To enable convenient genetic crossing, the G418 resistant marker cassette *kanMX* was inserted ~500 bases downstream of GFP-atb2 to preserve the *atb2* terminator region (**Fig. 11.2A**). The resulting strain was named P*atb2*-GFP-atb2 (P*atb2* stands for Promoter of the *atb2*$^{+}$ gene). We also constructed P*atb2*-CFP-atb2 with *nat* and P*atb2*-mRFP-atb2 with *hph*.

3.2.3. Expression Level of GFP-Atb2

In order to confirm that the GFP-Atb2 protein was expressed at similar levels to the endogenously expressed Atb2, expression levels were examined by Western blotting. In one of the widely used strains P*nmt1*-GFP-atb2, GFP-Atb2 is expressed artificially under the *nmt1* promoter (P*nmt1*) *(11)*. The *nmt1* promoter is repressed in the presence of thiamine (in media such as YE5S), and induced in the absence of thiamine (in media such as EMM2). P*nmt1*-GFP-Atb2 expression level was indeed highly up-regulated in EMM (-thiamine), which is toxic to the cells (**Fig. 11.2B** and **D**). In contrast, the level of GFP-Atb2 protein expressed under the *atb2* promoter (P*atb2*) remained constant at the endogenous level after shifting to thiamine-free EMM (**Fig. 11.2B**), suggesting that this newly constructed P*atb2*-driven GFP-Atb2 strain can also be utilized under conditions in which thiamine cannot be added for technical reasons.

3.2.4. Functional Test of Patb2-GFP-Atb2

To examine the function of GFP-Atb2 expressed under P*atb2*, cells were streaked on YE5S media containing TBZ, a drug that destabilizes microtubules. The *atb2*-deletion strain is known to show TBZ-hypersensitivity (**Fig. 11.2C**), indicating that Atb2 function is required to survive in the presence of TBZ. Note that the newly constructed P*atb2*-driven GFP-, mRFP-, and CFP-atb2 strains did not show TBZ-hypersensitivity (**Fig. 11.2C** and **D**). This suggests P*atb2*-driven GFP-Atb2, as well as mRFP-Atb2 and CFP-Atb2, is functional, at least in terms of TBZ resistance. Furthermore, whereas P*nmt1*-GFP-atb2 and P*nmt81*-GFP-atb2 (P*nmt81* is a weakened *nmt1* promoter) strains showed modest temperature sensitivity under our YE5S conditions, the newly constructed P*atb2*-driven GFP-, mRFP-, and CFP-atb2 strains did not (**Fig. 11.2D**). Therefore, a P*atb2*-GFP-atb2 strain is useful for live cell observation at high temperature *(8)*.

Other integrants such as P*nmt1*-GFP-atb2 and P*nmt81*-GFP-atb2 are synthetically lethal with some mutants of microtubule-associated proteins (MAPs), such as *alp7*Δ or *alp14*Δ (*(12)* and MS and TT, unpublished results), suggesting the protein amount of GFP-Atb2 might affect the viability of cells in these mutant strains. The P*atb2*-GFP-atb2 construct was, however, viable even with these mutations, although it showed slight synthetic growth delay, suggesting that GFP-Atb2 is not fully functional. Nonetheless, the P*atb2*-GFP-atb2 strain may therefore also be used for the observation of GFP-Atb2 in those microtubule-related mutant backgrounds.

3.3. Genetic Crossing with Genes for Other Marker Proteins

The crossing of strains containing different fluorescently tagged proteins enables the co-visualization of multiple cellular structures. A P*atb2*-GFP-atb2 strain was crossed with strains that contain other marker proteins, such as Cut12-CFP for an SPB marker *(13)*, Sfi1-CFP for SPB half-bridge *(14)*, Cut11-3mRFP for the nuclear envelope and the mitotic SPB *(15)*, and Mis6-2mRFP for the kinetochore *(16)*. Parental strains were mixed on EMM2 lacking nitrogen, and spores were separated by random spore analysis. Colonies showing appropriate multiantibiotic resistance were chosen as double- or triple-tagged strains.

3.4. Observation Under the Microscope

1. Cell culture is prepared in YE5S liquid media and placed on a lectin-coated glass-bottom dish.
2. Meiotic cells are taken from EMM-N plates to resuspend into EMM-N media before placing onto glass-bottom dish.
3. Approximately 10 min later, the dish is filled with the same media and placed on an inverted microscope (*see* **Note 3**).
4. Samples are filmed with 6–9 serial sections along the Z-axis in 1–15 s intervals, depending upon your purpose. For standard observation, images are taken every 15 s. To capture fast motion of fluorescent signal, especially in FRAP experiments (*see* **Section 3.6.1**), samples are filmed every second.
5. Images are subjected to deconvolution using SoftWoRx software followed by projection with the "maximum brightness" or "Sum" method depending upon your purpose. Images shown here were projected with the "maximum brightness." To quantify the total signal fluorescent intensity, "sum" projection is preferred.

3.4.1. Interphase Microtubule Structure

As an example of observation of cells expressing multiple fluorescently tagged proteins, GFP-Atb2 was observed with

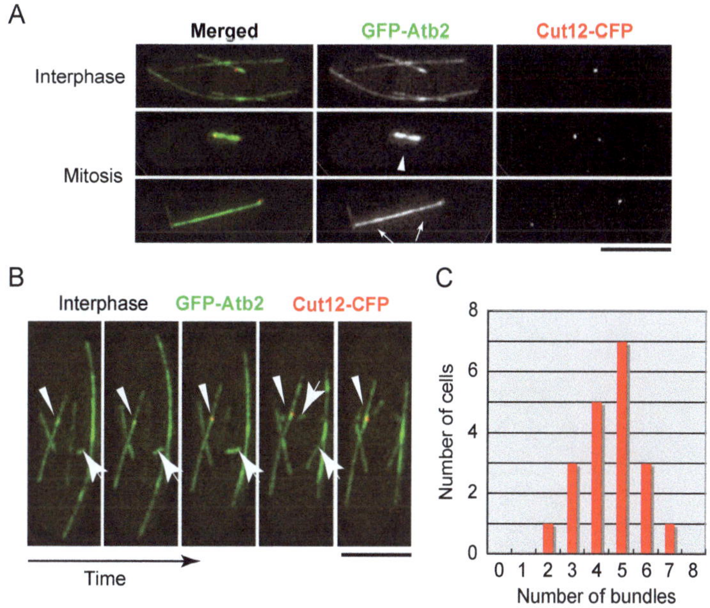

Fig. 11.3. Cytoplasmic microtubules visualized by GFP-Atb2. **(A)** P*atb2*-GFP-atb2 cells were observed with Cut12-CFP, an SPB marker. Cells in interphase (top), early mitosis (middle, the central dim region is marked with arrowhead), and late mitosis (bottom, the edges of central bright region are marked with arrows) are shown. **(B)** Dynamics of interphase microtubules was monitored. Images were taken every 15 s. Arrowheads: the position of the SPB. Arrows: the position of the iMTOC. Bars = 5 μm. **(C)** Distribution of the number of cytoplasmic microtubule bundles in the P*atb2*-GFP-atb2 strain.

Cut12-CFP, an SPB marker. As shown in many previous reports, an arraylike cytoplasmic microtubule structure was observed along the longitudinal cell axis (**Fig. 11.3A**). Cytoplasmic microtubules were nucleated from the SPB as well as around the nucleus (**Fig. 11.3B**). These non-SPB nucleation sites correspond to the iMTOCs (interphase microtubule organizing centers). Time-lapse recording revealed that there were 3.0 ± 0.47 iMTOC sites per cell on average. The number of cytoplasmic microtubule bundles in a cell was counted and is shown in **Fig. 11.3C**.

3.4.2. Meiosis and Horse-Tail Movement

The use of a GFP-Atb2 Sfi1-CFP Cut11-3mRFP strain was further validated by analyzing spore viability. Under nutrition starvation conditions, fission yeast cells of opposite mating types (h^+ and h^-, or h^{90} cells) undergo sexual conjugation (mating) and meiotic processes, such as karyogamy formation, premeiotic DNA synthesis, recombination, and meiosis I and II followed by sporulation *(17)*. To observe microtubule behavior in mating and meiosis, a single colony on a YE5S plate was taken to suspend in 2–3 μl of EMM-N liquid media and spotted onto EMM-N plates. Microtubule organization plays essential roles in many aspects

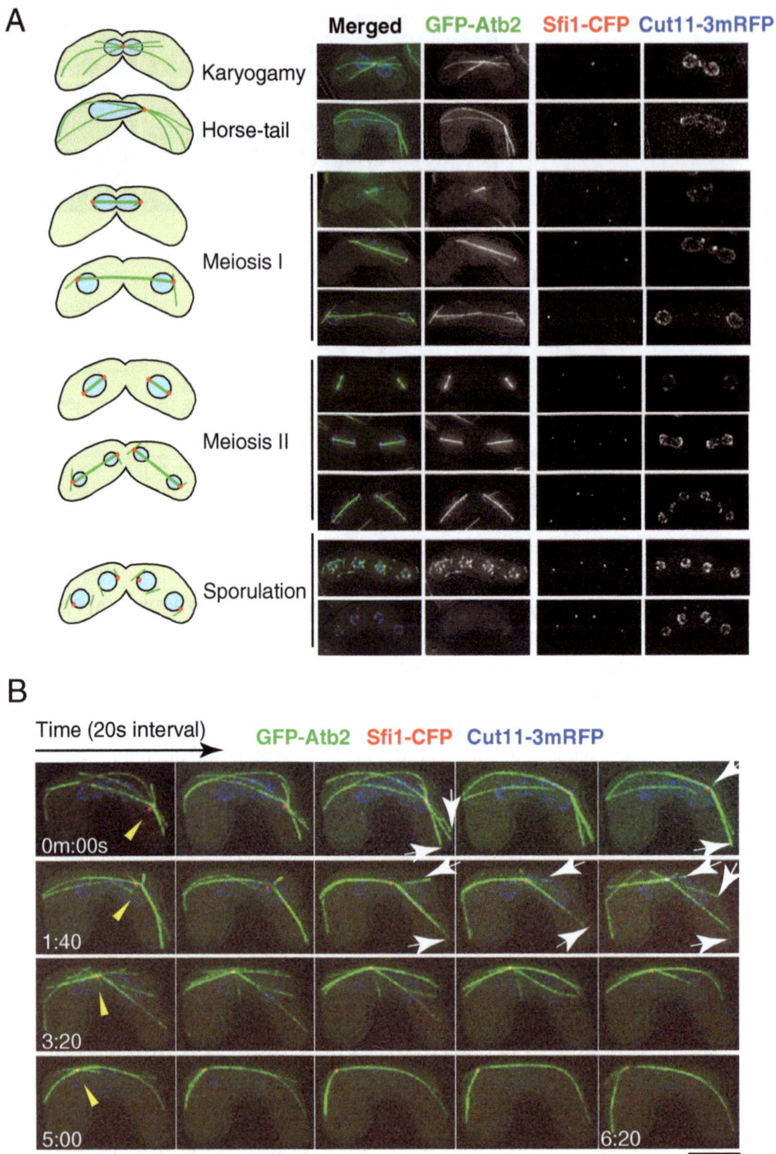

Fig. 11.4. Visualization of microtubule structures during meiosis. **(A)** Microtubule structures visualized by P*atb2*-GFP-atb2 together with Sfi1-CFP (an SPB half-bridge marker) and Cut11-3mRFP (a nuclear envelope marker) on each stage of meiotic cell cycle. Representative cells for each stage were chosen. Schematic images for each stage were depicted on the left. **(B)** Oscillatory movement of the nucleus driven by SPB–microtubule–cell cortex interaction. Images were taken every 20 s. Yellow arrowheads: the SPB, arrows: Ends of microtubules are anchored at the cell cortex. Bars = 5 μm.

of meiosis (**Fig. 11.4A**). In karyogamy formation, the microtubule array is focused at the center of the conjugated cells to promote nuclear fusion (**Fig. 11.4A**). During meiotic prophase, the nucleus shows characteristic back-and-forth movement in a cell (**Fig. 11.4A** and **B**). This oscillation is termed "horse-tail movement." *(18, 19)* Meiotic spindles are formed in meiosis I and II (**Fig. 11.4A**). After meiotic divisions, microtubules are fragmented and disappear upon spore wall formation (**Fig. 11.4A**). These experiments show that the use of a GFP-Atb2 Sfi1-CFP Cut11-3mRFP strain is valid, because spore viability of this strain was comparable to that of an untagged strain.

3.5. Observation of the Mitotic Spindle and the Kinetochore

3.5.1. Central Dim Region of the GFP-Atb2 Spindle

The P*atb2*-GFP-atb2 strain created in **Section 3.3** has been validated as a useful strain in which to observe dynamics and localization of tubulin in fission yeast. Interphase microtubules are reorganized into the nuclear mitotic spindle upon the G2-M transition. Because fission yeast undergoes a closed mitosis in which the nuclear envelope remains intact during mitosis, the mitotic spindle needs to be formed in the nucleus. **Fig. 11.3A** shows early and late mitotic spindles visualized in the P*atb2*-GFP-atb2 strain. The central region of the early mitotic spindle exhibited dim GFP signals compared to both edges of the spindle (arrowhead, middle panel, **Fig. 11.3A**), whereas in late mitosis this central domain became thick (anaphase B) due to the formation of the interdigitating spindle midzone (arrow, bottom panel, **Fig. 11.3A**). These characteristic GFP patterns were often hard to visualize when GFP-Atb2 was overexpressed from the pREP1 plasmid *(20)*, probably because an excess amount of GFP-Atb2 uniformly localized to the whole spindles, thereby masking the existence of dim or thick regions.

This result fits with previous studies of mitotic spindles using electron microscopy *(21, 22)* and immunostaining with anti-tubulin antibodies *(23)*, which revealed that the early mitotic spindle consists of (1) interpolar microtubules which connect the two SPBs and (2) kinetochore–microtubules (kMTs) which connect kinetochores to SPBs. This demonstrates the power of such a system expressing a fluorescently tagged gene of interest at endogenous levels.

3.5.2. Visualization of the Kinetochore in the GFP-Atb2 Dim Region

Previous FISH analysis demonstrated that the central dim region stained by anti-tubulin antibodies corresponds to the position of kinetochores *(24)*. To confirm this in our live observation system, a strain expressing the kinetochore protein Mis6 tagged with two copies of mRFP, GFP-Atb2, and Cut12-CFP was created using the gene targeting method outlined in **Section 3.2**. As expected, the bright GFP-Atb2 signal nucleating from SPBs ended at the position of Mis6-2mRFP signals (**Fig. 11.5A**). The position of the dim region changed during the oscillatory movement of kinetochores along the spindle during prometaphase, and note that its edge always corresponded to Mis6-2mRFP location. Mis6-2mRFP signals split to each pole upon anaphase onset (4:40, **Fig. 11.5A**), and simultaneously the

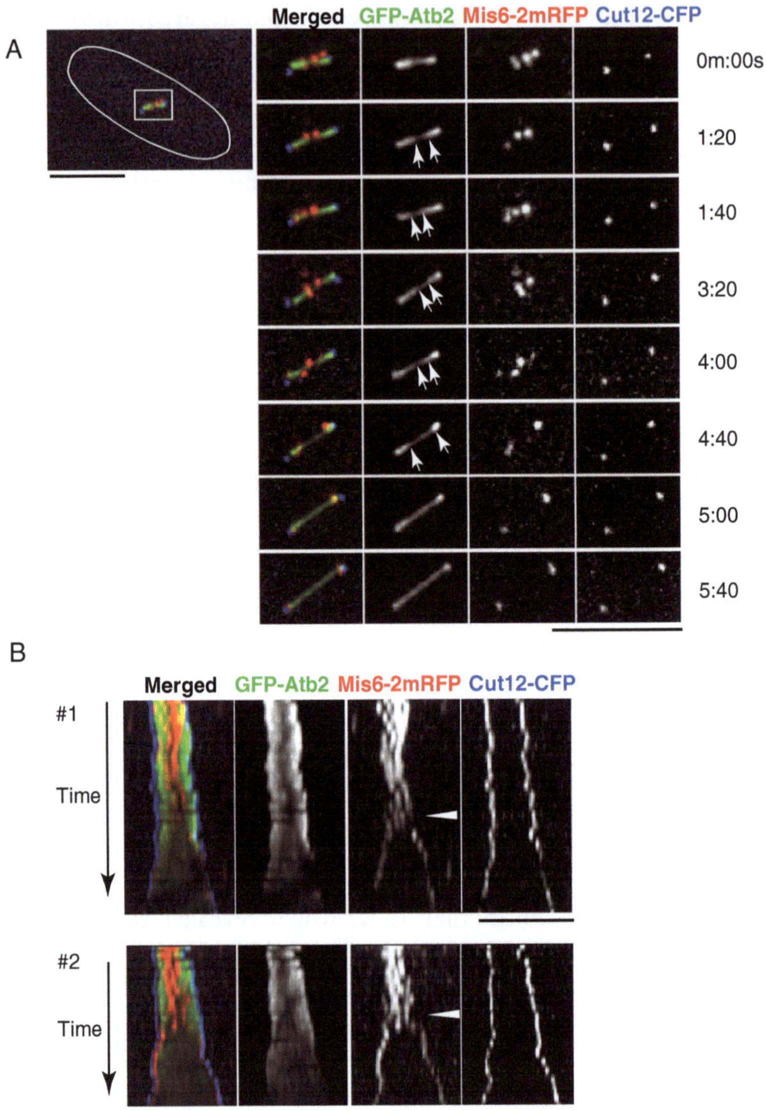

Fig. 11.5. GFP-tubulin and kinetochores during mitosis. **(A)** Live imaging of P*atb2*-driven GFP-Atb2 as well as Mis6-2mRFP (a kinetochore marker) and Cut12-CFP (SPB) in a preanaphase cell. Images were taken every 20 s. Arrowheads mark the edges of bright region of GFP-Atb2 signals, mostly co-localizing to the kinetochore position. **(B)** Kymographic view of (**A**). Two independent cells were recorded and processed for kymograph. Arrowheads indicate the timing of anaphase A. Bars = 5 μm.

bright GFP-Atb2 region shrank poleward and finally disappeared (~5:00; see also a kymographic view in **Fig. 11.5B**). This further confirms the existence of interpolar and kinetochore microtubules in fission yeast, as visualized by the P*atb2*-GFP-atb2 strain in live analysis with standard fluorescence microscopy (**Section 3.6.1** and **Fig. 11.3**).

3.6. Dynamics of Spindle Microtubules Using FRAP Analysis

3.6.1. FRAP Experiments

The creation of yeast strains expressing fluorescently tagged proteins also enables fluorescent recovery after photobleaching (FRAP) experiments to be performed, allowing investigation of spindle dynamics during prometaphase and anaphase. The FRAP experiments below were performed with a QLM-Laser bleaching system on the DeltaVision system.

1. Photobleach a circular region with an average radius of 0.5–1 μm encircling the central region of the GFP-Atb2 signal with a 488 nm laser for 0.05 s.
2. Record images of a single section along the z-axis every second after photobleaching.

3.6.2. Spindle Dynamics During Prometaphase

Prometaphase spindles of the P*atb2*-GFP-atb2 strain exhibit the central dim region where kinetochores are being captured by kMTs. Such prometaphase spindles were photobleached by laser exposure. Two representative examples are shown in **Fig. 11.6A,B**, ("pre" and "0"). Fluorescent recovery was monitored every second. Shortly after laser exposure, GFP-Atb2 signals of some spindles recovered from the SPB (region i, **Fig. 11.6A**), suggesting some microtubules were newly nucleated from the SPBs. GFP-Atb2 signals in other spindles emerged at sites apart from the SPB (5 s in region i, **Fig. 11.6B**), indicating that the existing bleached microtubules continued to polymerize at their plus-ends. Thus, the prometaphase microtubules are dynamic. Conversely, the central dim region remained dark after bleaching (**Fig. 11.6B**). Moreover, this blackout persisted even on the interpolar microtubules until anaphase B (the asterisk on 141 s in region iii, **Fig. 11.6B**). These observations indicate that the interpolar microtubules are stable; therefore the recovered GFP-Atb2 signal must belong to the kMTs, which had been newly polymerized after photobleaching. In line with this, the recovered GFP-Atb2 signal moved poleward and finally disappeared upon anaphase A onset (region ii, **Fig. 11.6A** and **B**). Thus, the P*atb2*-GFP-atb2 construct contributes to the visualization of kinetochore–microtubules polymerized *de novo* after laser exposure.

3.6.3. Spindle Dynamics During Anaphase B

Spindle dynamics in anaphase cells were then analyzed using FRAP. An anaphase B cell expressing P*atb2*-driven GFP-atb2 was captured and the central interdigitating region with bright GFP-Atb2 signals was photobleached (**Fig. 11.6C**). As the spindle elongated, the central region recovered its GFP signal whereas the rest of the spindle remained dark ("234" and the kymograph in **Fig. 11.6C**). This suggests (1) the plus-ends of overlapping microtubules at the spindle center are polymerizing, (2) other bleached parts of the spindle are not dynamic, and (3) the poleward flux at the SPBs seldom occurs in anaphase B

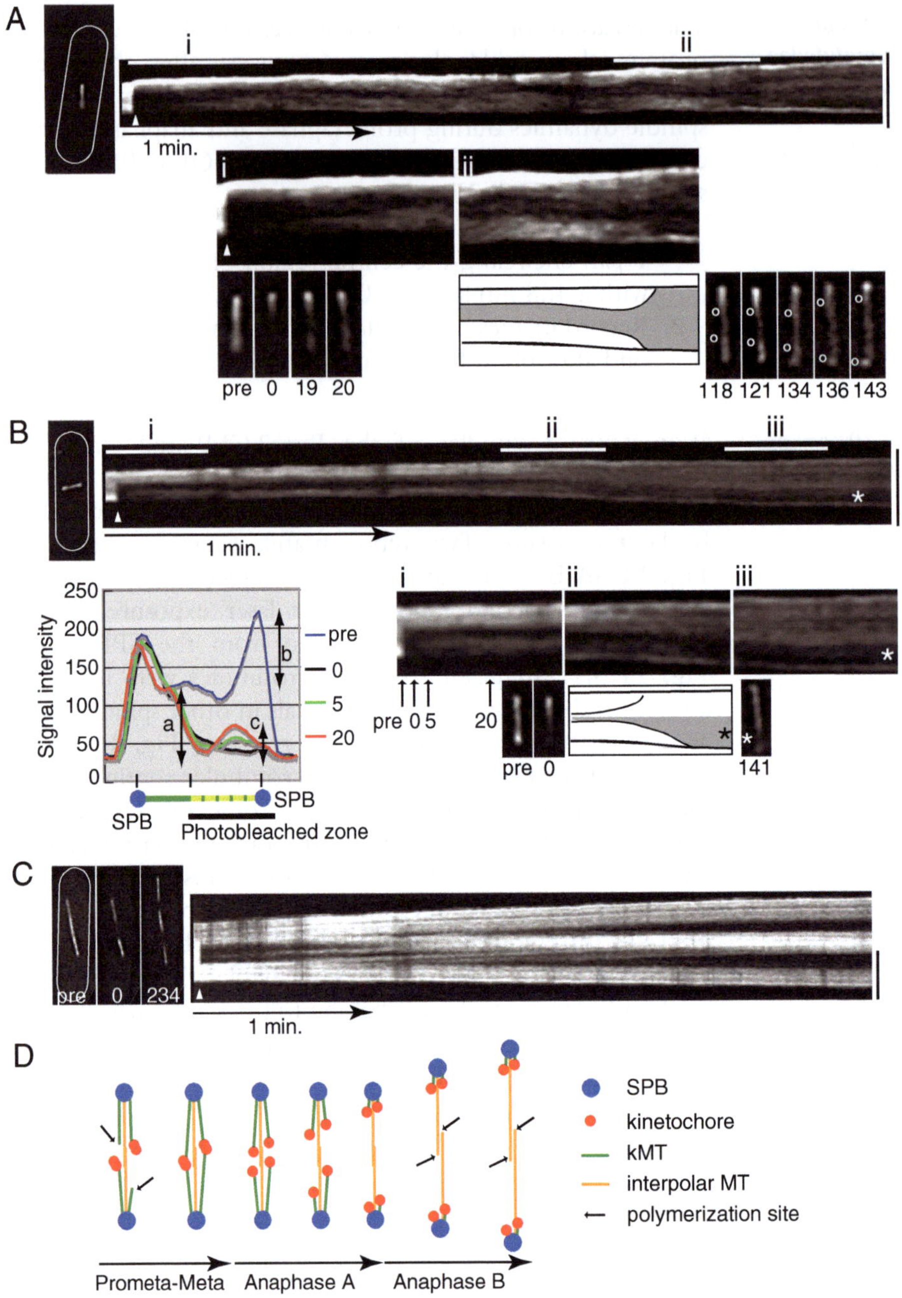

Fig. 11.6. Selective visualization of kinetochore–microtubules by FRAP analysis. P*atb2*-GFP-atb2 cells in prometaphase (**A**, **B**) and anaphase B (**C**) were photobleached. Time-lapse images before and after bleach were recorded and processed into the kymograph (from left to right). Length of horizontal arrows corresponds to 1 min. Arrowheads: timing of photobleach. Vertical bars = 5 μm. **(A)** An example of prometaphase cell. GFP-Atb2 signals recovered from bleached SPB (i). Representative timepoints are shown (prebleach, 0 s, 19 s, and 20 s). Recovered GFP-Atb2 signals split outward upon anaphase A onset (ii, shown with a schematic drawing). Examples at indicated times were also shown. Circles = the edges of bright GFP-Atb2 signals on microtubules upon anaphase A onset. **(B)** Another example of prometaphase cell. (i) The kymograph and the graph showing the recovery of GFP-Atb2 intensity measured at prebleach (pre) and indicated seconds after bleaching. Signal intensity presumably originated from interpolar microtubules

in fission yeast. This dynamic feature of the anaphase B spindle in the P*atb2*-GFP-atb2 stain is shared with other GFP-tubulin constructs as reported previously *(25–27)*. The dynamics of spindle microtubules from prometaphase to anaphase B visualized by P*atb2*-GFP-atb2 is summarized in **Fig 11.6D**. The central dim region in prometaphase contains interpolar microtubules, whereas the bright regions at both sides represent the coexistence of interpolar and kinetochore microtubules. Spindle elongation in anaphase B is driven by sliding of interdigitating interpolar microtubules which are continuously polymerized at their plus-ends.

4. Notes

1. One of the advantages of fission yeast is the ease of genetic manipulations. Recent cell biological studies of fission yeast are based on abundant historical accumulations in both classical and modern genetics. The development of gene-targeting methods of fission yeast will keep extending the field of the cell cycle and cell morphology onward.
2. Expression of GFP-tubulin in living cells has greatly contributed to the study of mitosis. However, we need to pay special attention to the expression level of GFP-tubulin which might be affected by several experimental factors, such as the promoter, the terminator, copy number, and so on. As a result, it is hard at the moment to judge which strain is the best as "wild-type", in as much as it is impossible to compare each construct with untagged wild-type without visualizing tubulin. We therefore consider each GFP-tubulin construct can be used depending upon your own purposes. The P*atb2*-GFP-atb2 construct introduced here would be useful for the study of kinetochore–microtubule interaction, because this construct made it possible to distinguish kinetochore microtubules from interpolar spindle microtubules. The P*atb2*-GFP-atb2 strain can also be tested to use under compromised conditions such as in some MAP mutant backgrounds. It may also be applied to the analysis

Fig. 11.6. (continued) (a) and kinetochore–microtubules (b) before photobleach, and estimated fluorescent recovery corresponding to newly polymerized kinetochore–microtubules (c). (ii) Anaphase A onset. (iii) The bottom-half of the spindle remained dim even during anaphase B (asterisks and 141 s). **(C)** A GFP-Atb2 cell in anaphase B was photobleached. Images before bleach (pre) and 0 s and 234 s after bleach are shown as examples. **(D)** A schematic of spindle microtubules visualized by the P*atb2*-GFP-atb2 strain. See **Section 3.6.3**. for details.

in nutritionally poor conditions: sexual conjugation, meiosis, and sporulation.

3. It is preferable to use an inverted microscope for live observation. Alternatively, the agar pad method may work with an upright microscope: for example, pour a drop of melted agar (YE5S, EMM, etc.) on a slide glass and immediately put a coverslip on it to flatten the agar. Remove the coverslip after the agar is solidified, mount cultured cells, and cover with a new coverslip prior to observation.

Acknowledgments

We thank Rafael E. Carazo-Salas for technical advice on microscopy and many invaluable discussions. We also thank Akira Yamashita for methods on meiosis, Miguel Angel Garcia and Kazuhide Asakawa for microscopy, Hiromi Maekawa and Elmar Schiebel for the 3GFP plasmid, Roger Tsien for providing mRFP and mCherry, and Kayoko Tanaka for transferring the mCherry plasmid. We are grateful to members of the Yeast Group on the third floor of the Lincoln's Inn Fields Laboratories and to Masayuki Yamamoto for continuous support. M.S. was a recipient of JSPS postdoctoral fellowship for research abroad. The London Research Institute is supported by Cancer Research UK.

References

1. Moreno, S., Klar, A. and Nurse, P. (1991) Molecular genetic analysis of fission yeast *Schizosaccharomyces pombe*. *Methods Enzymol.*, **194**, 795–823.
2. Bähler, J., Wu, J.Q., Longtine, M.S., Shah, N.G., McKenzie, A., 3rd, Steever, A.B., Wach, A., Philippsen, P. and Pringle, J.R. (1998) Heterologous modules for efficient and versatile PCR-based gene targeting in *Schizosaccharomyces pombe*. *Yeast*, **14**, 943–951.
3. Hentges, P., Van Driessche, B., Tafforeau, L., Vandenhaute, J. and Carr, A.M. (2005) Three novel antibiotic marker cassettes for gene disruption and marker switching in *Schizosaccharomyces pombe*. *Yeast*, **22**, 1013–1019.
4. Sato, M., Dhut, S. and Toda, T. (2005) New drug-resistant cassettes for gene disruption and epitope tagging in *Schizosaccharomyces pombe*. *Yeast*, **22**, 583–591.
5. Janke, C., Magiera, M.M., Rathfelder, N., Taxis, C., Reber, S., Maekawa, H., Moreno-Borchart, A., Doenges, G., Schwob, E., Schiebel, E. and Knop, M. (2004) A versatile toolbox for PCR-based tagging of yeast genes: new fluorescent proteins, more markers and promoter substitution cassettes. *Yeast*, **21**, 947–962.
6. Usui, T., Maekawa, H., Pereira, G. and Schiebel, E. (2003) The XMAP215 homologue Stu2 at yeast spindle pole bodies regulates microtubule dynamics and anchorage. *EMBO J.*, **22**, 4779–4793.
7. Tanaka, K., Mukae, N., Dewar, H., van Breugel, M., James, E.K., Prescott, A.R., Antony, C. and Tanaka, T.U. (2005) Molecular mechanisms of kinetochore capture by spindle microtubules. *Nature*, **434**, 987–994.
8. Sato, M. and Toda, T. (2007) Alp7/TACC is a crucial target in Ran-GTPase-dependent spindle formation in fission yeast. *Nature*, **447**, 334–337.
9. Grallert, A., Krapp, A., Bagley, S., Simanis, V. and Hagan, I.M. (2004) Recruitment of

NIMA kinase shows that maturation of the *S. pombe* spindle-pole body occurs over consecutive cell cycles and reveals a role for NIMA in modulating SIN activity. *Genes Dev.*, **18**, 1007–1021.

10. Radcliffe, P., Hirata, D., Childs, D., Vardy, L. and Toda, T. (1998) Identification of novel temperature-sensitive lethal alleles in essential b-tubulin and nonessential a2-tubulin genes as fission yeast polarity mutants. *Mol. Biol. Cell*, **9**, 1757–1771.
11. Maundrell, K. (1990) *nmt1* of fission yeast. A highly transcribed gene completely repressed by thiamine. *J. Biol. Chem.*, **265**, 10857–10864.
12. Garcia, M.A., Vardy, L., Koonrugsa, N. and Toda, T. (2001) Fission yeast ch-TOG/XMAP215 homologue Alp14 connects mitotic spindles with the kinetochore and is a component of the Mad2-dependent spindle checkpoint. *EMBO J.*, **20**, 3389–3401.
13. Bridge, A.J., Morphew, M., Bartlett, R. and Hagan, I.M. (1998) The fission yeast SPB component Cut12 links bipolar spindle formation to mitotic control. *Genes Dev.*, **12**, 927–942.
14. Kilmartin, J.V. (2003) Sfi1p has conserved centrin-binding sites and an essential function in budding yeast spindle pole body duplication. *J. Cell Biol.*, **162**, 1211–1221.
15. West, R.R., Vaisberg, E.V., Ding, R., Nurse, P. and McIntosh, J.R. (1998) *cut11⁺*: A gene required for cell cycle-dependent spindle pole body anchoring in the nuclear envelope and bipolar spindle formation in *Schizosaccharomyces pombe*. *Mol. Biol. Cell*, **9**, 2839–2855.
16. Saitoh, S., Takahashi, K. and Yanagida, M. (1997) Mis6, a fission yeast inner centromere protein, acts during G1/S and forms specialized chromatin required for equal segregation. *Cell*, **90**, 131–143.
17. Yamamoto, M., Imai, Y. and Watanabe, Y. (1997) Mating and sporulation in *Schizosaccharomyces pombe*. *The Molecular and Cellular Biology of the yeast Saccharomyces, Cold Spring Harbor Laboratory Press, Cold Spring Harbor, NY*, 1037–1106.
18. Yamamoto, A., West, R.R., McIntosh, J.R. and Hiraoka, Y. (1999) A cytoplasmic dynein heavy chain is required for oscillatory nuclear movement of meiotic prophase and efficient meiotic recombination in fission yeast. *J. Cell Biol.*, **145**, 1233–1249.
19. Ding, D.Q., Chikashige, Y., Haraguchi, T. and Hiraoka, Y. (1998) Oscillatory nuclear movement in fission yeast meiotic prophase is driven by astral microtubules, as revealed by continuous observation of chromosomes and microtubules in living cells. *J Cell Sci*, **111**, 701–712.
20. Maundrell, K. (1993) Thiamine-repressible expression vectors pREP and pRIP for fission yeast. *Gene*, **123**, 127–130.
21. Tanaka, K. and Kanbe, T. (1986) Mitosis in the fission yeast *Schizosaccharomyces pombe* as revealed by freeze-substitution electron microscopy. *J. Cell Sci.*, **80**, 253–268.
22. Ding, R., McDonald, K.L. and McIntosh, J.R. (1993) Three-dimensional reconstruction and analysis of mitotic spindles from the yeast, *Schizosaccharomyces pombe*. *J. Cell Biol.*, **120**, 141–151.
23. Hagan, I.M. and Hyams, J.S. (1988) The use of cell division cycle mutants to investigate the control of microtubule distribution in the fission yeast *Schizosaccharomyces pombe*. *J. Cell Sci.*, **89**, 343–357.
24. Uzawa, S. and Yanagida, M. (1992) Visualization of centromeric and nucleolar DNA in fission yeast by fluorescence *in situ* hybridization. *J. Cell Sci.*, **101**, 267–275.
25. Mallavarapu, A., Sawin, K. and Mitchison, T. (1999) A switch in microtubule dynamics at the onset of anaphase B in the mitotic spindle of *Schizosaccharomyces pombe*. *Curr. Biol.*, **9**, 1423–1426.
26. Khodjakov, A., La Terra, S. and Chang, F. (2004) Laser microsurgery in fission yeast; role of the mitotic spindle midzone in anaphase B. *Curr. Biol.*, **14**, 1330–1340.
27. Drummond, D.R. and Cross, R.A. (2000) Dynamics of interphase microtubules in *Schizosaccharomyces pombe*. *Curr. Biol.*, **10**, 766–775.

Chapter 12

Analysing Kinetochore Function in Human Cells: Spindle Checkpoint and Chromosome Congression

Christiane Klebig, Alberto Toso, Satyarebala Borusu and Patrick Meraldi

Abstract

During cell division microtubules of the mitotic spindle segregate the duplicated chromosomes into the two daughter cells. Chromosome–microtubule attachment is mediated by kinetochores, multiprotein complexes assembled on specialized regions of the DNA. Kinetochores modulate microtubule dynamics to generate the forces necessary to power chromosome movement and regulate the spindle checkpoint. Errors in kinetochore function can cause aneuploidy, a hallmark of 80% of solid tumors in humans, suggesting a fundamental link to tumorigenesis. Human kinetochores are complex protein machines with over 100 different proteins. Here we present fixed- and live-cell-based assays used to functionally categorize kinetochore proteins with regard to spindle checkpoint activity and kinetochore–microtubule attachment.

Key words: Kinetochore, microtubules, nuclear envelope breakdown, spindle checkpoint, mitosis, chromosome congression.

1. Introduction

During mitosis a cell divides into two daughter cells, each of them inheriting the same set of chromosomes. To achieve chromosome segregation, chromosomes bind to microtubules emanating from the spindle poles via kinetochores, multiprotein complexes located on centromeric DNA. Kinetochores have three fundamental functions during mitosis: (1) They act as the microtubule attachment platform, (2) they generate the forces necessary for chromosome movement during chromosome segregation, and (3) they regulate the spindle checkpoint *(1–4)*.

Bipolar kinetochore–microtubule attachment is a fundamental step not only for correct chromosome alignment on the

Andrew D. McAinsh (ed.), *Mitosis: Methods and Protocols*, Methods in Molecular Biology *545*,
DOI 10.1007/978-1-60327-993-2_12, © Humana Press, a part of Springer Science+Business Media, LLC 2009

metaphase plate but also for chromosome segregation during anaphase. In the case of attachment defects (e.g., unattached chromosomes, syntelic attachment), the spindle checkpoint is engaged and arrests the cell cycle at the metaphase/anaphase transition. After bipolar attachment of the chromosomes and alignment on the metaphase plate, the checkpoint is silenced and anaphase proceeds *(4)*. The spindle checkpoint depends on a set of conserved proteins such as Mad1, Mad2, Bub1, Bub3, Mad3/BubR1, and Mps1. Several studies have shown that failure of the spindle checkpoint can lead to aneuploidy, as segregation errors result in daughter cells with additional or missing chromosomes *(5–10)*. As the majority of cancers show aneuploidy this suggests that spindle checkpoint failure could be an early event in carcinogenesis *(11–14)*.

Based on biochemical and functional studies during the last few years, human kinetochore proteins have been clustered into several subcomplexes representing different functions *(3, 4, 15)*. The method of choice to study the function of human kinetochores is RNA-interference-mediated protein depletion *(16)*. Although this method is very powerful, there is an important caveat: the phenotype resulted from RNAi treatment does not necessarily reveal the function of the specific protein, but shows to which degree whole kinetochore function is impaired during cell division.

Here we describe several assays to analyze RNAi phenotypes according to kinetochore–microtubule attachment and the efficiency of the spindle checkpoint.

2. Materials

2.1. Cell Culture

1. Cells: Any adherent human cell line is appropriate to analyze kinetochore function (*see* **Note 1**). For live cell imaging, cells have to stably express a fluorescent marker, for example, a histone or kinetochore protein, to render chromosome or kinetochore movements visible by fluorescence microscopy.
2. Growth medium: For normal cell culture use standard mammalian medium appropriate for the cells. For live cell imaging, the cells either have to be cultured in CO_2-independent medium, for example, Leibovitz's L-15 medium (GIBCO) containing 10% FCS, or the microscope has to be equipped with a CO_2 incubator.
3. For all immunofluorescent staining, sterile coverslips should be prepared in advance: Treat them with 1 N HCl for 30 min followed by a washing step with distilled H_2O and a further treatment for 30 min with 100% ethanol. Then perform dry autoclavation of the coverslips (*see* **Note 2**).

2.2. Stock Solutions and Special Buffers

1. Phosphate buffered saline (PBS) made up as 0.144 g/l KH_2PO_4, 9 mg/l NaCl, 0.795 g/l $Na_2HPO_4*7H_2O$ in distilled H_2O, pH adjusted to 7.0 and autoclaved.
2. RNAi (20nM): HPLC-purified (Qiagen, Invitrogen, or other typical supplier).
3. Separate solutions of: 10 mM taxol, 10 mM nocodazole, 5 mM MG132 and 10 mM monastrol (all available from Sigma-Aldrich), all in DMSO; store at –20°C.
4. PHEM: 60 mM PIPES, 25 mM HEPES, 10 mM EGTA, 4 mM $MgSO_4$, 0.5% Triton X-100, 4% formaldehyde, 10% FCS, bovine serum albumin or normal donkey serum.
5. PTEMF: 0.2% Triton X-100, 20 mM PIPES, 1 mM $MgCl_2$, 10 mM EGTA, 4% formaldehyde.
6. Methanol/acetone (50/50%).
7. Fluorescence-labeled antibodies and anti-bleaching mounting medium for immunofluorescent-staining: for example, Alexa Fluorescence-labeled antibodies (Invitrogen) and Vectashield mounting medium with DAPI (Vector Laboratories).
8. Transfection reagent for RNAi treatment: for example, Oligofectamine or Lipofectamine RNAiMax (Invitrogen), OPTIMEM.
9. BSA-blocking solution: 3% BSA (bovine serum albumin), 0.02% sodium azide in 1× PBS); unlimited storage at 4°C.

2.3. Equipment

1. Microscope: Epi-fluorescence microscope equipped with a temperature-controlled incubator box as well as fast motorized excitation and emission filter wheels that allow for fast multicolor 3D live cell imaging (e.g., Olympus Deltavision (API) or alternatively a confocal microscope).

3. Methods

3.1. RNAi Treatment/Synchronizing Cells/Quantification of the Efficiency of RNA Depletion: Partial Versus Complete Depletion

The easiest way to analyze the function of a protein is to deplete the protein and investigate the effect in functional assays. RNAi treatment leads to depletion of the RNA level of the targeted gene, resulting in knockdown of protein expression *(16)*. Previous studies have shown that the depletion efficiency varies for different RNAs used against the same gene, thus potentially causing different phenotypes *(17)*. The efficiency of the RNA depletion can be quantified by performing Western blotting or immunofluorescence microscopy. Immunofluorescence is more quantitative than immunoblotting as there is a linear correlation between the output signal and the real protein level *(10)*. Immunofluorescent staining is based on the usage of antibodies against the protein of interest, a stable kinetochore reference, and DAPI-staining to label the DNA. For all experiments described below, a control sample treated with negative control RNAi without affecting any gene expression is processed in parallel.

Cell synchronization helps to study kinetochore protein RNA depletion: all cells in the population are treated with RNAi under the same conditions and the depletion occurs over a longer time (40 h) before the first mitosis starts. Due to this extended treatment a complete protein depletion can be reached. To obtain a synchronized cell population in mitosis we recommend performing a release from a double thymidine/aphidicoline block. Thymidine and aphidicoline inhibit DNA synthesis, causing a reversible G1/S or early S-phase arrest *(18, 19)*.

1. Inoculate HeLa cells or your cell line of interest in an appropriate dish containing sterile coverslips for immunofluorescent staining and standard growth medium (e.g., DMEM). Incubate at 37°C with 5% CO_2 overnight. Adjust the seeded cell number so that they reach 30% confluency the next day. Usually, 1×10^5–5×10^5 is the appropriate cell number for a 10 cm dish. When you do not synchronize the cells proceed directly with step 4 without adding thymidine and aphidicoline.
2. *For synchronization*: Add thymidine to a final concentration of 2 mM for 12 h.
3. Release the cells from thymidine by washing them twice with fresh medium. Incubate them for 5 h.
4. Perform your RNAi treatment according to manufacturer's protocol, for example:
   - Change the medium to MEM (*see* **Note 3**).
   - Tube A: Mix the appropriate volume of OPTIMEM and RNAi to a final concentration of 10–60 pmol.
   - Tube B: Mix the appropriate volume of OPTIMEM and transfection reagent.
   - Incubate each for 7–8 min at room temperature.
   - Mix tube A + B carefully and incubate again for 15–20 min at room temperature.
   - Apply the mixture to the cells and incubate them at 37°C for 48–72 h with a medium change after 24 h. Then, check the efficiency of the RNA depletion either by performing immunofluorescent staining or by harvesting the cells and performing Western blotting according to standard protocols. For immunofluorescent staining proceed with step 8.
5. *For synchronization*: 5 h after adding the RNAi to the cells, perform the second block by adding aphidicoline to a final concentration of 5 μg/ml for 24 h.

6. Release the cells from aphidicoline by washing them twice with fresh medium.
7. After further 14 h incubation, analyze the cells by Western blotting, immunofluorescent staining, or live cell imaging.
8. *For immunofluorescent staining*: Remove the medium and wash the cells gently with 1 × PBS to prevent losing mitotic cells.
9. Fix the cells with one of the following fixation solutions (*see* **Note 4**). These fixation procedures might give differences in the quality of your signal, depending on your antigen, and have to be evaluated for every protein analysed.
   a. Methanol-acetone (1:1): incubate for 3 min.
   b. 4% formaldehyde-solution (in PBS): incubate for 10 min, then wash the cells three times with 1 × PBS and permeabilize them by adding 0.5% Triton X-100 in PBS for another 10 min.
   c. PTEMF: incubate for 10–12 min.
   d. PHEM: Cells should be first rinsed with PHEM buffer (without Triton X-100 and formaldehyde). Then add PHEM with Triton X-100 and incubate for 5 min at 30°C. Aspirate the buffer, add PHEM with formaldehyde and incubate 20 min at 37°C. Aspirate the fixative buffer and permeabilize the cells with 0.05% Triton X-100 for 10 min.
   e. 100% Methanol: Use precooled (–20°C) methanol and incubate for 6 min at –20°C.
   f. Optional: Dipping coverslip into acetone for 0.5–1 min can improve extraction of epitopes.
10. Wash the samples three times with 1 × PBS.
11. Block the samples by adding 3% BSA-solution for 30 min.
12. Perform hybridization with the first specific antibodies against
    - Protein of interest
    - Kinetochore marker as a reference signal

    Dilute the antibody in blocking solution at the appropriate concentration. Be aware that both antibodies are synthesized in different animals. Incubate for 1 h.
13. Wash the samples three times with 1 × PBS.
14. Perform hybridization with second antibodies specific for the first ones. These antibodies must be labeled with different fluorescent markers. Dilute them also in blocking solution and incubate for 20 min in the dark.
15. Wash the samples three times with 1 × PBS.
16. Then, the coverslips are carefully inverted into a drop of mounting medium containing DAPI on a microscope slide. The excessive mounting medium should be removed by using filter paper and the coverslips fixed with nail polish.

Let the nail polish air-dry and store the samples at 4°C. The signal will be weakened in approximately four weeks.

17. The samples can be analyzed under a confocal microscope equipped with fluorescent filter wheels. Excitation at 490 nm induces the FITC fluorescence (green emission) for the protein of interest, excitation at 596 nm induces the Rhodamine fluorescence (red emission) for CREST, and excitation at 350 nm induces DAPI fluorescence (blue emission).
18. To quantify protein levels, determine signals with an appropriate software (e.g., softWoRx®, Applied Precision) using the formula:

$$ProteinLevel(\%) = \frac{\sum\limits_{1\to n} \frac{SRNAi(n)-b(n)}{rRNAi(n)-b(n)}}{\sum\limits_{1\to n} \frac{Sctrl(n)-b(n)}{rctrl(n)-b(n)}} \times 100$$

with s_{RNAi} (signal of protein of interest after RNAi treatment), b (background signal at the same wavelengths as for protein of interest and for reference protein, respectively), and r_{RNAi} (signal of reference protein, e.g., CREST, after RNAi treatment). The corresponding signals are also measured in the control RNAi treated samples (s_{ctrl} and r_{ctrl}). For each measurement, levels in at least five cells (50 kinetochores) should be determined to get statistically relevant results. An example of this analysis is shown in (**Fig. 12.1A, B**).

3.2. Spindle Checkpoint

Depletion of kinetochore proteins by RNAi treatment can lead either to activation or impairment of the mitotic spindle checkpoint. For example, knockdown of the microtubule motor CENP-E results in a mitotic arrest induced by an activated spindle checkpoint *(20)*. In contrast, depletion of checkpoint proteins such as Bub1 or Mad2 impairs the checkpoint *(10)*. Several methods exist to test whether the spindle checkpoint is impaired or activated by a particular RNAi depletion.

Fig. 12.1. (continued) Analysing kinetochore function in human cells: spindle checkpoint and chromosome congression. (**A**, **B**) Quantification of the efficiency of RNAi depletion. (**A**) Immunofluorescence images of prometaphase cells following RNAi treatment and stained with protein X and CREST antisera. Scale bar = 10 μm. (**B**) Quantification of specific protein levels. (**C**) Mitotic timing. Successive frames every 3 min from live-cell movies of H2B-GFP expressing HeLa cells. Nuclear envelope breakdown was set as T = 0 min. Full chromosome congression was determined at T = 30 min, anaphase onset at T = 45 min. Scale bar = 10 μm. (**D**) Cold stable assay. Cells treated with control RNAi or RNAi against protein X were subjected to cold treatment for 10 min before fixation and staining with anti-α-tubulin (*green*) and CREST antisera (*red*). Scale bar = 10 μm. (**E**) Chromosome congression. Chromosomes in metaphase cells were counted as unaligned if they were located outside the central 30% of the mitotic spindle (*dotted lines* in schematic) or if their kinetochores (KT) were aligned perpendicular to the spindle axis. Shown are representative immunofluorescence images of cells treated with RNAi and stained with DAPI (*blue*), α-tubulin (*green*) and kinetochore marker (*red*) antisera. Arrows indicate congression errors. Scale bar = 10 μm.

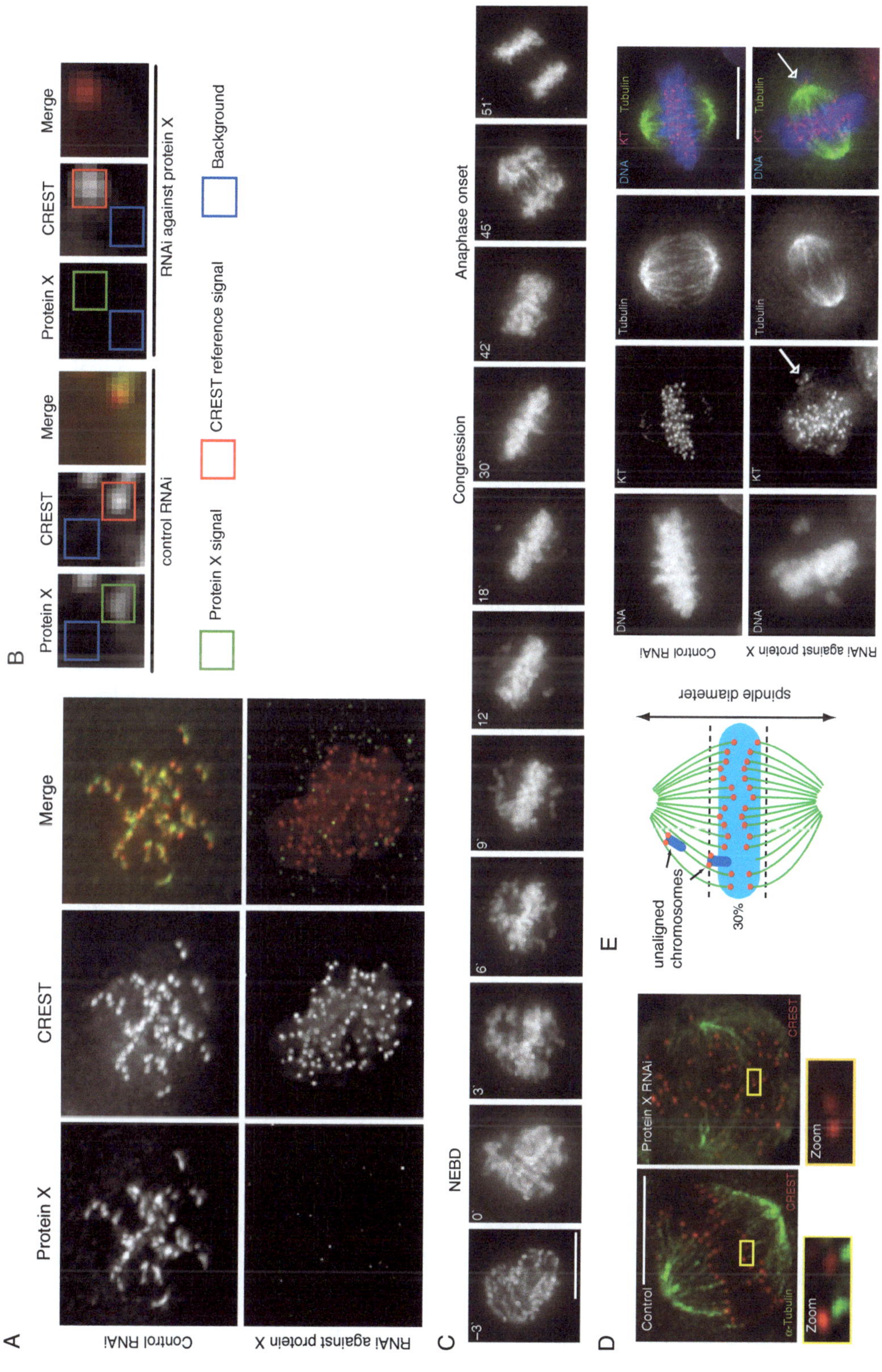

Fig. 12.1. (continued)

3.2.1. Analyzing Fixed Cells

3.2.1.1. Treatment with Nocodazole, Taxol, or Monastrol

One way to investigate the functionality of the spindle checkpoint is to treat cells with drugs that activate the checkpoint. Here we present three commonly used drugs.

(a) *Nocodazole*: The addition of the microtubule depolymerizing agent nocodazole leads to detachment of all chromosomes *(21)*. In cells with a functioning checkpoint this results in activation of the spindle checkpoint causing a mitotic arrest.

(b) *Taxol*: Paclitaxel (taxol) is a mitotic inhibitor used in cancer chemotherapy. The reagent interferes with normal microtubule growth during cell division. Taxol binds to the N-terminal region of β-tubulin and promotes the formation of highly stable microtubules that resist depolymerization. This interference of microtubule dynamics activates the spindle checkpoint *(22)*.

(c) *Monastrol*: Monastrol is a cell-permeable small molecule inhibitor of Eg5, a mitotic kinesin. It leads to monoastral or monopolar spindles that cause a spindle checkpoint dependent mitotic arrest *(23)*.

It is important always to use a negative and positive control sample in parallel with your experimental sample. Transfection with a control RNAi, such as Lamin A or scrambled oligos, that does not influence the spindle checkpoint is a common negative control. Depletion of the spindle checkpoint protein Mad2 represents a positive control as it abolishes the checkpoint control.

1. Inoculate HeLa cells or your cell line of interest in an appropriate dish containing standard growth medium. Incubate at 37°C with 5% CO_2 overnight. Adjust the seeded cell number so that they reach 20–30% confluency the next day. Perform RNAi treatment.
   a. *For nocodazole treatment*: Add nocodazole to a final concentration of 1 nM. Incubate at 37°C with 5% CO_2 for 16 h.
   b. *For taxol treatment*: Add taxol to a final concentration of 1 μM and incubate the cells for 16 h.
   c. *For monastrol treatment*: Add monastrol to a final concentration of 100 μM and incubate for 16 h.

For all three drugs, prepare a control dish under the same conditions but without adding the drug.

3.2.1.2. Analyzing the Spindle Checkpoint After Drug Treatment

(a) *Rounding-up:* Cells with a functioning spindle checkpoint will arrest in mitosis after drug treatment and show a rounded-up shape in comparison to nonmitotic cells that stay flattened. Cells without a functioning checkpoint do not react to the drug treatment and flatten down again after exiting mitosis. By using a phase-contrast microscope the number of rounded-up cells and the total number of cells are counted. The percentage of rounded-up cells in comparison to the whole cell number represents the mitotic index. Untreated HeLa cells usually have a mitotic index of 10% whereas the mitotic index increases to 40–60% after activation of the spindle checkpoint. In cells without a functioning spindle checkpoint, the mitotic index will be 10–20%. The control dish represents the usual number of cells being in mitosis.

(b) *Phospho-Histone H3*: Instead of counting the number of rounded-up cells, the cells can be immuno-stained with anti-phospho-histone H3 antibody as a mitotic marker, followed by the detection of the percentage of M-phase cells by FACSCALIBUR flow cytometric sorting for ρ-H3 *(24)*.

1. Perform cell culturing, RNAi treatment, and drug treatment as described above.
2. Wash the cells with 1 × PBS.
3. Trypsinize the cells.
4. Transfer the cells into a 15 ml Falcon tube and spin them down (2000 rpm, 5 min).
5. Fix the cells by adding 100 μl methanol (ice-cold); mix gently; place at –20°C for 10 min.
6. Centrifuge, and wash twice with ice-cold 1% BSA (in PBS).
7. Centrifuge, discard the supernatant, and resuspend the cells in the remaining volume.
8. First hybridization: Add the first antibody (anti-phospho-histone H3) at the appropriate concentration in ice-cold 3% BSA (in PBS). Incubate for 1 h in the dark.
9. Wash the cells three times with ice-cold PBS. Discard the supernatant and resuspend them in the remaining volume.
10. Second hybridization: Add the fluorochrome-labeled secondary antibody in 3% BSA (in PBS) at the optimal dilution. Incubate for 30 min in the dark.
11. Wash the cells three times in PBS.
12. Discard the supernatant and resuspend the cells in ice-cold 3% BSA, 1% sodium azide (in PBS).
13. Store the cell suspension immediately at 4°C in the dark.
14. Analysis: For best results, analyze the cells on the flow cytometer as soon as possible.

3.2.1.3. Quantification of the Levels of Spindle Checkpoint Proteins Recruited to the Kinetochores

Another criterion for an impaired or weakened spindle checkpoint is the failure to load checkpoint proteins at the kinetochores. In wild-type cells checkpoint activation results in high levels of checkpoint proteins at kinetochores. After depletion of the protein of interest, cells are treated with nocodazole to activate the spindle checkpoint. In case the spindle checkpoint is not working, the cells are further treated with the proteasome inhibitor MG132, which will arrest cells at metaphase/anaphase transition *(25)*. Then, the protein levels of the checkpoint proteins Bub1, Bub3, BubR1, Mad1, and Mad2 can be quantified (*see* **Section 3.1**).

1. Inoculate HeLa cells or your cell line of interest in a 6-well plate containing sterile coverslips. Incubate at 37°C with 5% CO_2 overnight. Adjust the seeded cell number so that they reach 20–30% confluency the next day. Perform RNAi treatment.
2. Add MG132 to a final concentration of 1 μM and nocodazole to a final concentration of 1 nM and incubate for 1 h.
3. Perform immunofluorescent staining against the checkpoint proteins Bub1, BubR1, Bub3, Mad1, or Mad2, and a kinetochore marker as a reference.

3.2.2. Analyzing Living Cells

3.2.2.1. Live Cell Imaging

Live cell imaging represents an excellent tool to investigate cell division events over a defined period of time at single cell level. A heating chamber that ensures the maintenance of optimal growth temperature, and the usage of CO_2-independent medium or a CO_2-incubator allows cell culture under normal conditions. Moreover, reagents (such as nocodazole) can be added directly to the cells to investigate the immediate effect in living cells. Live cell imaging offers the opportunity to monitor mitotic timing, efficiency of the spindle checkpoint, segregation errors, and detecting apoptotic events.

1. Inoculate HeLa-H2B-mRFP cells or your cell line of interest in a LabTech II chamber with 1–2 ml DMEM. Be aware that your cell line must have a fluorescent marker for live cell imaging. Incubate at 37°C with 5% CO_2 overnight. Adjust the seeded cell number so that they reach 20–30% confluency the next day. Perform RNAi treatment.
2. Two hours before imaging, change the medium to CO_2-independent medium.
3. One hour before imaging, preheat the microscope to 37°C, so that the stage is already at the optimal temperature and the cells can go on dividing without getting a cold-shock.
4. Just before starting, fix the lid of the LabTech chamber with silicone grease.
5. Perform live cell imaging by setting the area points that have to be investigated, defining the objective (depends on cell number and resolution, e.g., for large cell number with low resolution use 20X long distance, for low cell

number with high resolution use 40–60X), exposure time (e.g., 100–200 ms), fluorescent filter (e.g., Cy3), time interval (e.g., every 3 min), and the duration of the movie by defining the timepoint amount that has to be done (e.g., a total of 160 timepoints every 3 min gives a movie duration of 8 h).

3.2.2.2. Ch-TOG1 Depletion

The major functions of ch-TOG1 (colonic-hepatic tumor-overexpressed gene) during mitosis are to focus microtubule minus ends at spindle poles, maintain centrosome integrity, and contribute to spindle bipolarity. It has been shown that cells lacking ch-TOG1 assembled multipolar spindles, and showed decreased microtubule length and density in the spindle. Furthermore, depletion of ch-TOG1 affected K-fiber dynamics, kinetochore–microtubule attachment, and moreover microtubule tension at the kinetochores *(26, 27)*. Because microtubule polymerization is severely inhibited in the absence of ch-TOG1 this leads to activation of the spindle checkpoint, whereas the treatment of co-RNAi against ch-TOG1 and Mad2 rescues the phenotype and overrides the spindle checkpoint. Here, we describe a functional test to investigate whether depletion of your protein of interest also overrides the ch-TOG1 dependent arrest indicating an impairment of the spindle checkpoint.

Depletion of ch-TOG1 has advantages compared to nocodazole treatment. The cells show a more natural state than those treated with nocodazole, which disrupts the whole mitotic spindle causing an activation of an unknown number of stress kinases, and potentially several side effects.

1. Inoculate HeLa-H2B-mRFP cells in a LabTech II chamber with 1–2 ml DMEM. Incubate at 37°C with 5% CO_2 overnight. Adjust the seeded cell number so that they reach 20–30% confluency the next day.
2. Perform co-RNAi treatment against ch-TOG and your protein of interest.
3. After 48 h start live cell imaging (*see* **Section 3.2.2.1**.).
4. Analyze the movies by detecting whether the spindle checkpoint is activated initiating a mitotic delay or whether the cells go through mitosis.

3.2.2.3. Mitotic Timing

Mitotic timing is defined by the time that a cell needs to go through mitosis. In normal living cells this takes approximately 1 h. Depending on the experiment, not only the time of the whole mitosis can be measured but also the time from nuclear envelope breakdown until full chromosome congression or anaphase onset (**Fig. 12.1C**). This assay allows us to determine the state of the spindle checkpoint: An accelerated or normal timing in the presence of a high rate of segregation errors, such as lagging chromosomes or chromosomal bridges, denotes a weakened or impaired spindle checkpoint. In contrast, a delay of anaphase onset in the presence of unaligned chromosomes indicates an active spindle checkpoint. To make sure that this delay is caused by the spindle checkpoint one can co-deplete Mad2. An accelerated timing after co-depletion of Mad2 will indicate a spindle checkpoint dependency.

1. Perform live cell imaging as described above.
2. Define the nuclear envelope breakdown as time point = 0.
3. Determine the time of anaphase onset.
4. Analyze at least 100 cells to get statistically relevant results.

3.3. Kinetochore–Microtubule Attachment Defects

RNAi treatment against kinetochore proteins can lead to kinetochore–microtubule attachment defects that result in congression errors. Congression errors represent chromosomes that are not correctly aligned at the metaphase plate. Chromosome misalignment can arise from several mechanistically distinct defects in chromosome–microtubule binding, including a complete absence of attachment, monopolar attachment (binding of only one kinetochore to microtubules), syntelic attachment (both sister kinetochores are bound to microtubules emanating from the same pole), or merotelic attachment (one kinetochore shows microtubule binding from both poles) *(17, 28–30)*. No direct biochemical assays exist for these defects, but they can be at least partly distinguished by morphology and interkinetochore distances.

3.3.1. Analyzing Fixed Cells

3.3.1.1. Cold Stable Assay

When cells are cooled down to 6–8°C for several minutes the free, polar, and astral spindle microtubules disassemble, whereas the microtubules bound to kinetochores cluster together and persist as bundles *(31)*. Thus, this assay allows us to detect whether kinetochores are still capable of binding and mobilizing microtubules. An example is shown in **Fig. 12.1D**. This assay might also allow us to detect an enrichment of attachment defects, such as syntelic or merotelic attachments.

1. Inoculate HeLa cells or your cell line of interest in a 6-well plate containing sterile coverslips. Incubate at 37°C with 5% CO_2 overnight. Adjust the seeded cell number so that they reach 20–30% confluency the next day. Perform RNAi treatment.
2. Transfer the plate onto a 4°C surface (ice) for 10 min.
3. Perform fixation with precooled fixation solution at 4°C for 10 min.
4. Proceed with immunofluorescent staining (*see* **Section 3.1**). Use primary antibodies against α-tubulin to mark the microtubules and CREST as a kinetochore marker.

3.3.1.2. Detecting Congression Errors in "Metaphase" Cells

The rate of congression errors can be detected by counting the chromosomes that localize outside a defined 30% area around the metaphase plate. To overcome the problem that metaphase cells with unaligned chromosomes resemble those in prometaphase state, treatment with MG132 is recommended. This treatment for a duration of 1 h results in an accumulation of metaphase cells independent of differences in mitotic timing and checkpoint efficiency. In wild-type cells usually 10–15% of the cells show congression errors, whereas depletion of the key regulator of kinetochore–microtubule attachment, Bub1, results in 50–60% of cells with congression errors *(10)*. An example for congression errors is shown in **Fig. 12.1E**.

1. Inoculate HeLa cells in a 6-well plate containing sterile coverslips. Incubate at 37°C with 5% CO_2 overnight. Adjust the seeded cell number so that they reach 20–30% confluency the next day. Perform RNAi treatment.
2. Add MG132 to a final concentration of 1 μM and incubate for 1 h at 37°C.
3. Perform immunofluorescent staining (*see* **Section 3.1.**).
4. The first specific antibodies should be against a kinetochore marker (e.g., CREST) and α-tubulin to visualize the kinetochores and the microtubules.
5. Detect cells with metaphase plates by immunofluorescence and count the numbers of misaligned chromosomes (congression defects).

3.3.1.3. Interkinetochore Distances

The detection of interkinetochore distances presents a readout of the net force exerted on chromatid pairs during mitosis. Usually, interkinetochore distances average 2.0 ± 0.3 μm in control cells with bioriented chromatids and 0.6 ± 0.1 μm in nocodazole-treated cells with detached chromatid pairs *(17)*. Reduced interkinetochore distances reflect force generation defects that can also lead to congression errors.

1. Inoculate HeLa cells or your cell line of interest in a 6-well plate containing sterile coverslips. Incubate at 37°C with 5% CO_2 overnight. Adjust the seeded cell number so that they reach 20–30% confluency the next day. Perform RNAi treatment.
2. Add MG132 to a final concentration of 1 μM (for the control sample, also add nocodazole to a final concentration of 1 nM to get unattached chromatids) and incubate for 1 h at 37°C.
3. Perform immunofluorescent staining (*see* **Section 3.1.**).
4. The first specific antibodies should be against an outer kinetochore marker (e.g., Cenp-E) and α-tubulin to visualize the kinetochores and the microtubules.
5. Analyze the interkinetochore distances using softWoRx.

3.3.2. Analyzing Living Cells

3.3.2.1. Detection of Congression Efficiency

Investigating congression efficiency in living cells gives us the opportunity to monitor the amount of uncongressed chromosomes over a defined period of time. Furthermore, not only the dynamics of the unaligned chromosomes can be analyzed, but also their kinetics and long-term fate.

1. Perform live cell imaging as described above.
2. Define the nuclear envelope breakdown as time point = 0.
3. Determine the timepoint when all chromosomes are correctly aligned at the metaphase plate.
4. Analyze at least 100 cells to get statistically relevant results

4. Notes

1. Cell lines: The best studied cell line is the HeLa cell line due to well-documented and successful RNAi treatment during the last several years *(10, 17)*. Moreover, these cells do not move too much during live cell imaging. However, HeLa is a cancer cell line and has several disadvantages because it has an unstable karyotype and is therefore not suited to the investigation of failure of kinetochore function with regard to development of a malignant phenotype.
2. Preparing sterile coverslips: The procedure mentioned here helps to create a surface that ensures better binding of the cells.
3. RNAi treatment (*see* **Section 3.1.**): A medium shift from DMEM to MEM during RNAi treatment has been shown in our hands to result in higher efficiency of RNA depletion in HeLa cells.
4. Immunofluorescent staining (*see* **Section 3.1.**): All fixation solutions and antibody dilutions should be prepared freshly on the day of usage. Be aware that the fluorescence-dye-labeled secondary antibody is light-sensitive and should be stored in the dark.

Acknowledgments

We thank the Light Microscopy Center of ETH Zurich for technical support. We are grateful to members of the Meraldi group for helpful discussions. Work was supported by the Deutsche

Forschungsgemeinschaft (CK), Swiss National Science Foundation grant 3100A0-107912/1 (PM, AT, SB), and the Swiss National Science Foundation Förderungsprofessur (PM).

References

1. Cleveland, D. W., Mao, Y., and Sullivan, K. F. (2003) Centromeres and kinetochores: from epigenetics to mitotic checkpoint signaling. *Cell* **112**, 407–21.
2. Chan, G. K., Liu, S. T., and Yen, T. J. (2005) Kinetochore structure and function. *Trends Cell. Biol.* **15**, 589–98.
3. Westermann, S., Drubin, D. G., and Barnes, G. (2007) Structures and functions of yeast kinetochore complexes. *Annu. Rev. Biochem.* **76**, 563–91.
4. Musacchio, A., and Salmon, E. D. (2007) The spindle-assembly checkpoint in space and time. *Nat. Rev. Mol. Cell Biol.* **8**, 379–93.
5. Michel, L. S., Liberal, V., Chatterjee, A., Kirchwegger, R., Pasche, B., Gerald, W., Dobles, M., Sorger, P. K., Murty, V. V., and Benezra, R. (2001) MAD2 haplo-insufficiency causes premature anaphase and chromosome instability in mammalian cells. *Nature* **409**, 355–9.
6. Hernando, E., Nahle, Z., Juan, G., Diaz-Rodriguez, E., Alaminos, M., Hermann, M., Michel, L., Mittal, V., Gerald, W., Benezra, R., Lowe, S. W., and Cordon-Cardo, C. (2004) Rb inactivation promotes genomic instability by uncoupling cell cycle progression from mitotic control. *Nature* **430**: 797–802.
7. Warren, C. D., Brady, D. M., Johnston, R. C., Hanna, J. S., Hardwick, K. G., and Spencer, F.A. (2002) Distinct chromosome segregation roles for spindle checkpoint proteins. *Mol. Biol. Cell* **13,** 3029–41.
8. Draviam, V. M., Xie, S., and Sorger, P. K. (2004) Chromosome segregation and genomic stability. *Curr. Opin. Genet. Dev.* **14**, 120–5.
9. Dai, W., Wang, Q., Liu, T., Swamy, M., Fang, Y., Xie, S., Mahmood, R., Yang, Y. M., Xu, M., and Rao, C. V. (2004) Slippage of mitotis arrest and enhanced tumor development in mice with BubR1 haploinsuffiency. *Cancer Res.* **64**: 440–5.
10. Meraldi, P., and Sorger, P. K. (2005) A dual role of Bub1 in the spindle checkpoint and chromosome congression. *EMBO J.* **24**, 1621–33.
11. Lengauer, C., Kinzler, K. W., and Vogelstein, B. (1997) Genetic instability in colorectal cancers. *Nature* **386**, 623–7.
12. Masuda, A., and Takahashi, T. (2002) Chromosome instability in human lung cancers: possible underlying mechanisms and potential consequences in the pathogenesis. *Oncogene* **21**, 6884–97.
13. Lengauer, C., and Wang, Z. (2004) From spindle chechpoint to cancer. *Nature Genet.* **36**, 1144–5.
14. Yuen, K. W., Montpetit, B., and Hieter, P. (2005) The kinetochore and cancer: what's the connection? *Curr. Opin. Cell Biol.* **17**, 576–82.
15. Meraldi, P., McAinsh, A. D., Rheinbay, E., and Sorger P. K. (2006) Phylogenetic and structural analysis of centromeric DNA and kinetochore proteins. *Genome Biol.* 7, R23.
16. Elbashir, S. M., Harborth, J., Lendeckel, W., Yalcin, A., Weber, K., and Tuschl, T. (2001) Duplexes of 21-nucleotide RNAs mediate RNA interference in cultured mammalian cells. *Nature* **411**: 494–8.
17. McAinsh, A. D., Meraldi, P., Draviam, V. M., Toso, A., Sorger, P.K. (2006) The human kinetochore proteins Nnf1R and Mcm21R are required for accurate chromosome segregation. *EMBO J.* **25**, 4033–49.
18. Harper, J. V. (2005) Synchronization of cell populations in G1/S and G2/M phases of the cell cycle. *Methods Mol. Biol.* **296**, 157–66.
19. Uzbekov, R., Chartrain, I., Philippe, M., and Arlot-Bonnemains, Y. (1998) Cell cycle analysis and synchronization of the Xenopus cell line XL2. *Exp. Cell. Res.* **242**, 60–8.
20. Tanudji, M., Shoemaker, J., L'Italien, L., Russell, L., Chin, G., and Schebye, X. M. (2004) Gene silencing of CENP-E by small interfering RNA in HeLa cells leads to mis-segregation of chromosomes after a mitotic delay. *Mol. Biol. Cell.* **15**, 3771–81.
21. De Brabander, M. J., Van de Veire, R. M. L., Aerts, F. E. M., Borgers, M., and Janssen, P. A. J. (1976) The effects of methyl [5 -(2-Thienylcarbonyl)-1*H*-benzimindazol-2-yl]carbamate, (R 17934; NSC 238159), a new synthetic antitumoral drug interfering with microtubules, on mammalian cells cultures *in vitro*. *Cancer Res.* **36**: 905–916.
22. De Brabander, M., Geuens, G., Nuydens, R., Willebrords, R., and De Mey, J. (1981) Taxol induces the assembly of free microtubules in

living cells and blocks the organizing capacity of the centrosomes and kinetochores. *Proc. Natl. Acad. Sci. USA* **78**: 5608–12.
23. Mayer, T. U., Kapoor, T. M., Haggarty, S. J., King, R. W., Schreiber, S. L., and Mitchison, T. J. (1999) Small molecule inhibitor of mitotic spindle bipolarity identified in a phenotype-based screen. *Science* **286**, 971–4.
24. Tapia, C., Kutzner, H., Mentzel, T., Savic, S., Baumhoer, D., and Glatz, K. (2006) Two mitosis-specific antibodies, MPM-2 and phospho-histone H3 (Ser28), allow rapid and precise determination of mitotic activity. *Am. J. Surg. Pathol.* **30**, 83–9.
25. Rock, K. L., Gramm, C., Rothstein, L., Clark, K., Stein, R., Dick, L., Hwang, D., and Goldberg, A. L. (1994) Inhibitors of the proteasome block the degradation of most proteins and the generation of peptides presented on MHC class I molecules. *Cell* **78**: 761–71.
26. Gergely, F., Draviam, V. M., and Raff, J. W. (2003) The ch-TOG/XMAP215 protein is essential for spindle pole organization in human somatic cells. *Genes Dev.* **17**, 336–41.
27. Cassimeris, L., and Morabito, J. (2004) TOGp, the human homolog of XMAP215/Dis1, is required for centrosome integrity, spindle pole organization, and bipolar spindle assembly. *Mol. Biol. Cell.* **15**, 1580–90.
28. DeLuca, J. G., Moree, B., Hickey, J. M., Kilmartin, J. V., and Salmon, E.D. (2002) hNuf2 inhibition blocks stable kinetochore-microtubule attachment and induces mitotic cell death in HeLa cells. *J. Cell. Biol.* **159**, 549–55.
29. Hauf, S., Cole, R. W., LaTerra, S., Zimmer, C., Schnapp, G., Walter, R., Heckel, A., van Meel, J., Rieder, C. L., and Peters, J. M. (2003) The small molecule Hesperadin reveals a role for Aurora B in correcting kinetochore-microtubule attachment and in maintaining the spindle assembly checkpoint. *J. Cell. Biol.* **161**, 281–94.
30. Salmon, E. D., Cimini, D., Cameron, L. A., and DeLuca, J. G. (2005) Merotelic kinetochores in mammalian tissue cells. *Philos. Trans. R. Soc. Lond. B. Biol. Sci.* **360**: 553–68.
31. Salmon, E. D., and Begg, D. A. (1980) Functional implications of cold-stable microtubules in kinetochore fibers of insect spermatocytes during anaphase. *J Cell Biol.* **85**, 853–65.

Chapter 13

Probing Kinetochore Structure and Function Using *Xenopus laevis* Frog Egg Extracts

Michael J. Emanuele and P. Todd Stukenberg

Abstract

Kinetochores are multiprotein machines that initiate mitotic checkpoint signaling and control chromosome movement through interactions with microtubules. Our lab has utilized *Xenopus laevis* frog egg extracts to investigate the requirements for kinetochore assembly and disassembly in vertebrates. Egg extracts support the assembly of functional kinetochores that are capable of binding microtubules, aligning and segregating chromosomes, and sending spindle checkpoint signals. This is the only in vitro system that assembles functional kinetochores, making it particularly well suited for these types of studies. Probing kinetochore assembly using the biochemically tractable egg extract system has elucidated the intricate assembly requirements for numerous vertebrate kinetochore proteins. The following techniques have been used to characterize kinetochore assembly requirements. In addition, we describe assays that we utilized to identify factors that promote maintenance of preassembled kinetochores and those that induce kinetochore disassembly.

Key words: Kinetochore, mitosis, aurora B, protein phosphatase 1, *Xenopus* extracts: Ndc80.

1. Introduction

The addition of sperm nuclei to Cytostatic Factor (CSF) arrested frog egg extracts leads to the rapid construction of kinetochores onto the centromeres of unreplicated chromosomes. This can be visualized using standard immunofluorescent techniques, with kinetochores appearing as 18 distinct dots on each chromatin mass. Using polyclonal antibodies raised against a large panel of *Xenopus* kinetochore proteins we have dissected outer kinetochore assembly requirements using frog egg extracts. Antibodies are essential to deplete specific proteins from extracts and are

Andrew D. McAinsh (ed.), *Mitosis: Methods and Protocols*, Methods in Molecular Biology *545*,
DOI 10.1007/978-1-60327-993-2_13, © Humana Press, a part of Springer Science+Business Media, LLC 2009

also used to immunolocalize proteins to centromeres on mitotic nuclei. Thus, generating high-quality antibodies to various kinetochore proteins has been critical to these studies. In addition, the ability to examine the localization of a large panel of kinetochore proteins from a single assembly reaction has provided a semi-high-throughput method for dissecting the kinetochores' intricate assembly map.

Our lab has concentrated on assembly and function of the outer kinetochore. It is clear that the entire outer kinetochore rapidly assembles in these extracts (<12 min). It is likely that the inner kinetochore is also assembled under these conditions. The starting material for these reactions is demembranated sperm. Protamines and histones assemble onto the sperm DNA within 2–3 min of its introduction into the egg extract. The centromeres on sperm nuclei act as a template for kinetochore assembly. We have observed the assembly of inner centromere components and an increase in CENP-A earlier than outer kinetochore components in the periods after histone assembly. These data suggest that the *Xenopus* system may be a very useful system for studying the structure and function of the inner kinetochore and deposition of CENP-A nucleosomes.

In this chapter, we describe the generation of high-quality, polyclonal antibodies to kinetochore proteins. Because most kinetochore proteins are insoluble when expressed recombinantly, this has been a great challenge. We have devised reproducible methods to generate high quality antibodies and antigen affinity columns under denaturing conditions, which has made antibody production routine. We next describe our standard kinetochore assembly reaction, as we perform it in frog egg extracts. We have improved upon conventional techniques used for preparing mitotic nuclei from extracts for immunofluorescence. These improvements allow us to prepare dozens of coverslips for immunofluorescence, from a single assembly reaction in a quick, easy, inexpensive, and reproducible manner. We present methods used to probe maintenance requirements of preassembled kinetochores. Finally, we present methods used to examine kinetochore disassembly, as it occurs after inhibition of kinetochore maintenance factors, at exit from M-phase, and on isolated nuclei. The production and special handling techniques associated with generating CSF extracts has been described in detail in several excellent reviews and methods chapters, and is therefore not discussed herein *(1–3)*.

2. Materials

All chemicals used for making buffers were purchased from Sigma-Aldrich, unless otherwise stated.

2.1. Antibody Production

1. 6 × His protein purification, lysis buffer: 20 mM Tris, 500 mM NaCl, 5 mM imidizole, pH 7.9 (prepared as a 8 × stock, stored at room temperature)
2. 6 × His protein purification, wash buffer: 20 mM Tris, 500 mM NaCl, 30 mM imidizole, pH 7.9 (prepared as a 8 × stock, stored at room temperature)
3. 6 × His protein purification, elution buffer: 20 mM Tris, 200 mM NaCl, 300 mM imidizole, pH 7.9 (prepared as a 4 × stock, stored at room temperature)
4. Isopropyl β-D-1-thiogalactopyranoside (IPTG): prepared in ddH2O as a 1 M stock, and stored at –20°C in 1 ml aliquots
5. Ni-NTA Agarose (Qiagen): stored at 4°C
6. Econopac columns (Bio-Rad)
7. Dialysis tubing 12,000–14,000 MWCO (Spectra/Por): stored at 4°C
8. Cyanogen Bromide Activated Sepharose 4B (GE Healthcare): stored at 4°C
9. Guanidine hydrochloride
10. Bradford reagent: Stored at 4°C

2.2. Probing Kinetochore Assembly

1. *Xenopus laevis* cytostatic factor arrested frog egg extracts, made from the laid eggs of female frogs, are prepared as described in *(1–3)*.
2. Sperm nuclei isolated from male *Xenopus laevis* frogs, prepared as described in *(1)*.
3. CSF-XB: 10 mM HEPES, 100 mM KCl, 2 mM MgCl2, 0.1 mM $CaCl_2$, 50 mM sucrose, 5 mM EGTA, pH 7.8.
4. Nocodazole: stock prepared in DMSO at 5 mg/ml, stored at –20°C in 5–10 μl aliquots.
5. BRB80: 80 mM PIPES, 1 mM $MgCl_2$, 1 mM EGTA, pH 6.8 (prepared as a 5 × solution and stored at 4°C).
6. 80% Glycerol: diluted from 100% glycerol (glycerin) stock. Stored at room temperature protected from light.
7. 20% Triton X-100: Diluted from Triton-X 100, stored at room temperature, protected from light.
8. 37% Formaldehyde (by weight; Fisher Scientific).
9. Extract fixative: BRB80, 30% glycerol, 0.1% Triton X-100, 3.7% formaldehyde.
10. Extract glycerol cushion: BRB80, 40% glycerol.
11. Calcium cycling solution: 10 mM HEPES, 100 mM KCl, 1 mM $MgCl_2$, 5 mM $CaCl_2$, 50 mM Sucrose, pH 7.8.
12. Poly-l-lysine, 0.1% w/v in H_2O.
13. PBSt: 10 mM phosphate buffer, 137 mM NaCl, 0.1% Tween-20, pH 7.5.
14. TBSt: 10 mM Tris-HCl, 137 mM NaCl, 0.1% Tween-20, pH 7.5.
15. Bovine serum albumin, Fraction V (lyophilized).

16. Affi-Prep Protein-A Support (Bio-Rad).
17. 12-well tissue culture dishes (COSTAR).
18. Antibodies, used for immunodepletion or immunofluorescent analysis of kinetochore assembly.
19. No. 1 glass coverslips, 18 mm, precleaned (Fisher Scientific).

2.3. Effect of Protein Phosphatase 1 on Kinetochore Disassembly

1. Okadaic acid (Alexis Biochemicals).
2. Purified protein phosphatase 1 enzyme, purified as described in *(7)*.
3. Phosphatase buffer: 50 mM MOPS–NaOH, 0.1 mM EGTA 1 mg/ml BSA, 1 mM dithiothreitol (DTT), 20 mM 2-mercaptoetanol, 0.02% Brij-35, pH 7.0 (DTT and 2-mercaptoethanol are added immediately prior to use).

3. Methods

3.1. Antibody Production

1. *Xenopus* kinetochore proteins are identified by searching Expressed Sequence Tag (EST) databases using BLAST analysis *(4)*. The Gene index databases (http://compbio.dfci.harvard.edu/tgi/) of Xenopus EST tags are much more complete than the nonredundant (NR) databases at PubMed (http://www.ncbi.nlm.nih.gov/).
2. Genes corresponding to proteins of interest are commonly cloned into a pET-28 vector (Novagen) using standard PCR and molecular biology techniques. PCR amplification is preformed from either Stage 11.5–14 *Xenopus* cDNA libraries or directly from ESTs. When possible, cloning is done placing an NdeI site upstream of the gene coding region and an NotI site downstream of the stop codon (*see* **Note 1**).
3. Protein expression in BL21 (pLysS) bacteria is induced at 20°C by the addition of between 0.1 mM to 1 mM isopropyl β-D-1-thiogalactopyranoside (IPTG).
4. Bacterial pellets are frozen in liquid nitrogen, and stored at –80°C overnight. These pellets are stable at –80°C for several weeks.
5. Pellets are thawed and resuspended into ~30 ml of 6 × His protein purification lysis buffer using a dounce homogenizer and a loose-fitting pestle.
6. The bacterial suspension is lysed 2–3 times on a French Press and 1000–1500 psi.
7. The lysed solution is centrifuged at 17,500 rpm and 4°C in a Sorvall SS-34 rotor for 60–90 min.

8. The soluble material (supernatant) is transferred to a 50 ml conical tube and stored on ice.
9. The insoluble pellet is resuspended in lysis buffer containing 6 M guanidine hydrochloride at room temperature. The insoluble material is clarified by centrifugation at 17,500 rpm and 23°C in a Sorvall SS-34 rotor for 60–90 min
10. Ni-NTA agarose (1–2 ml of beads per binding reaction) is washed into the appropriate lysis buffer (insoluble containing 6 M guanidine hydrochloride) by resuspension in lysis buffer and centrifugation at 1000 rpm, for 2–3 min in a Jouan CR-412 centrifuge.
11. The 6 × His tagged proteins are absorbed onto Ni-NTA agarose from both soluble and insoluble material (soluble purifications are performed at 4°C and insoluble at room temperature). Binding is done in batch in a capped, 50-ml conical tube for 2–4 h.
12. The beads are washed in batch two times with wash buffer (with or without 6 M guanidine hydrochloride for insoluble and soluble purifications, resp.). Washing is performed by resuspending beads in 10–20 bed volumes of appropriate buffer, isolating beads by centrifugation (see step 10), and aspirating buffer from the beads.
13. Beads from the insoluble prep are exchanged into wash buffer containing 6 M urea, instead of guanidine hydrochloride.
14. Beads are transferred to Econopac columns (soluble and insoluble separately) and 6 × His tagged proteins are released from the beads with elution buffer (with or without 6 M urea for insoluble and soluble purifications, resp.). Six to eight 1 ml elutions are normally collected. Soluble protein is kept on ice, and insoluble at room temperature.
15. Protein concentration is determined by Bradford analysis.
16. Eluted protein is analyzed by SDS-PAGE to confirm that the purified product is free of contaminating proteins and is of the appropriate molecular weight.
17. Small aliquots are frozen in liquid nitrogen at 250 μg/aliquot.
18. Purified proteins are injected into rabbits, 250 μg at a time, for antibody production (Covance). We obtain equally good antibodies from soluble and insoluble protein preps.
19. Whole rabbit serum is affinity purified over columns of recombinant protein covalently coupled to cyanogen bromide activated sepharose, as described in *(5)*. Insoluble preps can be linked to sepharose as long as the concentration is greater than 0.5 mg/ml. Before coupling to the beads, insoluble proteins need to be dialyzed into coupling buffer containing fresh 6 M urea.

20. Affinity purified antibodies are exchanged into PBS or XB buffer using several rounds of dialysis in standard dialysis tubing.
21. Dialyzed, affinity purified polyclonal antibodies are snap-frozen in liquid nitrogen in 100 μl aliquots.

3.2. General Kinetochore Assembly Reaction

The following procedure describes our standard kinetochore assembly reaction. In short, CSF arrested extracts are supplemented with sperm nuclei, onto which kinetochores assemble. After assembly, the reactions are fixed and nuclei are pelleted onto coverslips and processed using immunofluorescence. Proteins and enzymes are inhibited in these reactions using immunodepletion, antibody addition, or drug addition.

1. *Xenopus* CSF egg extracts are prepared as described (*(1–3)*; *see* **Notes 2 and 3**).
2. To examine a protein's role in the kinetochore assembly, CSF arrested extracts are immunodepleted or supplemented with antibodies or drugs to specific proteins or complexes (*see* **Note 4**).
   a. For immunodepletion: polyclonal, affinity purified antibodies are bound to Affi-prep Protein A beads in batch, as follows.
      i. Affi-prep Protein A beads are first washed with PBSt. Antibodies are bound to beads in a reaction containing 10 bed volumes of PBSt plus antibody (*see* **Note 5**). We commonly use 25 μl of beads and 2–5 μg of antibody per μl of beads.
      ii. Binding is done on a rotator in a 1.5 ml microcentrifuge tube at room temperature or 4°C for at least 1 h. Reactions can be left overnight at 4°C.
   b. For small molecule inhibitors or antibody addition: these solutions can be added directly to the extracts to inactivate enzymes, proteins, or complexes.
   i. Drugs or antibodies are mixed into the extract, which is then stored on ice for ~20 min before initiating assembly (*see* **Note 6**).
3. For immunodepletion, antibody–bead conjugates are transferred to a tube of minimal volume (normally a 200 μl PCR tube), and are washed 3 times with CSF-XB. Immediately prior to immunodepletion, all residual liquid is aspirated from the bead bed with a 27 gauge needle. The bead to extract ratio should not exceed 1:4 at most.
4. Depletions are routinely performed in 1–3 rounds of 20–60 min, while rotating at 4°C. The extract is transferred to fresh antibody-coated beads after each round. It is critical that control extracts be mock-depleted in parallel over similar volumes of beads bound to control IgG (*see* **Notes 7 and 8**).

5. Optional: Prior to initiating kinetochore assembly extracts are supplemented with nocodazole (final concentration 5 μg/ml) that has first been diluted in CSF extract (depleted extract should be used when appropriate for dilution; *see* **Note 9**).
6. Kinetochore assembly reactions, 50 μl in volume, are warmed to room temperature (~22°C) for 1–2 min (this and the remaining steps are irrespective of the extract having been immunodepleted, supplemented with drug or antibodies, or treated with nocodazole).
7. Sperm nuclei are added to a final concentration of approximately 1000 per μl.
8. Reactions are incubated at room temperature for 45 min.
9. Assembly reactions are stopped by diluting 20-fold in approximately 1 ml of extract fixative (BRB80 + 30% glycerol + 0.5% Triton-X + 3.7% formaldehyde). Reactions are mixed by gently pipetting with 1 ml pipette tip and tubes are then inverted 10 times.
10. Reactions are fixed at room temperature for 15 min.
11. Fixed and diluted reactions are layered onto a glycerol cushion (BRB80 + 40% glycerol) overlaying a poly-l-lysine coated 18 mm No.1 coverslip (*see* **Note 10**). Cushions (~3 ml) and coverslips are set up in 12-well culture dishes.
    a. This allows us to prepare dozens of coverslips from the same reaction simultaneously. Reactions can be split over multiple coverslips. We have found that increasing the concentration of sperm in the extract has no noticeable affect on kinetochore assembly, but allows reactions to be split over increasing numbers of coverslips. Using tissue culture dishes for this purpose is an alternative to that performed by most groups in the past (*see* **Note 11**).
12. Culture dishes containing the assembly reactions, layered over cushions and coverslips, are spun on swinging platforms in a Jouan CR-412 centrifuge at 3000 rpm for 30 min at 18°C.
13. After centrifugation the tops of cushions are washed thoroughly with BRB80.
14. Coverslips are rinsed gently, once with BRB80, and are then postfixed with ice-cold methanol for 5 min at 4°C (*see* **Note 12**).
15. Coverslips are rehydrated with three 10-min incubations in PBSt or TBSt.
16. Coverslips are blocked, in their culture dish wells, in TBSt + 5% BSA (w/v).
17. Coverslips are processed using standard immunofluorescent staining techniques.

3.3. Probing Kinetochore Maintenance

Using a staggered start approach, we have added function blocking antibodies and drugs that affect kinetochore maintenance. An example of a kinetochore maintenance experiment, examining the kinetics of disassembly after the addition of hesperadin, an Aurora B inhibitor that induces kinetochore disassembly, is described below. This experiment has been complemented by adding function-blocking antibodies to the Aurora B complex, ensuring specificity of the drug. Any antibody, or drug, could be used in place of hesperadin to examine its affect on kinetochore maintenance.

1. Four-50 μl reactions (labeled A, B, C, and D) of a fresh CSF extract (that has been supplemented with nocodazole) are aliquoted into 1.5 ml microcentrifuge tubes and are stored on ice.
2. Kinetochore assembly is initiated with a staggered start approach, in 5 min time intervals. Reaction A is started at time zero minutes, B at time five minutes, C at time 10 min, and D at time 15 min.
3. The small molecule Aurora B kinase inhibitor, hesperadin, is added to reaction A at 45 min after time zero (*see* **Note 13**). Subsequently, hesperadin is added to reaction B at 50 min and to reaction C at 55 min.
   a. Thus, hesperadin has been added to each reaction 45 min after initiating kinetochore assembly.
4. At 60 min, all four reactions (A, B, C, and D) are stopped by dilution in 1 ml of extract fixative.
5. Reactions are fixed at room temperature for 15 min and are layered onto cushions, spun onto coverslips, and prepared for immunofluorescence, as described above.

3.4. Probing Kinetochore Disassembly at M-Phase Exit

In higher eukaryotes, outer kinetochore proteins dissociate from centromeres at the end of mitosis. This kinetochore disassembly reaction can be examined at the M-phase exit in frog egg extracts, through the addition of exogenous calcium which destroys CSF. The kinetics of kinetochore disassembly can be followed using the staggered start approach described above. An experiment examining disassembly at the M-phase exit in egg extracts is described.

1. Four 50 μl reactions (labeled A, B, C, and D) of a fresh CSF extract (that has been supplemented with nocodazole) are aliquoted into 1.5 ml microcentrifuge tubes and are stored on ice.
2. Kinetochore assembly is initiated with a staggered start approach, in 5 min time intervals, as described above. Reactions are first removed from ice and warmed to room temperature for 1 min, after which 1000 sperm nuclei per μl are added to initiate assembly.

a. Reaction A is started at time zero minutes, B at time five minutes, C at time 10 min, and D at time 15 min.

3. Forty-five minutes after initiating the first assembly reaction, 0.1 volume of calcium cycling solution is added to reaction A to drive it out of M-phase.

   a. At 50 and 55 min, reactions B and C are supplemented with 0.1 volume of calcium cycling solution.

4. 60 min after initiating assembly in reaction A, all four reactions (A, B, C, and D) are stopped by dilution in 1 ml of extract fixative.

5. Reactions are fixed at room temperature for 15 min and are layered onto cushions, spun onto coverslips, and prepared for immunofluorescence, as described above.

3.5. Effects of Protein Phosphatase 1 on Kinetochore Disassembly

Using the techniques described below, we have shown that Protein Phosphatase 1 (PP1) is sufficient to disassemble kinetochores on nuclei isolated from *Xenopus* extracts. These experiments utilize purified PP1 enzyme (*see* **Note 14**). Nuclei with preassembled kinetochores are isolated from extracts, onto coverslips, which are subsequently blocked, treated with PP1, and processed for immunofluorescence.

1. A 50 μl reaction of fresh CSF extract (that has been supplemented with nocodazole) is pipetted into a 1.5 ml microcentrifuge tube and stored on ice.
2. The reaction is thawed to room temperature for 1 min, and kinetochore assembly is initiated by adding 1000 sperm per μl.
3. Forty-five minutes after initiating the kinetochore assembly reaction, the extract is diluted with 5 ml (100 × dilution) of CSF-XB + 1 mM Okadaic Acid. The diluted reaction is layered onto a cushion of BRB80 + 40% glycerol + 0.5 μM okadaic acid, overlaying poly-l-lysine coated coverslips, in a 12-well culture dish. The reaction is split over the 12 cushions/coverslips.
4. Nuclei are pelleted out of the extract and onto the coverslips in a Jouan CR-412 centrifuge at 3000 rpm for 30 min at 18°C.
5. After centrifugation, the cushions are washed thoroughly with CSF-XB, and after removing the cushion, the coverslips are gently rinsed two times with CSF-XB + 0.1 μM okadaic acid.
6. Coverslips are blocked at room temperature for 45 min, in the 12-well dish, with 250 μl per coverslip of CSF-XB + 1 mM okadaic acid + 2.5% BSA (w/v).
7. During the blocking step, solutions of PP1 or BSA in phosphatase dilution buffer are prepared and stored on ice.

8. The BSA or PP1 solutions are aliquoted in 50 μl spots onto parafilm, and set up in a humidified chamber containing wetted paper using a covered, plastic kitchen container.
9. After the 45 min block, coverslips are overturned into a solution of PP1 or BSA, at various time intervals, between 0 and 20 min. Thus, those placed in enzyme (or BSA last) spend the additional time in blocking solution.
10. After incubation with enzyme (or BSA), coverslips are placed in a fresh 12-well dish containing CSF-XB.
11. Coverslips are rinsed two times with CSF-XB buffer.
12. Fixative of CSF-XB + 4% formaldehyde + 0.1% Triton X is added to the coverslips.
13. Coverslips are fixed at room temperature for 15 min.
14. Coverslips are washed 3 times in PBSt, and are then post-fixed in ice-cold methanol, for 5 min at 4°C.
15. Coverslips are rehydrated in PBSt (see above), and processed using standard immunofluorescent techniques.

4. Notes

1. The minimal tags used for expressing recombinant proteins minimize the generation of nonspecific antibodies.
2. *Xenopus laevis* frog egg extracts, arrested in M-phase by CSF, are very sensitive to handling technique. They should be handled with the utmost care at all times; this includes gently flicking for mixing, and pipetting through cut-off, or wide-bore, pipette tips. These handling procedures are discussed more extensively in *(1–3)*.
3. We have found that CSF extracts that have been frozen and stored at –80°C perform kinetochore assembly as well as fresh extract. CSF extracts are frozen in liquid nitrogen immediately after being prepared fresh and supplemented with an additional 250 mM sucrose.
4. High concentrations of antibodies are normally required to induce inactivation of specific proteins, when being added directly to extracts. We commonly concentrate an antibody to greater than 5 mg/ml, and add it at a dilution of 1:20 to the extract.
5. For immunodepletion prior to kinetochore assembly, Affi-Prep beads are used because they are less expensive, easier to use, and bind more antibody relative to comparable magnetic bead reagents.
6. Care should be taken to dilute the extract with as minimal a volume as possible. Diluting extract in excess of 10% of

the starting volume can be detrimental to maintenance of CSF arrest.

7. Prior to performing a depletion and kinetochore assembly reaction, it is important to empirically determine the depletion requirements for each antibody independently. Depletions are done as described above, and aliquots of extract are examined by immunoblot for the removal of specific proteins from the extract supernatant after each round of depletion. The minimal time needed to fully deplete the extract should be used (*see* **Note 8**).
8. The depletion process can degrade the ability of extracts to perform kinetochore assembly or to maintain their mitotic arrest, making it essential that control extracts are treated in an identical manner to those being immunodepleted, diluted with antibodies or drugs, and so on.
9. Because microtubule attachments can strip proteins from kinetochores, we normally include nocodazole in assembly reactions, at a final concentration of 5–10 μg/ml.
10. Poly-l-lysine coated coverslips are prepared by mixing 1 box of 100–18 mm, no. 1 cover glass in a plastic Petri dish with poly-l-lysine for 1 h at room temperature on a orbital rotator. The glass coverslips are washed at least 5 times in ddH2O and are stored in ethanol at room temperature.
11. Standard procedures for spinning nuclei out of extract and onto glass coverslips have utilized glass tubes, with a chock inserted into the bottom of the tube with a cover glass lying atop the chock. Using a standard HB-6 rotor accommodates the preparation of 6 coverslips at a time. Using our methods, we can simultaneously prepare upwards of 48 coverslips. We have used 24-well dishes, together with 12 mm coverslips, to prepare hundreds of coverslips at one time.
12. When adding and removing buffers to and from coverslips in multiwell dishes, care should be taken not to dispense liquid directly onto the coverslips. Solutions are added slowly down one sidewall of the well and are removed by aspiration from the opposing corner of the well.
13. We have found that many kinase inhibitors need to be used at higher concentrations in egg extracts than they are normally used in cells. For example, to inhibit Aurora B, hesperadin is used at 2 μM in extracts. However, IP-kinase assays of Aurora B suggest that hesperadin has an IC50 value of ~250 nm *(6)*.
14. PP1 enzyme purification was done as first described in *(7)* and was a generous gift of David Brautigan at the University of Virginia.

Acknowledgments

Special thanks to David Brautigan (University of Virginia) for the generous gift of purified PP1 enzyme and its associated buffers. This work was supported by funding from the American Cancer Society and the Jim Craig Scholars Foundation to PTS. MJE is the Philip O'Bryan Montgomery Jr. MD fellow of the Damon Runyon Cancer Research Foundation (DRG-1996–08).

References

1. Murray, A.W. (1991). Cell cycle extracts. *Methods Cell Biol.* **36**, 581–605.
2. Desai, A., Murray, A., Mitchison, T.J., and Walczak, C.E. (1999). The use of Xenopus egg extracts to study mitotic spindle assembly and function in vitro. *Methods Cell Biol.* **61**, 385–412.
3. Maresca, T.J. and Heald, R. (2006). Methods for studying spindle assembly and chromosome condensation in Xenopus egg extracts. *Methods Mol.Biol.* **322**, 459–474.
4. Altschul, S.F., Gish, W., Miller, W., Myers, E.W., and Lipman, D.J. (1990). Basic local alignment search tool. *J. Mol. Biol.* **215**, 403–410.
5. Harlow, E. and Lane, D. (1988). Antibodies: A Laboratory Manual. (Cold Spring Harbor, NY: Cold Spring Harbor Laboratory).
6. Hauf, S., Cole, R.W., LaTerra, S., Zimmer, C., Schnapp, G., Walter, R., Heckel, A., van Meel, J., Rieder, C.L., and Peters, J.M. (2003). The small molecule Hesperadin reveals a role for Aurora B in correcting kinetochore-microtubule attachment and in maintaining the spindle assembly checkpoint. *J. Cell Biol.* **161**, 281–294.
7. Brautigan, D.L., Shriner, C.L., and Gruppuso, P.A. (1985). Phosphorylase phosphatase catalytic subunit. Evidence that the Mr = 33,000 enzyme fragment is derived from a native protein of Mr = 70,000. *J. Biol. Chem.* **260**, 4295–4302.

Chapter 14

Live Cell Imaging of Kinetochore Capture by Microtubules in Budding Yeast

Kozo Tanaka and Tomoyuki U. Tanaka

Abstract

For high-fidelity chromosome segregation, kinetochores must be properly captured by spindle microtubules, but the mechanisms of initial kinetochore capture have remained elusive. Observation of individual kinetochore–microtubule interaction has been difficult, because multiple kinetochores are captured by microtubules during a short period and within a small space. By isolating one of the kinetochores from others through regulation of the activity of a centromere, we could visualize individual kinetochore–microtubule interactions in *Saccharomyces cerevisiae*. This technique, which we have called the 'centromere reactivation system', allowed us to dissect the process of kinetochore capture and transport on the mitotic spindle into several steps, thus enabling us to identify genes involved in each step. Kinetochores are captured by the side of microtubules extending from a spindle pole, and subsequently transported poleward along them. This process is evolutionarily conserved from yeast to vertebrate cells. Therefore, our system has proved useful in elucidating the underlying mechanisms of kinetochore capture by spindle microtubules.

Key words: *Saccharomyces cerevisiae*, kinetochore, centromere, spindle, microtubule, mitosis, live cell imaging, fluorescence microscopy

1. Introduction

The proper sister chromatid segregation to opposite poles of the cell during mitosis is crucial for maintenance of genetic integrity in eukaryotic cells. For high-fidelity chromosome segregation, kinetochores must be properly captured by spindle microtubules *(1)*. In animal cells, kinetochores are captured by microtubules in prometaphase after nuclear envelope breakdown *(2–4)*. Because the kinetochore capture and transport can only be visualised in a few cell types and very rarely, mechanisms of kinetochore capture by microtubules have remained elusive.

Andrew D. McAinsh (ed.), *Mitosis: Methods and Protocols*, Methods in Molecular Biology *545*,
DOI 10.1007/978-1-60327-993-2_14, © Humana Press, a part of Springer Science+Business Media, LLC 2009

In budding yeast, *Saccharomyces cerevisiae*, the nuclear envelope remains intact throughout the cell cycle. Microtubule organising centres, called spindle pole bodies (SPBs), are embedded in the nuclear envelope, and kinetochores are tethered to SPBs by microtubules during G1 phase. We found that centromeres detach from microtubules during S phase upon their replication which results in kinetochore disassembly. Soon afterwards kinetochores are reassembled, leading to their capture by microtubules *(5)*. However, in normal S phase, it is difficult to analyze the individual kinetochore–microtubule interaction in detail, because all kinetochores interact with microtubules in the vicinity of a spindle pole where microtubules frequently overlap each other.

To circumvent this problem, we spatially separated one of the 16 chromosomes of budding yeast from the others; this was achieved by regulating the centromere function *(6)*. A budding yeast centromere spans approximately 130 base pairs and we can inactivate the centromere function by activating transcription from an adjacently inserted promoter, which inhibits kinetochore assembly *(7, 8)*.

We replaced *CEN3* on chromosome III with *CEN3* under control of the *GAL1-10* promoter *(7)*, to conditionally inactivate and activate the centromere by turning on and off transcription in the presence of galactose and glucose, respectively. For live cell imaging by fluorescence microscopy, we labelled the *CEN3*-adjacent sequence and α-tubulin (*TUB1*) with fluorescent proteins to visualise *CEN3* and microtubules, respectively *(9)*. We inactivated the *CEN3* and simultaneously arrested cells in metaphase by depleting Cdc20, which is required for sister chromatid separation and for anaphase onset *(10)*. In this situation, *CEN3* localised away from the spindle and well separated from all other centromeres locating on the spindle. Then, we reactivated *CEN3* by turning off the adjacent *GAL1-10* promoter while cells were still in metaphase, and followed the behaviour of *CEN3*.

In this system, which we have called the 'centromere reactivation system', we found that kinetochores were captured by the lateral surface of a single microtubule extended from a spindle pole, and were subsequently transported polewards along the microtubule *(6)*. This process of initial kinetochore capture and subsequent transport along the microtubule is evolutionarily conserved from yeast to vertebrate cells. The centromere reactivation system allowed us to observe kinetochore capture in a large number of cells, and dissect the process of kinetochore capture and transport on the mitotic spindle into several steps, thus enabling us to identify genes involved in each step. Therefore, our system is useful in elucidating the underlying mechanisms of kinetochore capture by spindle microtubules, a process which may also be conserved throughout evolution *(6)*.

2. Materials

2.1. Yeast Strain

We use the yeast strain T3531 for live cell imaging of kinetochore capture *(6)*. T3531 has the following genotype with the W303 strain background: *MATa*, *cdc20::PMET3-CDC20::TRP1*, *cen3::PGAL-CEN3-tetOs::URA3*, *leu2::TetR-GFP::LEU2*, *trp1::YFP-TUB1::TRP1*. Each allele of T3531 was obtained as follows.

2.1.1. Replacement of the Authentic CEN3 with PGAL-CEN3-tetOs

First, *PGAL-CEN3 (7)*, *CYC1* transcription terminator (350 bp amplified by PCR) and 112 tandem copies of *tetOs (9)* (spanning 5.6 kb) were cloned into YIplac211 (National Centre for Biotechnology Information X75462) in the above order; second, the left and right *CEN3*-flanking regions (about 1 kb; not containing *CEN3* itself) were amplified by PCR and cloned next to the *GAL1-10* promoter in the above plasmid (at the opposite side from *CEN3*) with the opposite orientation (joining their 3' ends together) (pT389); third, pT389 was cut between two para-*CEN3* DNA fragments and used for transformation of yeast cells (**Fig. 14.1**).

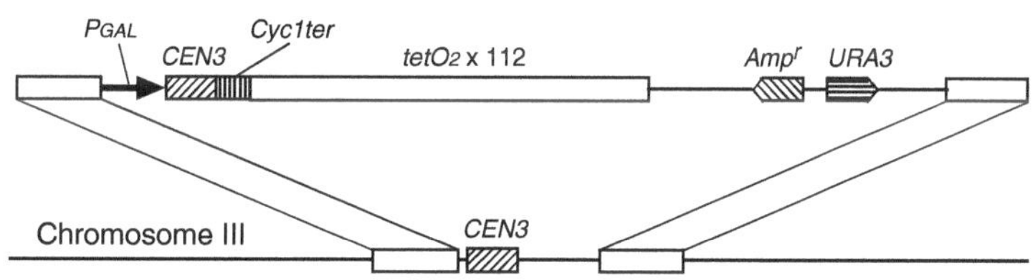

Fig. 14.1. Replacement of *CEN3* with the *PGAL-CEN3* construct for centromere reactivation system. The linearised plasmid pT389 (*see* **Section 2**) was integrated into the *CEN3* locus through recombination at para-*CEN3* regions. *CEN3*, *GAL1-10* promoter, *CYC1* transcription terminator, an array of *tetOs*, ampicillin-resistant gene, and *URA3* gene are indicated.

2.1.2. Other Alleles

TetR-GFP (9) and *PMET3-CDC20* (11) were previously reported. *YFP-TUB1* plasmid (pDH20) was obtained from the Yeast Resource Center (Seattle, USA) and used for yeast cell transformation (integrated at *TRP1* locus).

2.2. Yeast Cell Culture

1. YPA medium: Dissolve 20 g of peptone (Becton Dickinson 211677), 10 g of yeast extract (Becton Dickinson 212750), and 10 ml of 1% adenine (Sigma A8751) in 950 ml of deionised water. Autoclave and depending on requirements, add an appropriate carbon source, such as glucose, raffinose, and galactose, usually at 2%.
2. Synthetic complete (SC) and methionine dropout (SC-met) medium: Dissolve 6.7 g of yeast nitrogen base without amino acids (Becton Dickinson 291940), 1 ml of 1M

NaOH, 10 ml of 1% adenine, and amino acid mix (790 mg of CSM complete [Foremedium DCS0011] for SC medium, or 770 mg of CSM-Met [Foremedium DCS0111] for SC-met medium) in 950 ml of deionised water. Autoclave and depending on the purpose, add an appropriate carbon source to 2%.

3. α factor (13 amino acid peptide: WHWLQLKPGQPMY): Dissolve in methanol to make 5 mg/ml stock solution. Store at –80°C.
4. Methionine (Sigma M9625): Dissolve in distilled water to make 100 mM stock solution. Filter to sterilize and keep at room temperature.

2.3. Sample Preparation

1. Filtration device.
2. Membrane filter (Whatman ME28 10 400 812; diameter 47 mm, pore size 1.2 μm).
3. Microscope slides (VWR Superfrost 631-0103).
4. Microscope cover slips (VWR 631–0125; 22 × 22 mm, thickness no. 1.5).
5. Agarose (Invitrogen Ultra pure for molecular biology 15510-027).
6. Glass bottom dishes: MatTek dish (MatTek Corporation P35G-1.5-10-C; dish diameter 35 mm, glass thickness no. 1.5, microwell diameter 10 mm), Delta T culture dishes (Bioptechs 04200415C; 0.17 mm thick clear).
7. Concanavalin A (Sigma C7275): Dissolve in distilled water to 0.2% (w/v). Make aliquots and keep at –20°C.

2.4. Fluorescence Microscopy

1. Fluorescence microscope (Applied Precision Deltavision microscope)
2. Objective heater system (Bioptechs)
3. Dish heater system (Bioptechs Delta T open culture dish system)

3. Methods

3.1. Yeast Cell Culture

1. Culture *PMET3-CDC20 PGAL-CEN3-tetOs TetR-GFP YFP-TUB1* cells (T3531) in SC-met medium + 2% raffinose at 25°C overnight. We usually prepare 30 ml of culture in a 100 ml flask, and culture cells in a shaking water bath with rotation at 160 rpm.
2. Adjust OD600 of cell culture to 0.2–0.3. Treat cells with α factor to arrest cells in G1 phase. Add 5 mg/ml α factor stock solution in methanol to a final concentration of 0.5 μg/ml at t = 0, 1, and 2 h. Check for proper arrest

under a microscope 20–30 min after the last treatment (*see* **Note 1**).

3. Pour cell culture onto a membrane filter set in a vacuum filtration device to remove α factor. Wash cells three times with YPA medium or distilled water. Take the membrane filter with cells on it from the filtration device and place into YPA medium + 2% raffinose + 2% galactose + 2 mM methionine in a flask. We usually use 40 ml medium in a 100 ml flask. Resuspend cells by shaking.
4. Release cells from G1 arrest in a shaking water bath at 25°C. Cdc20 is depleted in the presence of methionine and *CEN3* is inactivated in the presence of galactose. Cells arrest at metaphase with large buds in 2 h. When cells are arrested at metaphase for longer (>3 h), nuclei are often elongated between buds and mother cell bodies (**Figs. 14.3** and **14.4**; *see* **Note 2**). If *CEN3* and the spindle are located on the opposite sides of the elongated nucleus, *CEN3* capture by long microtubules can be observed in detail.

3.2. Sample Preparation for Microscopy

For microscopy, cells are mounted either on microscope slides or on glass-bottom dishes depending on the purpose. Cells on microscope slides become somewhat flattened under the coverslip, enabling most of the cells to be kept within a single microscope field and on the same focal plane. On the other hand, cells sometimes change their location during imaging when the mounting medium dries up. In addition, this method may negatively affect cell viability, especially when images are collected over an extended time period. Mounting cells on glass-bottom dishes has the opposite attributes; cells within a single microscope field are not usually on the same focal plane, but their viability is maintained for longer. When observing temperature-sensitive cells at their restrictive temperature, the objective lens and glass-bottom dishes need to be heated.

3.2.1. Live Cell Imaging on Microscope Slides

1. Prepare microscope slides with agarose pads. Boil 5% agarose in water in a microwave and then allow to cool to 68°C in a heat block. Heat also SC medium (×1.5 concentration) + 3% glucose to 68°C in a heat block. Mix 1 volume of 5% agarose to 2 volumes of SC medium + 2% glucose. To make the pad, prepare 2 slides covered by a piece of catering wrap (taped to the slide; **Fig. 14.2**). Place a clean slide between these 2 slides. Drop 3 μl of medium/agarose onto the center of the slide, and quickly place another slide on top, perpendicular to the first slide so that its ends are supported by the catering wrap (*see* **Fig. 14.2**). The thickness of the catering wrap prevents the pad being too thin between the two slides. Remove the top slide soon after agarose becomes hard (~10 min), leaving the pad on the bottom slide. These

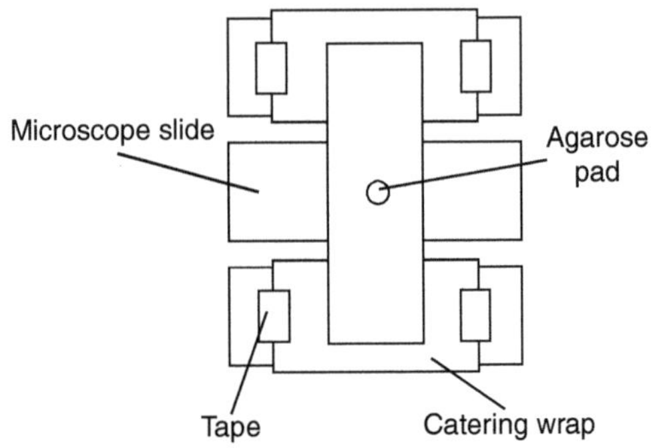

Fig. 14.2. Making agarose pads on microscope slides. Arrangement of slides is schematically shown. See **Section 3** for details.

microscope slides with agarose pads can be used for a couple of days.

2. Take 0.5–1.0 ml of cell culture. Spin down by centrifugation at 5000 rpm for 30 s at room temperature. Wash twice with 1 ml of SC medium. Resuspend in a small volume of SC medium + 2% glucose + 2 mM methionine. Now *CEN3* is reactivated by the addition of glucose and begins to be captured by microtubules. Therefore, the following procedures should be done as quickly as possible (within 5 min).
3. Mount 2–3 μl of cell suspension on the agarose pad on a microscope slide. Cover cells with a coverslip. Gently tap edges of the coverslip to spread cell suspension. Cells tend to be trapped on the agarose pad whereas medium spreads to the edge of the coverslip.
4. Set the microscope slide on the stage of a microscope and find a field suitable for imaging. In our experience, cells at the edge of the agarose pad are well concentrated and do not move even when medium dries up in other areas.

3.2.2. Live Cell Imaging on Glass-Bottom Dishes

1. Coat a glass-bottom dish with Concanavalin A. Apply 100 μl of 0.2% Concanavalin A to the microwell of a MatTek dish. Place in the dark for 10 min. Remove liquid and dry the dish in the dark.
2. Take 0.5–1.0 ml of cell culture. Spin down by centrifugation at 5000 rpm for 30 s at room temperature. Wash with 1 ml of SC medium + 2% raffinose + 2% galactose. Resuspend in 0.2–0.5 ml of SC medium + 2% raffinose + 2% galactose.
3. Mount 100 μl of cell suspension in the microwell of a Concanavalin A-coated MatTek dish on the stage of a microscope. Wait 7 min; then remove medium. To wash out galactose and remove floating cells, wash the dish with 1 ml of SC medium twice (carefully add and then immediately remove the medium by pipetting).

4. Mount 1 ml of SC medium + 2% glucose + 2 mM methionine to reactivate *CEN3*.

3.2.3. Live Cell Imaging of Temperature-Sensitive Mutant Strains

3.2.3.1. On Microscope Slides

1. Heat objective lens to 35°C with an objective heater system (Bioptechs) following the manufacturer's instructions. Heat microscope slides with agarose pads to 35°C on a heating block. Also heat SC medium for washing and SC medium + 2% glucose + 2 mM methionine to 35°C on a heating block.
2. Transfer the cell culture in YPA + 2% raffinose + 2% galactose + 2 mM methionine from a shaking water bath at 25°C to another water shaking bath at 35°C prior to imaging.
3. After incubation at 35°C for a period required to inactivate a mutant (typically 30 min), take samples, and then wash and mount cells as in **Section 3.2.1**, except for using preheated medium and microscope slides.

3.2.3.2. On Glass-Bottom Dishes

1. Use Delta T culture dishes (Bioptechs) for which a heating device is available, instead of MatTek dishes. To coat the glass surface of the dishes with Concanavalin A, mount 150 μl of 0.2% Concanavalin A (sufficient to cover the middle of the glass surface). Incubate for 10 min in the dark, remove liquid, and dry the dishes.
2. Heat the objective lens with the objective heater system and heat a Concanavalin A-coated Delta T culture dish with the dish heater system to 35°C. Also heat SC medium + 2% raffinose + 2% galactose and SC medium without carbon source for washing and SC medium + 2% glucose + 2 mM methionine to 35°C in a heating block.
3. Transfer the cell culture in YPA + 2% raffinose + 2% galactose + 2 mM methionine from a shaking water bath at 25°C to one at 35°C prior to imaging.
4. After incubation in a shaking water bath at 35°C for a required period to inactivate a mutant, take samples; then wash the cells as in **Section 3.2.2**, but use preheated medium.
5. Mount 150 μl of cell suspension in the microwell of a Concanavalin A-coated Delta T culture dish on the stage of a microscope. Wait 7 min and remove medium. To wash out galactose and remove floating cells, wash the dish twice using 1 ml of preheated SC medium, carefully adding the medium and removing immediately by pipetting.
6. Mount 2 ml of preheated SC medium + 2% glucose + 2 mM methionine to reactivate *CEN3*.

3.3. Microscopy

3.3.1. Image Collection

Images are taken using an inverted microscope (Olympus IX-71 in Deltavision microscope) with a 100 × 1.4 numerical aperture optical lens, a cooled CCD camera (Photometrics CoolSNAP

HQ), and softWoRx software (Applied Precision). In typical experiments, time-lapse images are collected every 10–15 s for 20–40 min with 3–7 (0.5–0.7 μm apart) z-sections at 23°C (ambient temperature) or 35°C (for temperature-sensitive cells). To distinguish GFP and YFP signals in time-lapse fluorescence microscopy, the JP3 filter set (Chroma) is used. In typical experiments, exposure time to excitation light is 0.1 s for both GFP/JP3 and YFP/JP3 channels with appropriate neutral density filters, but this depends on the signal intensity of target samples.

3.3.2. Image Analysis

Acquired images are deconvoluted and projected with maximum intensity signals to two-dimensional images with the softWoRx software.

Cells with unattached *CEN3*s localising distant from spindles in elongated nuclei are suitable for analysis of kinetochore capture by long microtubules (**Figs. 14.3** and **14.4**, 0 s). *CEN3*s are

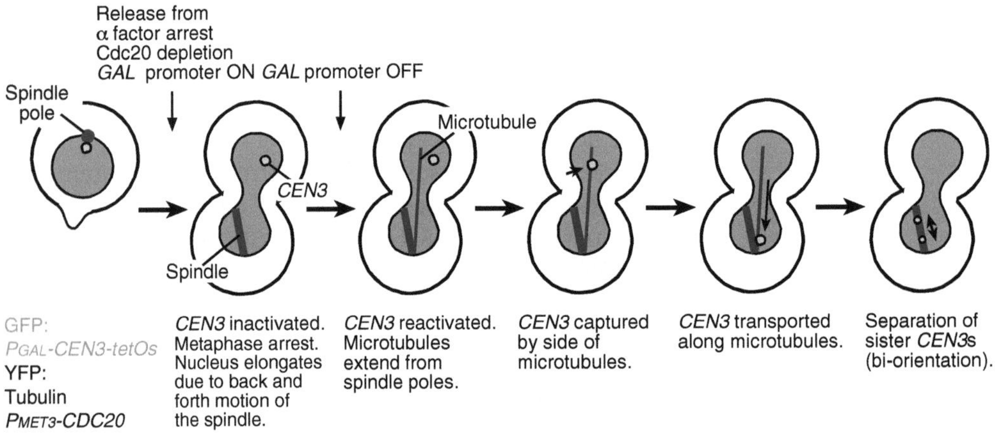

Fig. 14.3. Diagrams of the experimental system for observing kinetochore capture by microtubules (centromere reactivation system). See **Section 3** for details.

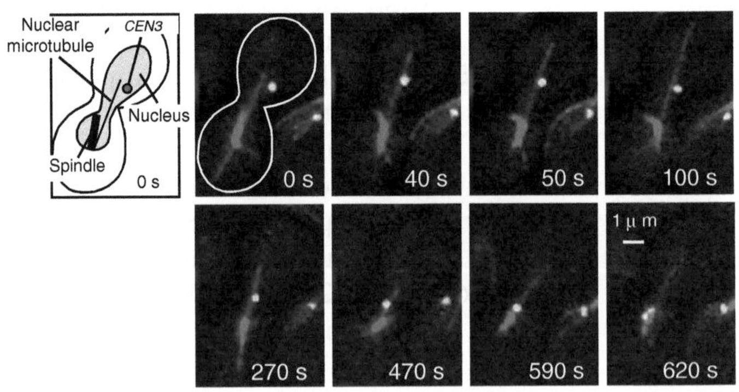

Fig. 14.4. Visualizing kinetochore capture by microtubules. *PMET3-CDC20*, *PGAL-CEN3-tetOs*, *TetR-GFP*, *YFP-TUB1* cells (T3531) were treated and observed as described in **Section 3**. Zero time is set arbitrarily for the first panel, in which the cell shape is outlined. Scale bar, 1 μm.

captured by microtubules within 20 min after addition of glucose in most of the cells. *CEN3*s are usually captured by the side of microtubules (**Fig. 14.4**, 40 s), and subsequently transported along the microtubules (**Fig. 14.4**, 100–590 s). Shortly after *CEN3*s reach a spindle pole, the majority of *CEN3* GFP signals split on the spindle, indicating that sister *CEN3*s biorient on the metaphase spindle (**Fig. 14.4**, 620 s). Molecular mechanisms of these processes are described in *(6, 12)*.

4. Notes

1. Depending on the purpose, cells can be directly arrested at metaphase from asynchronous culture. As it is not necessary to wash out α factor in this case, dilute culture in SC-met medium + 2% raffinose into more than 2 volumes of YPA medium + 2% raffinose + galactose (final 2%).
2. This elongation is probably due to earlier back-and-forth motions of the spindle between the two cell bodies, and is not due to cells leaking into anaphase, because the spindle length stays short. Typically, inactivated *CEN3* remains in the mother and the short spindle is found in the bud in the elongated nucleus (**Figs. 14.3** and **14.4**).

Acknowledgments

We thank L. Clayton and other members of TUT laboratory for discussions and reading the manuscript; C. Allan and S. Swift for technical help for microscopy/computing; and R. Ciosk, F. Uhlmann, K. Nasmyth, K. Bloom, and the Yeast Resource Centre for reagents. We learned the use of Concanavalin A for yeast-cell immobilization at the Y. Hiraoka lab website. This work was supported by Cancer Research UK, the Wellcome Trust, the EMBO Young Investigator Program, Human Frontier Science Program, Lister Research Institute Prize, and Association for International Cancer Research.

References

1. McIntosh, J. R., Grishchuk, E. L. and West, R. R. (2002) Chromosome-microtubule interactions during mitosis. *Annu. Rev. Cell. Dev. Biol.* **18**, 193–219.
2. Rieder, C. L. and Alexander, S. P. (1990) Kinetochores are transported poleward along a single astral microtubule during chromosome attachment to the spindle in newt lung cells. *J. Cell Biol.* **110**, 81–95.
3. Hayden, J. H., Bowser, S. S. and Rieder, C. L. (1990) Kinetochores capture astral microtubules during chromosome

attachment to the mitotic spindle: direct visualization in live newt lung cells. *J. Cell Biol.* **111**, 1039–45.
4. Merdes, A. and De Mey, J. (1990) The mechanism of kinetochore-spindle attachment and polewards movement analyzed in PtK2 cells at the prophase-prometaphase transition. *Eur. J. Cell Biol.* **53**, 313–25.
5. Kitamura, E., Tanaka, K., Kitamura, Y. and Tanaka, T. U. (2007) Kinetochore-microtubule interaction during S phase in *Saccharomyces cerevisiae*. *Genes Dev.* **21**, 3319–30.
6. Tanaka, K., Mukae, N., Dewar, H., van Breugel, M., James, E. K., Prescott, A. R., Antony, C. and Tanaka, T. U. (2005) Molecular mechanisms for kinetochore capture by spindle microtubules. *Nature* **434**, 987–994.
7. Hill, A. and Bloom, K. (1987) Genetic manipulation of centromere function. *Mol. Cell. Biol.* 7, 2397–405.
8. Collins, K. A., Castillo, A. R., Tatsutani, S. Y. and Biggins, S. (2005) De novo kinetochore assembly requires the centromeric histone H3 variant. *Mol. Biol. Cell* **16**, 5649–5660.
9. Michaelis, C., Ciosk, R. and Nasmyth, K. (1997) Cohesins: chromosomal proteins that prevent premature separation of sister chromatids. *Cell* **91**, 35–45.
10. Nasmyth, K., Peters, J. M. and Uhlmann, F. (2000) Splitting the chromosome: cutting the ties that bind sister chromatids. *Science* **288**, 1379–1385.
11. Uhlmann, F., Wernic, D., Poupart, M. A., Koonin, E. V. and Nasmyth, K. (2000) Cleavage of cohesin by the CD clan protease separin triggers anaphase in yeast. *Cell* **103**, 375–386.
12. Tanaka, K., Kitamura, E., Kitamura, Y. and Tanaka, T. U. (2007) Molecular mechanisms of microtubule-dependent kinetochore transport towards spindle poles. *J. Cell Biol.* **178**, 269–281.

Chapter 15

The Spindle Checkpoint: Assays for the Analysis of Spindle Checkpoint Arrest and Recovery

Josefin Fernius and Kevin G. Hardwick

Abstract

The spindle checkpoint is a surveillance mechanism that ensures the fidelity of chromosome segregation by inhibiting anaphase onset until all chromosomes have established stable bipolar attachments. Here we describe a number of protocols that can be used to assay the ability of budding and fission yeast cells to (1) establish and maintain a spindle checkpoint arrest, and (2) segregate chromosomes efficiently upon recovery from mitotic arrest. We focus on experimental detail of the budding yeast protocols, but also point out important differences between budding and fission yeast assays.

Key words: Checkpoint, segregation, recovery, biorientation.

1. Introduction

Genetic screens in budding yeast identified components of the spindle checkpoint (Mad1-3, Bub1-3) using drugs that inhibit microtubule polymerisation *(1, 2)*. High levels of these drugs (e.g., benomyl or nocodazole) result in depolymerised spindle microtubules and therefore in lack of kinetochore–microtubule attachments. Wild-type cells respond to these unattached kinetochores and halt the cell cycle in metaphase. In the presence of low levels of microtubule drugs, wild-type cells are viable due to their ability to delay anaphase onset until all kinetochores have been properly attached to microtubules. However, spindle checkpoint (*mad* and *bub*) mutants ignore the unattached kinetochores induced by the microtubule poison, and undergo anaphase prematurely. This precocious segregation of sister-chromatids gives

Andrew D. McAinsh (ed.), *Mitosis: Methods and Protocols*, Methods in Molecular Biology *545*,
DOI 10.1007/978-1-60327-993-2_15, © Humana Press, a part of Springer Science+Business Media, LLC 2009

rise to high levels of aneuploidy and cell death in these mutants. Although loss of viability in the presence of these drugs can be a useful method to assess mutants *(3)*, there are limited conclusions that can be drawn from these assays. This chapter describes further methods that can be used to analyse the ability of cells to initiate and maintain a spindle checkpoint arrest, and to recover accurately from spindle damage.

2. Materials

2.1. Method 1 (Securin Immunoblot)

2.1.1. Yeast Media, Mating Pheromone, and Anti-Microtubule Drugs

1. All yeast strains are derivatives of W303 (*see* **Table 15.1**).
2. Rich yeast media (YPDA): 1% (w/v) yeast extract, 2% (w/v) Bacto-peptone, 2% (w/v) glucose, 0.003% (w/v) adenine sulphate.
3. α-factor (Peptide Protein Research Ltd) is stored at –20°C as a stock solution of 10 mg/ml, dissolved in DMSO. Avoid repetitive freeze–thawing.
4. Nocodazole (Sigma) is stored at –20°C as a stock solution of 10 mg/ml, dissolved in DMSO.

2.1.2. Trichloroacetic Acid (TCA) Precipitation, and Immunoblotting

1. 5% TCA is made up from a 100% stock (Fisher Scientific) and stored in the dark at 4°C. Care should be taken because TCA is an acid and is corrosive.
2. Lysis buffer: 50 mM Tris pH 7.5, 1 mM EDTA, 2 mM DTT with 1X protease inhibitors made up from tablets (Roche).
3. A Hybaid Ribolyser is used to smash open yeast cells, according to the manufacturer's instructions.
4. 3X protein sample buffer: 187 mM Tris pH 6.8, 6% β-mercaptoethanol, 30% glycerol, 9% SDS, 0.05% bromophenol blue.
5. Blotto: 5% Marvel dry milk, 0.1% Tween-20 in PBS (13.7 mM NaCl, 0.27 mM KCl, 1 mM Na_2PO_4, 0.176 mM KH_2PO_4, pH 7.2).
6. A14 anti-myc rabbit primary antibody (Santa Cruz Biotechnology).
7. 9E10 mouse monoclonal antibody (Covance).
8. Anti-Pgk1 mouse antibody (Invitrogen).

2.2. Method 2 (Sister-Chromatid Cohesion Assay)

1. Rich yeast media (YPDA): 1% (w/v) yeast extract, 2% (w/v) Bacto-peptone, 2% (w/v) glucose, 0.003% (w/v) adenine sulphate.
2. α-factor (Peptide Protein Research Ltd) is stored at –20°C as a stock solution of 10 mg/ml, dissolved in DMSO. Avoid repetitive freeze–thawing.

Table 15.1
Budding yeast strain list (all are *MATa*)

Strain	Genotype
KH186	*ade2-1 his3-11,15 leu2-3,112 trp1-1 ura3-1 bar1*
JF004	*trp1-1:lacO::URA GFPlacI::HIS3 Pds1-Myc18::LEU ade2-1 leu2-3,112*
EK013	*mad3Δ2::URA3 his3-11:pCUP-GFP12-lacI12:HIS3 trp1-1:lacO::TRP1 leu2,3-112 ade2-1 bar1*
JF125	*bub1ΔK::HPH LacO::URA GFP-LacI::HIS Pds1-Myc18::LEU ade2-1 leu2-3,112 trp1-1*
JF140	*bub1Δ::HPH LacO::URA GFP-LacI::HIS Pds1-Myc18::LEU ade2-1 leu2-3,112 trp1-1*
SBY1646	*mtw1-1 his3-11pCUP1GFP12-lacI12:HIS3 trp1-1 256 lacO::TRP1 PDS1-myc18::LEU2 bar1 ade2-1 ura3-1*
SBY1724	*mtw1-1 his3-11pCUP1GFP12-lacI12:HIS3 trp1-1 256 lacO::TRP1 PDS1-18myc18::LEU2 ipl1-321 bar1 ade2-1 ura3-1*
JF100	*mtw1-1 bub1ΔK::HPH his3-11pCUP1GFP12-lacI12:HIS3 trp1-1 256 lacO::TRP1 PDS1-myc18::LEU2 ade2-1 ura3-1 bar1*
JF152	*pURA3-tetR::GFP::LEU2 cenIV::tetO(x448)::URA3 METprom-CDC20::URA Spc42-Tomato::NAT ade2-1 leu2-3,112 ura3-3 trp1-1 his3-11,15*
JF154	*pURA3-tetR::GFP::LEU2 cenIV::tetO(x448)::URA3 METprom-CDC20::URA Spc42-Tomato::NAT bub1ΔK::HPH bar1 ade2-1 leu2-3 ura3 trp1-1 his3-11,15*
JF156	*pURA3-tetR::GFP::LEU2 cenIV::tetO(x448)::URA3 METprom-CDC20::URA Spc42-Tomato::NAT sgo1Δ::KAN ade2-1 leu2-3,112 ura3-3 trp1-1 his3-11,15*

3. Nocodazole (Sigma) is stored at –20°C as a stock solution of 10 mg/ml, dissolved in DMSO.
4. 37% formaldehyde solution.
5. 80% ethanol.
6. 1 μg/ml DAPI solution.

2.3. Method 3 (Lack of Tension, GAL-MCD1, Assay)

1. YP media: 1% (w/v) yeast extract, 2% (w/v) Bacto-peptone, 0.003% (w/v) adenine sulphate
2. Filter sterilised 20% galactose and 20% raffinose solutions.
3. α-factor (Peptide Protein Research Ltd) is stored at –20°C as a stock solution of 10 mg/ml, dissolved in DMSO. Avoid repetitive freeze–thawing.
4. Securin (Pds1) immunoblotting; see **Section 2.1**.

2.4. Method 4 (Kinetochore Attachment Defects)

1. Rich yeast media (YPDA): 1% (w/v) yeast extract, 2% (w/v) Bacto-peptone, 2% (w/v) glucose, 0.003% (w/v) adenine sulphate.
2. α-factor (Peptide Protein Research Ltd) is stored at –20°C as a stock solution of 10 mg/ml, dissolved in DMSO. Avoid repetitive freeze–thawing.
3. Securin (Pds1) immunoblotting; see **Section 2.1**.

2.5. Method 5 (Checkpoint Recovery Assay)

As above (**Section 2.4**)

2.6. Method 6 (Chromosome Segregation After Spindle Damage: Monitored by Spindle Staining and GFP Chromosome Analysis)

1. Rich yeast media (YPDA): 1% (w/v) yeast extract, 2% (w/v) Bacto-peptone, 2% (w/v) glucose, 0.003% (w/v) adenine sulphate.
2. α-factor (Peptide Protein Research Ltd) is stored at –20°C as a stock solution of 10 mg/ml, dissolved in DMSO. Avoid repetitive freeze–thawing.
3. Securin (Pds1) immunoblotting, see **Section 2.1**.
4. Anti-tubulin immunofluorescence:
   a. 0.1 M potassium phosphate, pH 7.5.
   b. Zymolyase 100,000 (ICN Pharmaceuticals Inc.) is stored at –20°C as a stock solution of 5 mg/ml in dH_2O.
   c. Blotto (5% Marvel dry milk, 0.1% Tween-20 in PBS) is well dissolved and centrifuged at 8000 g to remove any milk particles.
   d. 37% formaldehyde (Sigma).
   e. Anti-GFP rabbit primary antibody (Invitrogen).
   f. Anti-tubulin rat primary antibody, YOL 1/34 (Abcam).

2.7. Method 7 (Chromosome Bi-Orientation Assay)

1. CSM-Met media (Formedium).
2. 200 mM methionine (Formedium) stock solution is prepared fresh for each experiment and sterile filtered.

3. α factor (Peptide Protein Research Ltd) is stored at –20°C as a stock solution of 10 mg/ml, dissolved in DMSO. Avoid repetitive freeze–thawing.
4. Nocodazole (Sigma) is stored at –20°C as a stock solution of 10 mg/ml, dissolved in DMSO.

3. Methods

3.1. Assays for Spindle Checkpoint Arrest in Budding Yeast

3.1.1. Biochemical Analysis of Securin (Pds1) Levels

The spindle checkpoint components ensure that anaphase is actively inhibited until all kinetochores have established stable bipolar attachments. The downstream target of the checkpoint is Cdc20 *(4, 5)*, which is an activator of the E3 ubiquitin ligase known as the Anaphase Promoting Complex/Cyclosome (APC/C). The APC/C promotes ubiquitination of target proteins such as cyclins and Securin (Pds1 in budding yeast). The ubiquitination and destruction of Pds1 releases the protease Separase (Esp1 in budding yeast), which proteolytically cleaves the cohesion subunit Mcd1 thereby initiating chromosome segregation. Therefore the level of Pds1 reflects the cell cycle stage: high (and maintained) levels of Pds1 indicate that the spindle checkpoint is active and APC/C is being inhibited and conversely, degradation of Pds1 reflects activation of APC/C. The levels of Pds1 in cells treated with microtubule depolymerising drugs can be monitored easily by using a budding yeast strain carrying Pds1-myc (made by PCR mediated C-terminal tagging *(6)*).

To allow the study of a synchronous cell culture, budding yeast cells are first arrested in G1 using the mating pheromone α-factor. These experiments are carried out in YPDA media.

1. Cells are grown overnight and diluted to OD_{600} 0.2 in the morning; then allowed to grow at 30°C for three hours.
2. Bar1 is a protease that degrades α-factor. Strains carrying the wild-type *BAR1* gene are arrested in YPDA media containing 10 μg/ml α-factor, and strains with the *bar1* mutation are arrested using 1 μg/ml α-factor at 30°C for three hours or until >90% of cells have formed schmoos, monitored by DIC light microscopy.
3. 2.5 μg/ml α-factor is readded to *BAR1* cell cultures after 90 min to prevent cells escaping from G1 (*see* **Note 1**).
4. The α-factor is then carefully removed from the cell cultures, either by washing the cells three times with YPDA media, or by filtering cells with 10 volumes of YPDA media. Washing by centrifugation is carried out by resuspending the cells in at least five volumes of YPDA, collecting the cells by centrifugation at 1000 g, and repeating three times.

5. The cells are then resuspended in YPDA media containing microtubule drugs. Nocodazole is more soluble than benomyl and is most commonly used in liquid media. 20 μg/ml nocodazole is added to the cultures, which are then incubated at 23°C for the duration of the time-course. After 60 min into the time-course, α-factor is added back to the cultures to prevent cells that exit from mitosis from entering another cell cycle.
6. 10 ml of yeast culture is harvested at each timepoint for immunoblotting of Pds1-myc.
7. To analyse the levels of Pds1, protein samples are prepared by TCA precipitation (*see* **Note 2**). The cell pellet is resuspended in 5 ml 5% TCA and kept on ice for 10 min.
8. The sample is then centrifuged at 1500 g for three minutes at 4°C.
9. The pellet is then snap-frozen in liquid nitrogen and stored at –80°C until processed.
10. An equal amount of glass beads is added to the pellet which is then lysed in 100 μl of lysis buffer, using a Ribolyser for 20 s (on setting 4), and kept on ice.
11. 50 μl of 3× sample buffer are added and the samples are then boiled for five minutes, then run on 10% SDS-PAGE, and blotted for Pds1-myc using anti-myc antibodies (e.g., A14 anti-myc rabbit antibody), and a loading control such as Pgkl can be used (**Fig. 15.1**).

3.1.2. Analysis of Spindle Checkpoint Arrest by Monitoring Sister-Chromatid Cohesion

In addition to monitoring the levels of Securin in budding yeast cells, analysis of single cells using fluorescence microscopy can be an informative indicator of spindle checkpoint arrest. Spindle checkpoint arrest induced by microtubule depolymerisation can be visualised using fluorescence microscopy to monitor sister-chromatid cohesion. In wild-type cells sister-chromatids remain cohesed during spindle checkpoint arrest, whereas spindle checkpoint mutants precociously separate their sister-chromatids. This can be directly visualised, using budding yeast strains expressing lac (or tet) repressor-GFP (lacI-GFP) that contain a lac-operator array integrated on a chromosome, thereby marking that chromosome with GFP (originally described in *(7)*). Cells in which the GFP marked sister-chromatids remain cohesed display one GFP focus, whereas cells in which the sister-chromatids have separated display two GFP foci. This enables quantitative analysis of spindle checkpoint status by single cell analysis.

1. Cells are synchronised in G1 using α-factor (as described above), released into YPDA media containing 20 μg/ml nocodazole, and incubated at 23°C for four hours.
2. Samples for analysis by fluorescence microscopy are taken (450 μl) at time point zero (G1) and after two, three, and four hours of nocodazole treatment.

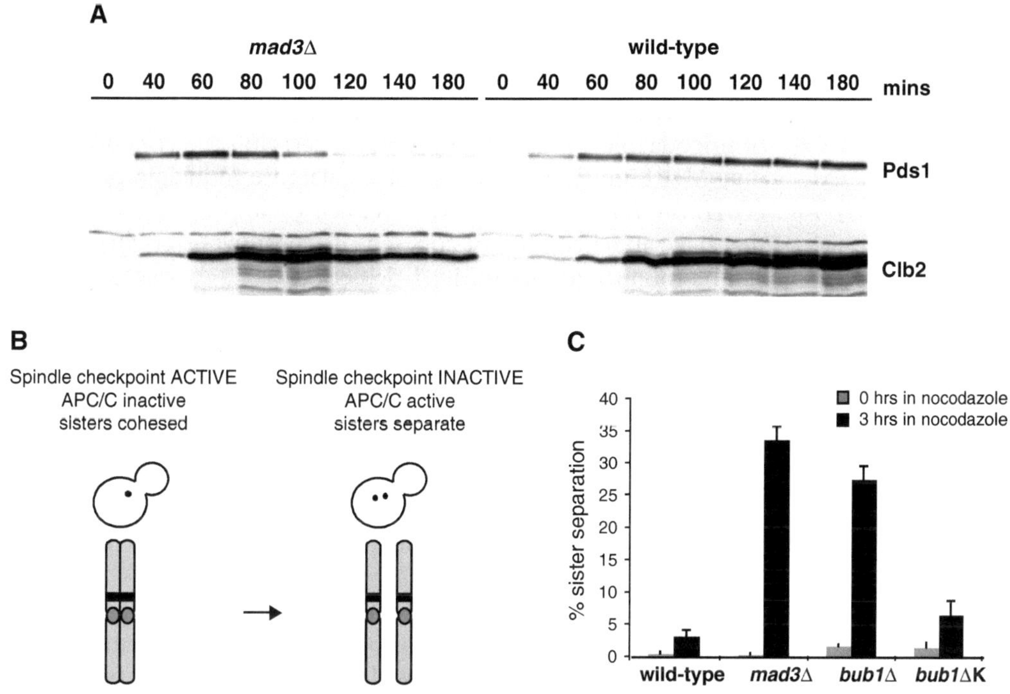

Fig. 15.1. Spindle checkpoint mutants fail to arrest in mitosis in response to the microtubule depolymerising drug nocodazole. (**A**) Cells are synchronised in G1 and released into media containing nocodazole. Wild-type cells stabilise Pds1 in response to unattached kinetochores induced by the microtubule drugs, whereas *mad3Δ* mutant prematurely degrades Pds1 and the mitotic cyclin Clb2. (**B**) Schematic diagram of GFP labelled sister-chromatids. Wild-type cells maintain sister-chromatid cohesion (one GFP focus) as they arrest in mitosis in response to nocodazole, whereas spindle checkpoint mutants separate their sister-chromatids prematurely (two GFP foci). (**C**) Spindle checkpoint mutants (*mad3Δ*, *bub1Δ*) fail to maintain sister-chromatid cohesion in cultures containing nocodazole, as judged by a single GFP focus. In contrast, cells lacking the Bub1 C-terminal kinase domain (*bub1Δkinase*) are able to keep most sister-chromatids cohesed for three hours.

3. 450 μl of cells are then fixed with 50 μl 37% formaldehyde at room temperature for 10 min, then collected by centrifugation at 14,000 g for 2 min.
4. The supernatant is removed and the cells are resuspended in 1 ml of 80% ethanol (*see* **Note 3**).
5. Collect cells by centrifugation at 14,000 g and remove supernatant.
6. Centrifuge again and remove any residual ethanol.
7. Resuspend cells in 20 μl of 1 μg/ml DAPI.
8. GFP dots are analysed by fluorescence microscopy. Samples can be stored for a couple of days, but the signal will be stronger if analysed immediately.

3.1.3. Spindle Checkpoint Arrest Induced by Lack of Tension at Kinetochores

The assays to monitor efficient spindle checkpoint arrest described above use microtubule depolymerising drugs to produce unattached kinetochores. There are some proteins which are not required for such an arrest, but are required for the delay in response to kinetochore attachments that lack tension. For example, the Aurora B homologue (Ipl1), Sgo1, and the Bub1 kinase domain are all required for cells to respond to reduced cohesion at centromeres, but not the defects induced by high levels of nocodazole *(8–10)*. Using a strain carrying the cohesion subunit Mcd1 under the control of a galactose-inducible promoter (*GAL-MCD1*) and also Pds1-myc, cell cycle progression in cells that have attached but not cohesed (no tension) sister-chromatids can be monitored, upon Mcd1 depletion.

1. Cells are synchronised in G1 using α-factor (as described above) for 3 h in YP media containing 2% galactose and 2% raffinose to allow expression of Mcd1.
2. To switch off the expression of Mcd1, the cells are incubated for 2 h in YPDA media containing α-factor (to synchronise the cells in G1).
3. The α-factor is then carefully washed out (as described above) and cells are released into the cell cycle in YPDA media.
4. Samples are taken at 10 min timepoints and α-factor is added back after 60 min (*see* **Note 4**). The samples are then TCA precipitated (as described above) and processed for immunoblotting with anti-myc antibodies to analyse the levels of Pds1-myc. Pgk1 can be used as a loading control (~45kD).

3.1.4. Spindle Checkpoint Arrest Induced by Kinetochore Defects

Mtw1 is an essential kinetochore component, and some proteins (e.g., Ipl1) are required to delay cells in metaphase in response to the kinetochore defects induced by a temperature-sensitive mutation of Mtw1 (*mtw1-1*) *(11)*. Ipl1 kinase is required to break the defective kinetochore attachments, and the Mad and Bub proteins are then required to respond to the unattached kinetochores. The requirement of a protein for the metaphase delay induced by such kinetochore defects can be assayed by monitoring a mitotic delay as measured by stabilised Securin levels. This experimental procedure is based on methods from *(12)*.

1. Budding yeast cells are grown in YPDA media overnight at the permissive temperature of 23°C.
2. The cells are diluted to OD_{600} 0.2 in the morning, and grown at 23°C for 2 h.
3. The cells are then treated with α-factor (as above, to synchronise the cells in G1) for 2 hours and 30 min. At this point the cultures are switched to the restrictive temperature of 36°C and incubated for another 30 min with α-factor.
4. The cells are washed three times in YPDA media and released from G1 into a three-hour time course at 36°C. α-factor is added back after 60 min. Pds1-myc levels are then monitored by immunoblotting, as above.

3.1.5. Assays for Spindle Checkpoint Recovery in Budding Yeast

There are several reasons why a mutant can be sensitive to microtubule depolymerising drugs. The original spindle checkpoint deletions mutants are sensitive due to their inability to arrest in response to these drugs. Other mutants may be sensitive to microtubule drugs due to an inability to silence the spindle checkpoint and recover from the arrest. Thirdly, some mutants may be able to arrest, and to release from the arrest, but fail to release in an appropriate manner. For example, they may display chromosome mis-segregation upon release. In this section we describe assays that can be used to distinguish such phenotypes.

3.1.5.1. Analysis of the Ability of Cells to Recover from a Spindle Checkpoint Arrest

To analyse whether cells have the ability to silence a prolonged spindle checkpoint arrest, Pds1 levels can be monitored as cells are recovering from the arrest.

1. The cells are first synchronised in G1 and then arrested with microtubule depolymerising drugs as described in **Section 3.1.1**
2. The nocodazole is carefully removed, and α-factor is added immediately to prevent entry into the next cell cycle.
3. Samples are taken from the time of release and then every 10 min for one hour. Levels of Pds1 monitored by immunoblotting as described above.

3.1.6. Analysis of Chromosome Segregation Following Spindle Damage

The fidelity of chromosome segregation upon spindle checkpoint recovery can be monitored by analysing GFP-marked sister chromatids (*see* **Section 3.1.2**) and mitotic spindles during anaphase. Segregation defects are easily identified, and quantified, as daughter cells containing two GFP foci (*see* **Fig. 15.2**).

1. Cells are synchronised in G1 using α-factor (as described above), released into YPDA media containing 20 μg/ml

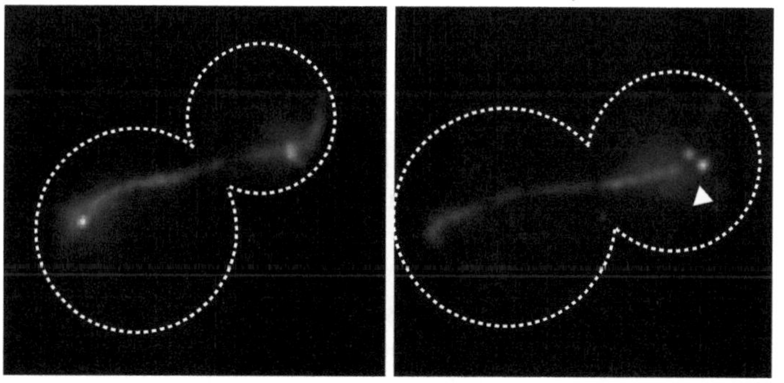

Fig. 15.2. Analysis of chromosome segregation during nocodazole release. Budding yeast cells are analysed by fluorescence microscopy upon the recovery from a mitotic arrest induced by microtubule depolymerising drugs. Cells containing anaphase spindles are analysed and scored for either correct sister-chromatid segregation (*left panel*), or chromosome missegregation as judged by both sister-chromatids segregating to the same daughter cell (two GFP foci, *right panel*, white arrow head).

nocodazole, and incubated at 23°C for three hours to allow for the spindle to disassemble and the cells to arrest.

2. The microtubule drugs are then carefully washed out from the media (as described above) and the cells are released into YPDA media.
3. The cells are then incubated, shaking, at 30°C for 30 min to allow for spindle assembly and anaphase to occur.
4. The cells are then fixed in 3.7% formaldehyde, by adding 1 ml of 37% formaldehyde (Sigma) to 9 ml of cell culture, and incubated at room temperature for 1 h, then washed with 0.1 M potassium phosphate (pH 7.5).
5. To permeabilise the cell wall, the cells are treated with 50 μg/ml Zymolyase 100,000 in 0.1 M potassium phosphate/0.7 M sorbitol for 30 min at 30°C.
6. The permeabilised cells are then collected by centrifugation at 1000 g and resuspended in 1 ml PBS with 0.7 M sorbitol.
7. Multiwell microscope slides are prepared for immunofluorescence by pipetting 10 μl 0.1% polylysine (Sigma) onto the wells for 1 min, then removing the polylysine with an aspirator and letting the slides dry. Cells will stick to the polylysine.
8. 10 μl of cells are then added to the wells. After 3 min the suspension is aspirated off and the wells are washed three times with 15 μl PBS with 0.7 M sorbitol.
9. To further fix and to flatten the cells, the slides are immersed in ice-cold methanol for 5 min and then in ice-cold acetone for 30 s, then left to dry.
10. To block unspecific binding sites, 15 μl of blotto is added to each well and incubated in a humid chamber for 30 min.
11. The slides are then washed three times with PBS.
12. Primary antibodies (anti-GFP and anti-tubulin, YOL1/34, to detect mitotic spindles) are added to blotto, and 15 μl of this is added to each well and incubated in a humid chamber at 4°C overnight.
13. Steps 11 and 12 are then repeated with secondary antibodies
14. A small drop of Vectashield containing DAPI (Vector) is then added to each well and a coverslip placed on top.
15. The distribution of the GFP spots can then be analysed in cells with anaphase spindles.

Alternatively, a strain containing CFP-tubulin can be used, and the segregation of the GFP chromosome analysed by live-cell imaging.

3.1.7. Analysis of Chromosome Biorientation of Budding Yeast Chromosomes

The ability of chromosomes to achieve biorientation is crucial for chromosome segregation. Budding yeast metaphase chromosomes transiently separate their centromeres, due to the forces of the mitotic spindle upon biorientation *(13)*. This has been called centromeric breathing, and it can be visualised by microscopy when using a strain that carries an array of *tet*-operators integrated close to a centromere (in the JF152 strain they are 2 kb from the CEN on chromosome IV) and expresses a Tet-repressor-GFP fusion protein. This strain also has the spindle pole bodies (SPBs) marked with Spc42-tomato, in addition to a repressible promoter for *CDC20 (pMET-CDC20)*. Depletion of Cdc20, through addition of methionine, causes a metaphase arrest that is independent of microtubule drugs or spindle checkpoint activity. This is because the APC/C cannot be activated in mitosis if Cdc20 is depleted. With this strain, the ability of cells to biorient is directly visualised by red/green fluorescence microscopy, either each cell cycle or after spindle damage. Here we describe a procedure, based on methods from *(14)*, that analyses the ability of budding yeast cells to biorient their chromosomes on a metaphase spindle following spindle damage.

1. Budding yeast cells are grown overnight in CSM-Met media, diluted back to OD_{600} 0.2, then arrested in G1 with α-factor in medium lacking methionine (CSM-Met) for 3 h.
2. Cells are then transferred to YPDA media with 8 mM methionine, to deplete Cdc20, and α-factor and incubated for 2 h at 30°C.
3. The α-factor is washed out and cells are then incubated at 23°C for 3 h in YPDA media containing 30 μg/ml benomyl and 20 μg/ml nocodazole, to depolymerise microtubules, and 8 mM methionine to deplete Cdc20.
4. The microtubule drugs are then carefully washed out, and the spindle is allowed to reform by incubating the cells in YPDA media at 30°C, with 8 mM methionine to maintain the metaphase arrest.
5. 4 mM methionine is added back every hour to prevent anaphase. The cells are fixed at the indicated times in 3.7% formaldehyde for 5 min, then washed once in PBS and stored in a small volume of PBS at 4°C.
6. To analyse biorientation the GFP-dots are scored as one or two GFP dots in cells that have a short bipolar spindle (i.e., two red SPBs, *see* **Fig. 15.3**). *See* **Note 5.**

3.2. Assays for Spindle Checkpoint Arrest in Fission Yeast

Cellular pools of securin (Cut2) are not completely degraded in fission yeast *(15)*, so its degradation upon anaphase onset is difficult to monitor by immunoblotting. Instead its levels, and those of the mitotic cyclin (Cdc13), are best monitored at the single cell level, using fluorescent tags such as cdc13-GFP and cut2-GFP. This analysis is aided by the fact that both proteins are enriched on mitotic spindles. Alternatively, in vitro H1 kinase assays can be used to measure the levels of active Cdc2 kinase. Microtubule-

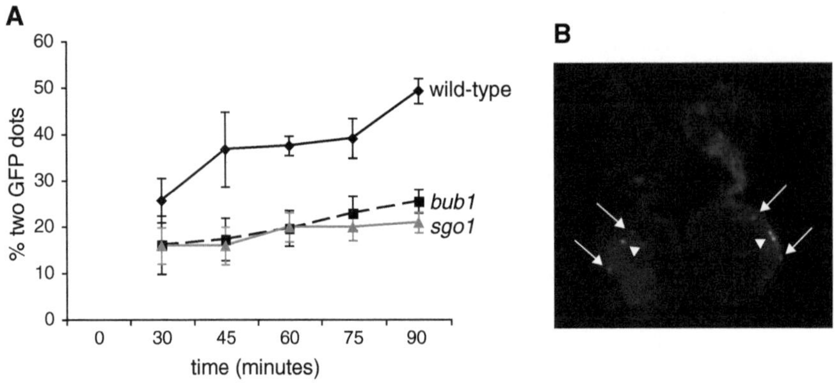

Fig. 15.3. Analysis of chromosome biorientation following spindle damage in budding yeast. The efficiency of sister-chromatid biorientation is scored by analysing cells containing a bipolar spindle (these have two separated SPBs, labelled with tdTomato-Spc42, *white arrows*) and determining whether their GFP marked sister-chromatids are in one or two GFP foci (white *arrow heads*). In the experiment shown, both the *bub1*Δ*kinase* and the *sgo1* mutant display a clear defect in biorientation.

destabilising drugs, such as nocodazole, benomyl, thiabendazole, or carbendazim, have limited ability to arrest liquid cultures of fission yeast. Where necessary we would recommend 25 μg/ml carbendazim addition to liquid cultures (from a fresh 5 mg/ml stock in DMSO). Several drugs are effective in solid media, to test for anti-microtubule drug sensitivity. We recommend benomyl (~6–12 μg/ml in YES, with yeast grown on plates for 2–3 days at 30°C). Thiabendazole is popular, but also perturbs the actin cytoskeleton *(16)*.

The most popular assays for mitotic arrest in fission yeast are the response of cells due to specific, conditional mitotic mutations.

3.2.1. The Response to Unattached Kinetochores: Nda3-KM311 (Cold-Sensitive β Tubulin Mutant)

At their restrictive temperature *nda3-KM311* mutants have no microtubules and are therefore unable to build a mitotic spindle *(17)*. Cells are grown at their permissive temperature (32°C) and then switched to 18°C to induce a mitotic arrest. For an efficient *nda3* arrest one should use a fresh yeast stock, and shift the cells at a relatively low density (~0.3–0.5). The arrest can be monitored by chromosome condensation, which becomes very pronounced after 6–8 h at 18°C, and lack of septation (scored with calcofluor staining). Cells which fail to maintain the mitotic arrest frequently septate with unevenly segregated chromosomes, which can be monitored by DAPI staining. Sometimes the septa physically cut through the nucleus producing the classic Cells Untimely Torn (*cut*) phenotype. *Checkpoint, nda3* double mutants die much faster then the *nda3* single mutant, when tested for colony-forming ability through an 8–10 h time-course at 18°C (*see (18)* for further experimental details, and *see* **Note 6**).

3.2.2. The Response to Deformed Spindles: Cut7-24 Mutant

Cut7 is a kinesin-related microtubule motor protein required for bipolar spindle assembly. *Cut7-24* mutants form V-shaped spindles at their restrictive temperature, having failed to interdigitate microtubules from their two half-spindles *(19)*. They typically arrest as overcondensed chromosomes and septated cells. The arrest is frequently monitored by FACS analysis, where the *cut7* single mutant maintains a 2C peak for several hours, but *cut7, checkpoint* double mutants re-replicate their chromosomes in the absence of cytokinesis producing 4C cells *(18)*.

3.2.3. The Response to Kinetochores with Reduced Cohesion ('Lack of Tension' Assays): Mis4-242 (Cohesin Mutant)

Because mitosis is only extended by a few minutes in a *mis4-242* strain, single-cell movies are needed to monitor the effect of reduced cohesion. Using intense Bub1-GFP association with kinetochores as a mitotic marker, mitosis was extended from 2 min (wild-type) to around 11 min in the *mis4* mutant *(20)*, and this delay would be abolished in checkpoint mutants. Note that checkpoint mutants may perturb Bub1 localisation directly *(21)*.

3.2.4. psc3-1T (Another Temperature-Sensitive Cohesion Mutant)

A similar assay was recently described *(22)* using Cut2-GFP and Sad1-GFP (spindle pole body marker) to monitor prometaphase duration (the time from SPB separation to Cut2-GFP disappearance) in *psc3-1T* mutants. This is extended in 40% of cells by up to 100 min (extended from 22 to 45 min on average).

3.2.5. Chromosome Segregation Assays

Fission yeast is extremely well suited to live-cell analysis of mitosis and chromosomne segregation. Here we simply highlight a few key reports in the literature. For a general description of live-cell GFP studies in fission yeast and in particular the duration of different mitotic phases see *(23)* and *(24)*.

Fission yeast has 3–4 microtubule attachment sites per kinetochore *(25)*, compared to only one in budding yeast. This added complexity means that cells must co-ordinate these microtubule-binding sites, such that they attach to the same spindle pole, otherwise merotely will occur. Many kinetochore and some checkpoint mutants display 'lagging chromosomes' on anaphase spindles *(26, 27)*. This is often interpreted as being due to merotelic attachment of kinetochores, leading to a 'tug-of-war' on the spindle, with a single sister chromatid being pulled to both spindle poles. Stretching and splitting of single kinetochores was recently visualised during live-cell analysis of lagging chromosomes in both heterochromatin and kinetochore mutants *(28)*. Many of the above are well suited to quantitative analysis, and GFP marked chromosomes (e.g., cen2-GFP *(29)*) are now very popular in this regard, as one can simply score the presence of nuclei containing two GFP-marked chromosomes within binucleate cells. Colony sectoring assays are also frequently used to quantitatively monitor the loss rates of artificial marker chromosomes (both circular, e.g., CM3112 *(30)* and linear, e.g., Ch16 *(31)*). These rely on the pink/red colony colour phenotypes of different *ade6* alleles. Overall the relative disadvantages of fission yeast studies, due to poor synchronisation of the cell cycle, are balanced by improved cytology and an ability to study complex segregation defects, such as merotelic attachments.

4. Notes

1. Cells expressing the wild-type Bar1 protease can escape from α-factor induced G1 arrest. Readdition of α-factor after 90 min will help prevent this, and allows for a more efficient arrest.
2. Protein extracts can also be prepared by addition of silica beads and protein sample buffer straight to the snap-frozen protein pellet, followed by bead beating. However, the Pds1 protein is degraded to a lesser extent when extracted following TCA precipitation, and it is then easier to compare protein levels between different strains.
3. Care should be taken here as the cell pellet may be difficult to see.
4. Because the length of the metaphase delay in this 'lack of tension' assay may vary between mutants, any differences between yeast strains are easier to detect when taking 10 min timepoints following the release from G1.
5. Live cell analysis can be performed for more detailed studies of chromosome biorientation and segregation (*see (14)* and Chapter 9, Tanaka).
6. Loss of viability in *nda3* arrest: it is important to note that whilst checkpoint mutations will lead to a rapid loss of viability in the *nda3* background (when compared to the *nda3* single mutant), this is not in itself sufficient to demonstrate a checkpoint function. It is possible for a double mutant to arrest reasonably efficiently, yet die rapidly during release/recovery from the checkpoint arrest. Examples of such a phenotype include *sgo2* mutants and *bub1* kinase mutants *(32, 22, 10)*. Therefore, the arrest should also be analysed morphologically (mitotic cdc13-GFP levels, chromosome condensation, *cut* phenotype, septation).

References

1. Hoyt, M. A., Totis, L., and Roberts, B.T. (1991) *S.cerevisiae* genes required for cell cycle arrest in response to loss of microtubule function. *Cell* **66,** 507–17.
2. Li, R., and Murray, A. W. (1991) Feedback control of mitosis in budding yeast. *Cell* **66,** 519–31.
3. Straight, A. F., and Murray, A. W. (1997) The spindle assembly checkpoint in budding yeast. *Methods Enzymol* **283,** 425–40.
4. Hwang, L. H., Lau, L. F., Smith, D. L., Mistrot, C. A., Hardwick, K. G., Hwang, E. S., Amon, A., and Murray, A. W. (1998) Budding yeast Cdc20: a target of the spindle checkpoint. *Science* **279,** 1041–4.
5. Kim, S. H., Lin, D. P., Matsumoto, S., Kitazono, A., and Matsumoto, T. (1998) Fission yeast Slp1: an effector of the Mad2-dependent spindle checkpoint. *Science* **279,** 1045–7.
6. Longtine, M. S., McKenzie, A., 3rd, Demarini, D. J., Shah, N. G., Wach, A., Brachat, A., Philippsen, P., and Pringle, J. R. (1998) Additional modules for versatile and economical PCR-based gene deletion and modification in Saccharomyces cerevisiae. *Yeast* **14,** 953–61.
7. Straight, A. F., Belmont, A. S., Robinett, C. C., and Murray, A. W. (1996) GFP tagging of budding yeast chromosomes reveals that protein-protein interactions can mediate sister chromatid cohesion. *Curr Biol* **6,** 1599–608.
8. Biggins, S., and Murray, A. W. (2001) The budding yeast protein kinase Ipl1/Aurora allows the absence of tension to activate the spindle checkpoint. *Genes Dev* **15,** 3118–29.
9. Indjeian, V. B., Stern, B. M., and Murray, A. W. (2005) The centromeric protein Sgo1 is required to sense lack of tension on mitotic chromosomes. *Science* **307,** 130–3.
10. Fernius, J., and Hardwick, K. G. (2007) Bub1 kinase targets Sgo1 to ensure efficient chromosome bi-orientation in budding yeast mitosis. *PLoS Genetics* **3,** e213.
11. Pinsky, B. A., Kung, C., Shokat, K. M., and Biggins, S. (2006) The Ipl1-Aurora protein kinase activates the spindle checkpoint by creating unattached kinetochores. *Nat Cell Biol* **8,** 78–83.
12. Pinsky, B. A., Tatsutani, S. Y., Collins, K. A., and Biggins, S. (2003) An Mtw1 complex promotes kinetochore biorientation that is monitored by the Ipl1/Aurora protein kinase. *Dev Cell* **5,** 735–45.
13. He, X., Asthana, S., and Sorger, P. K. (2000) Transient sister chromatid separation and elastic deformation of chromosomes during mitosis in budding yeast. *Cell* **101,** 763–75.
14. Indjeian, V. B., and Murray, A. W. (2007) Budding yeast mitotic chromosomes have an intrinsic bias to biorient on the spindle. *Curr Biol.* 17, 1837–46.
15. Funabiki, H., Yamano, H., Kumada, K., Nagao, K., Hunt, T., and Yanagida, M. (1996) Cut2 proteolysis required for sister-chromatid seperation in fission yeast. *Nature* **381,** 438–41.
16. Sawin, K. E., and Snaith, H. A. (2004) Role of microtubules and tea1p in establishment and maintenance of fission yeast cell polarity. *J Cell Sci* **117,** 689–700.
17. Hiraoka, Y., Toda, T., and Yanagida, M. (1984) The nda3+ gene of fission yeast encodes beta-tubulin: a cold-sensitive nda3 mutation reversibly blocks spindle formation and chromosome movement in mitosis. *Cell* **39,** 349–58.
18. Millband, D. N., and Hardwick, K. G. (2002) Fission yeast Mad3p is required for Mad2p to inhibit the anaphase-promoting complex and localises to kinetochores in a Bub1p, Bub3p and Mph1p dependent manner. *Mol Cell Biol* **22,** 2728–42.
19. Hagan, I., and Yanagida, M. (1990) Novel potential mitotic motor protein encoded by the fission yeast cut7+ gene. *Nature* **347,** 563–66.
20. Toyoda, Y., Furuya, K., Goshima, G., Nagao, K., Takahashi, K., and Yanagida, M. (2002) Requirement of chromatid cohesion proteins Rad21/Scc1 and Mis4/Scc2 for normal spindle-kinetochore interaction in fission yeast. *Curr Biol* **12,** 347–58.
21. Vanoosthuyse, V., Valsdottir, R., Javerzat, J. P., and Hardwick, K. G. (2004) Kinetochore targeting of fission yeast Mad and Bub proteins is essential for spindle checkpoint function but not for all chromosome segregation roles of Bub1p. *Mol Cell Biol* **24,** 9786–801.
22. Kawashima, S. A., Tsukahara, T., Langegger, M., Hauf, S., Kitajima, T. S., and Watanabe, Y. (2007) Shugoshin enables tension-generating attachment of kinetochores by loading Aurora to centromeres. *Genes Dev* **21,** 420–35.
23. Nabeshima, K., Saitoh, S., and Yanagida, M. (1997) Use of green fluorescent protein for intracellular protein localization in living fission yeast cells. *Methods Enzymol* **283,** 459–71.
24. Nabeshima, K., Nakagawa, T., Straight, A. F., Murray, A., Chikashige, Y., Yamashita,

Y. M., Hiraoka, Y., and Yanagida, M. (1998) Dynamics of centromeres during metaphase-anaphase transition in fission yeast: Dis1 is implicated in force balance in metaphase bipolar spindle. *Mol Biol Cell* **9,** 3211–25.
25. Ding, R., West, R. R., Morphew, D. M., Oakley, B. R., and McIntosh, J. R. (1997) The spindle pole body of Schizosaccharomyces pombe enters and leaves the nuclear envelope as the cell cycle proceeds. *Mol Biol Cell* **8,** 1461–79.
26. Bernard, P., Hardwick, K., and Javerzat, J. P. (1998) Fission yeast bub1 is a mitotic centromere protein essential for the spindle checkpoint and the preservation of correct ploidy through mitosis. *J Cell Biol* **143,** 1775–87.
27. Pidoux, A. L., Uzawa, S., Perry, P. E., Cande, W. Z., and Allshire, R. C. (2000) Live analysis of lagging chromosomes during anaphase and their effect on spindle elongation rate in fission yeast. *J Cell Sci* **113 Pt 23,** 4177–91.
28. Gregan, J., Riedel, C. G., Pidoux, A. L., Katou, Y., Rumpf, C., Schleiffer, A., Kearsey, S. E., Shirahige, K., Allshire, R. C., and Nasmyth, K. (2007) The kinetochore proteins Pcs1 and Mde4 and heterochromatin are required to prevent merotelic orientation. *Curr Biol* **17,** 1190–200.
29. Ding, D. Q., Yamamoto, A., Haraguchi, T., and Hiraoka, Y. (2004) Dynamics of homologous chromosome pairing during meiotic prophase in fission yeast. *Dev Cell* **6,** 329–41.
30. Matsumoto, T., Murakami, S., Niwa, O., and Yanagida, M. (1990) Construction and characterization of centric circular and acentric linear chromosomes in fission yeast. *Curr Genet* **18,** 323–30.
31. Niwa, O., Matsumoto, T., Chikashige, Y., and Yanagida, M. (1989) Characterization of *Schizosaccharomyces pombe* minichromosome deletion derivatives and a functional allocation of their centromere. *Embo J* **8,** 3045–52.
32. Vanoosthuyse, V., Prykhozhij, S. and Hardwick, K.G. (2007) Shugoshin 2 regulates localisation of the chromosomal passenger proteins in fission yeast mitosis. *Mol Biol Cell* **18**, 1657–69.

Chapter 16

Measuring Proteolysis in Mitosis

Catherine Lindon and Barbara Di Fiore

Abstract

The targeted destruction of key regulators helps to drive the cell cycle. Here we describe a quantitative assay to measure destruction of different regulators in mitotic cells. This assay uses GFP-tagged substrates and time-lapse fluorescence microscopy of single cells to pinpoint the timing of destruction of different substrates at different stages in mitosis.

Key words: GFP, protein destruction, ubiquitin, microinjection, fluorescence time-lapse imaging.

1. Introduction

The targeted destruction of key regulators via the ubiquitin–proteasome pathway is one of the mechanisms that ensures orderly and irreversible passage between phases of the cell cycle *(1, 2)*. These transitions occur very rapidly during mitosis (on the order of minutes), such that the different phases of mitosis are not easily accessible to biochemical analysis. Developing quantitative assays for measuring events occurring in single cells is therefore essential for studying mitosis. Many mitotic regulators are known targets for proteolysis as cells enter and exit mitosis, but with few exceptions *(3–8)* the exact timing of their destruction is not known. However, this information can be critical to understanding how destruction is regulated, and with which mitotic events it is associated. For example, pioneering work from Jonathon Pines' group used live cell fluorescence assays to measure the destruction of cyclin B1, a critical target of the Anaphase Promoting Complex/Cyclosome (APC/C) ubiquitin ligase. Onset of cyclin

Andrew D. McAinsh (ed.), *Mitosis: Methods and Protocols*, Methods in Molecular Biology *545*,
DOI 10.1007/978-1-60327-993-2_16, © Humana Press, a part of Springer Science+Business Media, LLC 2009

B1 destruction was found to occur at precisely the moment that the spindle checkpoint was satisfied *(3)*, a result that has been of major importance in developing models of how the mitotic checkpoint works, and in refining our understanding of metaphase *(9)*.

This chapter describes an assay for measuring the proteolysis of substrates in single cells. The theory behind the assay is very simple: GFP-tagged versions of substrates are targeted at the same time and place as the endogenous protein. After poly-ubiquitination, the GFP-substrate is targeted to the 26S proteasome, where it is unfolded, passed into the barrel of the 20S proteasome, and destroyed. The loss of fluorescence that accompanies this destruction can be measured as a readout for the destruction of the protein. There are some assumptions inherent in this assay (*see* **Note 1**) that can limit the interpretation of data, but experience has shown that it is possible to obtain reliable information about timing and regulation of proteolysis, and about relative rates of proteolysis of different substrates (by simultaneously measuring substrates tagged with spectral variants of GFP; *see* **Note 2**). Much of this information can only be obtained by imaging live cells, because measurements from fixed cells cannot capture the timing of dynamic processes, nor take into account stochastic variation in protein levels between individual cells.

The assay was first developed for use in HeLa cells, which are readily synchronised and do not arrest before mitosis following nuclear injection with plasmid DNA. Here we describe an enhanced version of the assay, schematically illustrated in **Fig. 16.1**, that we have adapted for use with mRNAs in cytoplasmic injection. This assay performs better than nuclear injection assays in HeLa cells, and we have found that it is also suitable for use on more sensitive cells, such as the nontransformed cell line, hTERT-RPE1 (*see* **Note 3**). In principle, this assay can be used with any adherent cell type. After cytoplasmic microinjection with mRNA, cells are filmed by a combination of Differential Interference Contrast (DIC) and fluorescence microscopy. Fluorescence imaging allows us to monitor the localisation of the GFP-substrate (to confirm that it behaves as a suitable marker for the endogenous substrate), and to quantify the GFP fluorescence in the cell. Plotted against time, this produces

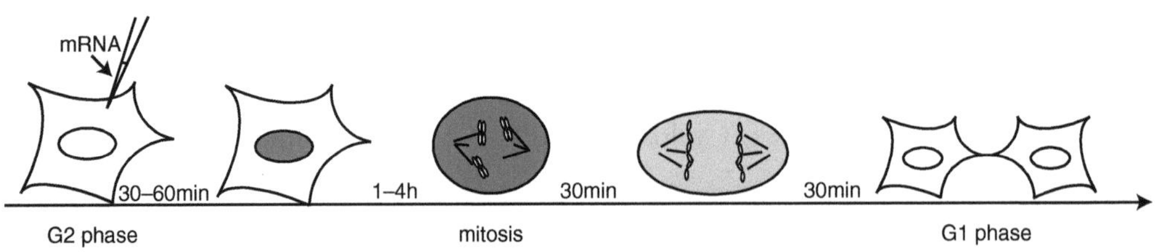

Fig. 16.1. Schematic representation of a single-cell proteolysis assay. The timings of the different steps are indicated. *Shading* represents GFP.

a degradation curve. DIC imaging allows optimal visualisation of different events in mitosis (chromosome condensation, nuclear envelope breakdown (NEBD), congression of chromosomes to the metaphase plate, segregation of chromosomes at anaphase), and when compared with the degradation curve, allows new insight into the timing and regulation of proteolysis at mitosis (*see* **Note 4**).

2. Materials

2.1. Cell Culture and Synchronisation

1. Delta T® glass-bottom tissue culture dishes (Bioptechs, Butler, PA).
2. HeLa cells.
3. Gibco® 10 × Trypsin-EDTA mix (Invitrogen Corporation, Carlsbad, CA) diluted to 1 × working solution (0.05% Trypsin, 0.5 mM EDTA) in phosphate-buffered saline (PBS).
4. Gibco Advanced Dulbecco's Modified Eagle's Medium (Advanced DMEM; *see* **Note 5**; Invitrogen) supplemented with 2% heat-inactivated Fetal Bovine Serum (FBS; PAA Laboratories GmbH, Pasching, Austria), 2 mM Gibco GlutaMAX™-1 and 1:100 Gibco Penicillin-Streptomycin mix of antibiotics (final concentrations 100 U/ml penicillin, 100 μg/ml streptomycin; Invitrogen).
5. 100 mM thymidine (Sigma-Aldrich, St Louis, MO) made up in water and sterilised by passing through a single-use 0.2 μm filter unit (Sartorius AG, Goettingen, Germany).
6. Aphidicolin dissolved at 5 mg/ml in dimethylsulfoxide (DMSO; both from Sigma).

2.2. Preparation of mRNA Encoding GFP-Tagged Substrate

1. Plasmid construct for expression of the substrate of interest tagged to GFP. A T7 RNA Polymerase promoter is required upstream of the coding sequence (*see* **Note 6**).
2. mMESSAGE mMACHINE® T7 Ultra Kit (Ambion, Austin, TX).
3. RNase-free filtered tips and microfuge tubes (Ambion).
4. DEPC-treated water.

2.3. Cell Injection and Microscopy

1. Diamond-tip pen.
2. Gibco Leibovitz L-15 medium (Invitrogen) supplemented with 10% FBS and antibiotics (*see* **Section 2.1, step 4**).
3. mRNA (prepared as in **Section 3.1**) at 0.5–1 mg/ml in DEPC-treated water (*see* **Note 7**).
4. Inverted fluorescence microscope (such as the Leica DM IRBE pictured in **Fig. 16.2**) fitted with the following features (*see also* **Note 8**):

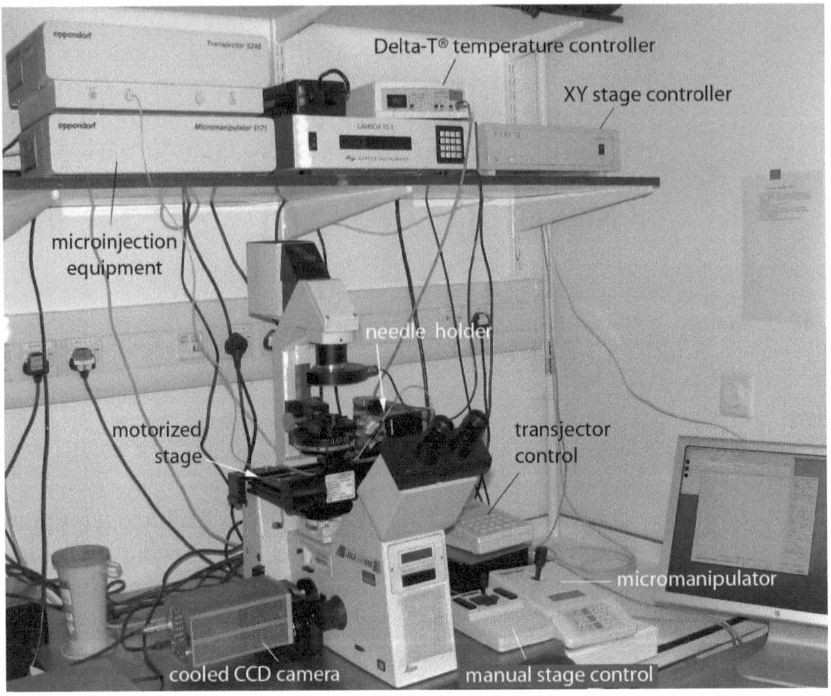

Fig. 16.2. Leica DMIRBE microscope fitted for microinjection and time-lapse imaging as described in **Section 2.3, step 4**.

- Eppendorf micromanipulator and transjector apparatus for microinjection of cells (Eppendorf AG, Hamburg, Germany).
- DIC optics.
- 40X objective with numerical aperture (NA) of about 1.0.
- Delta T stage adapter and temperature controller (Bioptechs).
- Cooled CCD camera.
- Acquisition software. We use Slidebook (Intelligent Imaging Innovations, Denver, CO) or Scanalytics IP Lab (BD Biosciences, Rockville, MD).

5. Microloaders (tips) and Femtotips® (needles) for injections (Eppendorf).

2.4. Analysis

1. Image analysis software such as Image J (available at http://rsb.info.nih.gov/ij/).
2. Data analysis software such as Microsoft Office Excel.

3. Methods

We describe how to prepare G2 cells and inject them with mRNA transcribed from plasmid templates encoding the substrate of interest. We find that this represents the most flexible approach, one that can be readily adapted to experimentation with different substrates. However, *see* **Note 9** for discussion of the possible alternatives to microinjection.

3.1. Preparing mRNA for Injection

The protocol for the mMESSAGE mMACHINE T7 Ultra Kit from Ambion can be downloaded at http://www.ambion.com/techlib/prot/fm_1345.pdf and is briefly as follows.

1. Linearise the plasmid using a restriction site downstream of the coding sequence, verify on an agarose gel that restriction is complete, and precipitate the DNA.
2. Resuspend the DNA at a concentration of 0.5 mg/ml, and set up the transcription reaction and subsequent polyA-tailing reaction using filter tips and sterile Eppendorf tubes as detailed in the manual.
3. Extract the mRNA with phenol/chloroform and precipitate with isopropanol.
4. Resuspend the mRNA in 30 μl of nuclease-free water.
5. Measure the concentration of mRNA using a spectrophotometer (the usual yield is 1–3 mg/ml), aliquot and freeze at –20°C. Each aliquot should be used only once.

3.2. Synchronising Cells in G2 Phase

Here we describe the synchronisation procedure that we use for HeLa cells. We have also developed a synchronisation procedure for hTERT-RPE1 cells (*see* **Note 3**).

1. HeLa cells are grown throughout the 5-day synchronisation procedure in supplemented Advanced DMEM culture medium at 37°C in 10% CO_2.
2. Harvest HeLa cells when approaching confluence, using trypsin/EDTA, and seed onto Delta T dishes at 3.6×10^4 cells per dish (*Day 1*). The Delta T dishes can be kept without their lids on, inside 60 mm tissue culture dishes.
3. The following day (*Day 2*) add 2.5 mM thymidine to the culture medium and incubate for 24 h.
4. Release from thymidine arrest (*Day 3*) by rinsing dishes twice with prewarmed medium. Add back culture medium.
5. After a further 12 h, add 5 μg/ml aphidicolin to the dishes and return cells to the incubator for 24 h.
6. Release from aphidicolin arrest (*Day 4*) by rinsing 3 times with prewarmed medium. Add back culture medium.
7. 9–10 h after release from aphidicolin (*Day 5*), cells will be in late G2 phase and ready for microinjection.

3.3. Injecting Cells and Time-Lapse Microscopy

1. Dilute the mRNA for injection if required (using DEPC-treated water), and spin for 10 min at top speed in a bench-top microfuge at 4°C to remove any insoluble matter.

2. Discard cell culture medium from a Delta T dish of G2 phase cells and gently mark the top surface of the dish with the diamond pen in order to mark the future site of injected cells.
3. Replace the culture medium with supplemented L-15 medium prewarmed to 37°C, and transfer the dish onto the heated stage of the microscope, set to 37°C.
4. Load a Femtotips needle with 1 ul of mRNA using an Eppendorf microloader and attach the needle to the microinjector. Set the back-pressure to about 40 (units) to prevent medium from entering the needle as it is lowered into the dish.
5. Focus on cells in the scored region under the microscope, and using the micromanipulator, gently lower the needle until it is in the field of vision.
6. Set the injection height (Z-limit) to that at which the needle is pressing down on the cell surface, in a region close to but not overlying the nucleus, without rupturing the plasma membrane. Providing the whole field of cells is in focus, it should be possible to inject several cells at a time without adjusting the Z-limit (*see* **Note 10**). Typical injection parameters are 100 hPa (1.5 PSI) for 0.4 s, but these should be adjusted so that on injection, a small wave is seen spreading from the injection site. This wave corresponds to an increase in volume of the cell of approximately 1/10th.
7. Inject 200–300 cells for a single experiment. Expect to obtain data from up to 30 cells using a motorised stage (*see* **Note 8**).
8. Cover dish with the heated lid to prevent evaporation of medium and wait for GFP-tagged substrate to become visible (30 min–1 h).
9. Find GFP-expressing cells and set-up time-lapse imaging for 5–6 h, using a 40× objective. Image DIC and GFP fluorescence in a single plane. As long as the objective does not have a high NA (and therefore a narrow depth of field), small alterations in the focus of the fluorescence (such as that which occurs when cells round up at mitosis) will not influence the quantification of fluorescence. Imaging every 2–4 min with exposure times of between 50 and 200 ms is optimal in terms of obtaining good data whilst minimising perturbation to the cells at mitosis. Do not be tempted to bin pixels during acquisition, even when the fluorescence is weak. A signal of only 20 pixels above background is enough to measure degradation. Save data for subsequent analysis as 12- or 16- bit images.

3.4. Analysis of Protein Levels

1. Assemble images in a series using image analysis software such as ImageJ. Draw a Region Of Interest (ROI) around

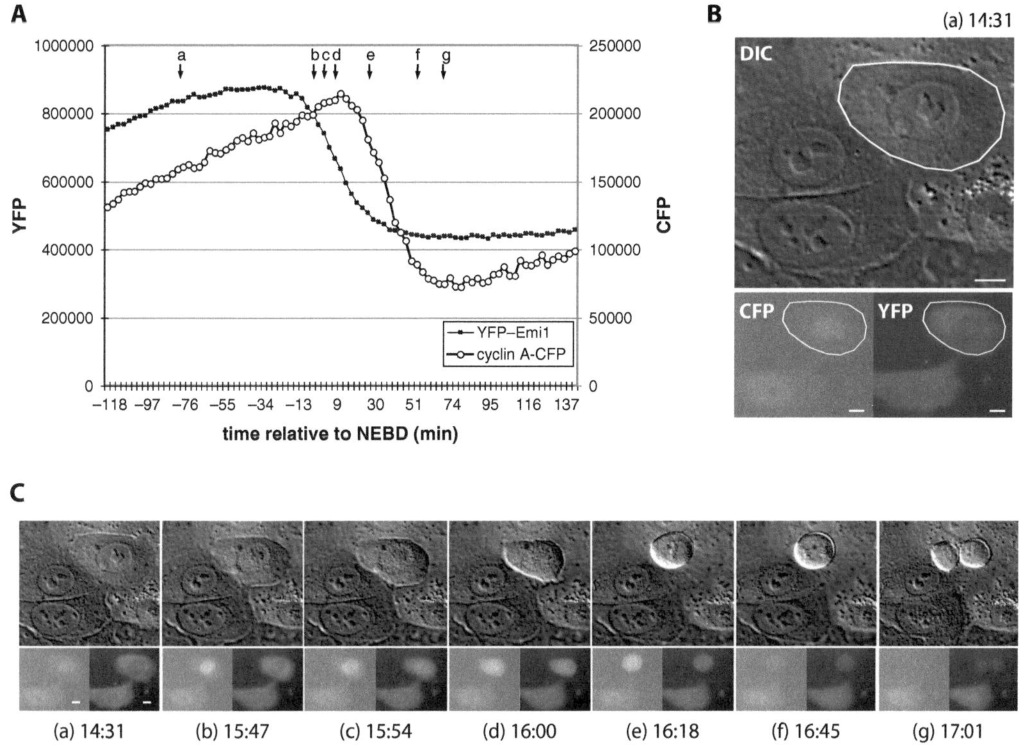

Fig. 16.3. Example of a proteolysis assay used to compare the targeting of two substrates. Proteolysis of the APC/C inhibitor Emi1, targeted by the SCF ubiquitin ligase during mitotic entry *(14, 15)*, is compared with the proteolysis of an early APC/C substrate, cyclin A *(4)*. The close timing of these two events does not allow them to be resolved by immunoblot analyses of cell populations. However, single-cell assays reveal that proteolysis of Emi1 is not sufficient for activation of the APC/C *(8)*. (**A**) Degradation curves for YFP-Emi1 and cyclin A-CFP in a single cell. Images acquired at 3 min intervals. (**B**) DIC (*top panel*) and CFP/YFP (*bottom panels*) images at the start of the time-lapse series, showing the ROI applied for quantifying fluorescence. (**C**) Selected images from the time-lapse series showing some of the major events of mitosis visible by DIC. These have been correlated with the degradation curves shown in (**A**). *Bars*, 10 μm.

the cell that is large enough to contain the boundaries of the cell throughout the series, but that does not contain any unnecessary background, and apply to all of the images in the series (**Fig. 16.3**).

2. Measure the average pixel values within the ROI (F) and the area of the ROI (A). Select a second ROI in a region where there are no GFP-expressing cells and measure the average pixel value within this ROI to give background fluorescence (F*).
3. Transfer all data into a spreadsheet application such as Microsoft Excel to calculate the total cell fluorescence at each timepoint as AF – AF*. Plot total cell fluorescence over time to give degradation curves such as those shown in **Figs. 16.3** and **16.4**.

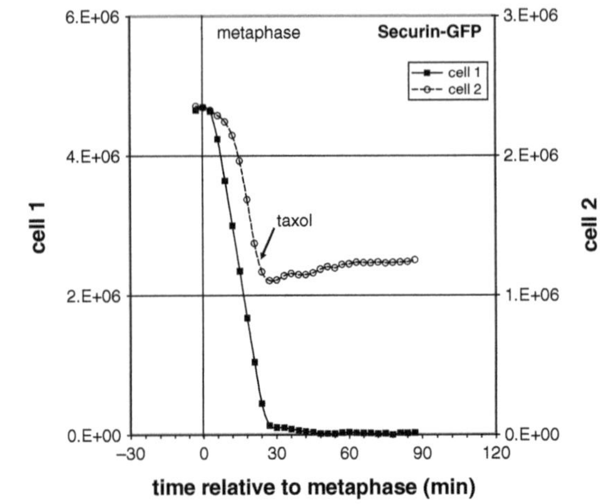

Fig. 16.4. Example of a proteolysis assay used to examine the regulation of proteolysis. Proteolysis of the APC/C substrate securin begins at metaphase once the spindle checkpoint is satisfied. In this experiment, proteolysis of securin-GFP in a control cell is allowed to go to completion (cell 1). Taxol is then added to the dish during filming of a second cell (cell 2), to show that proteolysis of this substrate is arrested by reactivation of the checkpoint during metaphase *(5)*. Images acquired every 3 min. Data provided by Dr. Anja Hagting.

3.5. Analysis of Degradation Timing

1. Establish the timing of degradation onset by reference to the sequence of DIC images of a mitotic cell, showing chromosome condensation, NEBD, chromosome congression to the metaphase plate, anaphase, cytokinesis, chromosome decondensation, and nuclear envelope reformation (**Fig. 16.3**).
2. Compare the onset of degradation of different substrates (**Fig. 16.3**).
3. Use degradation assays to analyse how substrate targeting is regulated; for example, after challenging the spindle checkpoint by treatment of mitotic cells with a microtubule poison such as taxol (**Fig. 16.4**).

4. Notes

1. This assay depends on the assumptions (a) that GFP-tagged substrates behave in a similar fashion to the endogenous substrate (b) that injection or overexpression of a particular substrate does not influence its pattern of degradation, and (c) that there is negligible synthesis of new protein during the time of the assay. With regard to these assumptions: (a) This assay should only be used for GFP-substrates

that show the same localisation and behaviour at mitosis as the endogenous substrate. In general, injecting mRNA into G2 phase cells maximises the chances of exogenously introduced substrates being incorporated into mitotic complexes (compared to, e.g., injecting proteins directly into mitotic cells). (b) Expression levels can be controlled by microinjection (*see* **Note 7**). Typical levels of expression achieved are 0.5–10× the level of the endogenous substrate, for moderately abundant proteins. It is possible to estimate relative expression levels from comparisons of the fluorescence measured in transiently transfected cell populations (where the transfection efficiency is known) with immunoblots of the same cell populations. (c) Suppression of cap-dependent translation at mitosis *(10)* makes this proteolysis assay well suited to mitotic cells. We note that substrate proteolysis in interphase cells could be examined using one of the following approaches: (i) treatment of cells with a protein synthesis inhibitor (such as cycloheximide) during time-lapse imaging in order to assess the contribution of fresh synthesis to the apparent degradation curve, (ii) injection of purified GFP-tagged protein instead of mRNA, and (iii) using photoactivatable GFP *(11)* to allow generation of a fixed population of fluorescent substrate at the start of the assay.

2. The development of distinct spectral variants of GFP (and indeed the discovery of alternative fluorescent proteins) make it possible – with the use of appropriate filter sets – to compare directly the degradation profiles of two or more substrates within the same cell. Yellow Fluorescent Protein (YFP) and Cyan Fluorescent Protein (CFP) have proved a useful combination. The brightest versions that we currently use are the Venus variant of YFP *(12)* and the Cerulean variant of CFP *(13)*.
3. Some aspects of cell cycle regulation are lost in cancer cell lines such as HeLa, but retained in nontransformed cells. Thus for some questions it would be advantageous to measure proteolysis in nontransformed cells, although use of these cells poses several practical difficulties. These difficulties extend to synchronising the cells (they tend not to recover from G1/S phase arrest procedures), injecting them (because of their flatter morphology), and obtaining fluorescent cells in mitosis (they are more sensitive to mechanical damage during injections and photodamage during filming). However, we have developed a synchronisation procedure for human telomerase-immortalised retinal pigment epithelial (hTERT-RPE1) cells that allows them to be used for single-cell proteolysis assays. Cell culture conditions for hTERT-RPE1

cells can be found in the Clontech discontinued manuals archive at http://clontech.com/images/pt/dis_manuals/PT3358-1.pdf. These cells can be synchronised for microinjection in G2 as follows.

- Seed 1.6×10^4 cells per Delta T dish.
- Serum starve for 24 h.
- Refeed with medium containing 25% FBS.
- Cells will be in late G2 22–26 h after adding back serum.

4. Phase contrast microscopy can be used instead of DIC to monitor progress through mitosis. However, it has the disadvantage that the phase ring reduces the transmission of fluorescence.
5. Gibco DMEM supplemented with 10% FBS (and GlutaMAX-1 and antibiotics as described in **Section 2.1, step 4**) can also be used.
6. In vitro transcription kits for use with SP6 and T3 promoters are also available, but we have consistently obtained the best yields using the T7 RNA polymerase-based kit, which includes a polyA-tailing step.
7. 0.5–1 mg/ml is the recommended 'starting concentration' of mRNA. The level of expression of GFP-substrate can be controlled by diluting mRNA as required in DEPC-treated water.
8. We strongly recommend using a microscope fitted with a motorised XY stage, to increase the number of cells that can be imaged.
9. Microinjection of mRNA in G2 phase is the recommended approach for introducing GFP-substrates into cells, because it minimises potential perturbation of the cell cycle by the activity of the GFP-substrate. In principle, the same result can be achieved by injecting plasmid DNA into the nucleus of a G2 cell, and we have successfully used this approach in HeLa cells for many years. However, in practice we find that injecting mRNA presents several advantages:
   - mRNA is injected into the cytoplasm, and this causes less stress for the cell than nuclear DNA injection. This results in cells more likely to enter mitosis.
   - Expression levels are more homogeneous in mRNA-injected cells.
   - Expression of mature protein from mRNA is faster, and therefore cells can be injected later in G2 phase (and therefore injection is less likely to trigger a G2 checkpoint arrest).

   These considerations are particularly relevant for cell lines that are easily stressed, and we have found use of mRNA to be essential for proteolysis assays in hTERT-RPE1 cells. The use of stable cell lines in which expression of GFP-tagged substrates can be induced during G2 phase is suitable for these assays (in which case, workers should skip to **Section 3.3, step 8** of the method), but are more limited in terms of the number of substrates that can be analysed.

10. Our experience suggests that injecting the cell cytoplasm far from the nucleus may be less stressful for cells. However, it is more likely to result in cells fusing if they are damaged by injection.

Acknowledgments

The authors would like to thank Jonathon Pines, and an earlier generation of researchers in the Pines group (in particular Paul Clute, Christina Karlsson-Rosenthal, Anja Hagting, and Nicole den Elzen) who first measured proteolysis in mitotic cells. Many thanks to Anja Hagting for the data shown in **Fig. 16.4** and to Anja Hagting and Jonathon Pines for their comments on this chapter.

References

1. Reed, S. I. (2006) The ubiquitin-proteasome pathway in cell cycle control. *Results Probl. Cell Differ.* **42**, 147–81.
2. Pines, J. (2006) Mitosis: a matter of getting rid of the right protein at the right time. *Trends Cell Biol.* **16**, 55 63.
3. Clute, P. and Pines, J. (1999) Temporal and spatial control of cyclin B1 destruction in metaphase. *Nat. Cell Biol.* **1**, 82–87.
4. den Elzen, N. and Pines, J. (2001) Cyclin A is destroyed in prometaphase and can delay chromosome alignment and anaphase. *J. Cell Biol.* **153**, 121–36.
5. Hagting, A., den Elzen, N., Vodermaier, H. C., Waizenegger, I. C., Peters, J.-M. and Pines, J. (2002) Human securin proteolysis is controlled by the spindle checkpoint and reveals when the APC/C switches from activation by Cdc20 to Cdh1. *J. Cell Biol.* **157**, 1125–37.
6. Lindon, C. and Pines, J. (2004) Ordered proteolysis in anaphase inactivates Plk1 to contribute to proper mitotic exit in human cells. *J. Cell Biol.* **164**, 233–41.
7. Hayes, M. J., Kimata, Y., Wattam, S. L., Lindon, C., Mao, G., Yamano, H. and Fry, A. M. (2006) Early mitotic degradation of Nek2A depends on Cdc20-independent interaction with the APC/C. *Nat. Cell Biol.* **8**, 607–14.
8. Di Fiore, B. and Pines, J. (2007) Emi1 is needed to couple DNA replication with mitosis but does not regulate activation of the mitotic APC/C. *J. Cell Biol.* **177**, 425–37.
9. Pines, J. and Rieder, C. L. (2001) Re-staging mitosis: a contemporary view of mitotic progression. *Nat. Cell Biol.* **3**, E3–E6.
10. Pyronnet, S. and Sonenberg, N. (2001) Cell-cycle-dependent translational control. *Curr. Opin. Genet. Dev.* **11**, 13–8.
11. Patterson, G. H. and Lippincott-Schwartz, J. (2002) A photoactivatable GFP for selective photolabeling of proteins and cells. *Science* **297**, 1873–7.
12. Nagai, T., Ibata, K., Park, E. S., Kubota, M., Mikoshiba, K. and Miyawaki, A. (2002) A variant of yellow fluorescent protein with fast and efficient maturation for cell-biological applications. *Nat. Biotechnol.* **20**, 87–90.
13. Rizzo, M. A., Springer, G. H., Granada, B. and Piston, D. W. (2004) An improved cyan fluorescent protein variant useful for FRET. *Nat. Biotechnol.* **22**, 445–9.

14. Margottin-Goguet, F., Hsu, J. Y., Loktev, A., Hsieh, H. M., Reimann, J. D. and Jackson, P. K. (2003) Prophase destruction of Emi1 by the SCF(betaTrCP/Slimb) ubiquitin ligase activates the anaphase promoting complex to allow progression beyond prometaphase. *Dev. Cell* **4**, 813–26.

15. Guardavaccaro, D., Kudo, Y., Boulaire, J., Barchi, M., Busino, L., Donzelli, M., Margottin-Goguet, F., Jackson, P. K., Yamasaki, L. and Pagano, M. (2003) Control of meiotic and mitotic progression by the F box protein beta-Trcp1 in vivo. *Dev. Cell* **4**, 799–812.

Chapter 17

An In Vitro Assay for Cdc20-Dependent Mitotic Anaphase-Promoting Complex Activity from Budding Yeast

Scott C. Schuyler and Andrew W. Murray

Abstract

Cell cycle transitions are controlled, in part, by ubiquitin-dependent proteolysis. In mitosis, the metaphase to anaphase transition is governed by an E3 ubiquitin ligase called the cyclosome or Anaphase-Promoting Complex (APC), and a WD40-repeat protein co-factor called Cdc20. In vitro Cdc20-dependent APC (APC^{Cdc20}) assays have been useful in the identification and validation of target substrates, and in the study of APC enzymology and regulation. Many aspects of the regulation of cell cycle progression have been discovered in the budding yeast *Saccharomyces cerevisiae*, and proteins purified from this model organism have been employed in a wide variety of in vitro assays. Here we outline a quantitative in vitro mitotic APC^{Cdc20} assay that makes use of a highly active form of the APC that is purified from budding yeast cells arrested in mitosis.

Key words: Anaphase-promoting complex, APC, cyclosome, mitosis, Cdc20, anaphase, metaphase, Pds1, E3 ubiquitin ligase, ubiquitin, budding yeast, *saccharomyces cerevisiae*.

1. Introduction

Ubiquitin-dependent proteolysis is widely employed within the eukaryotic cell to promote regulated protein degradation. In general, ubiquitinylation of a target protein substrate requires three enzyme activities; a ubiquitin-activating enzyme (E1), a ubiquitin-conjugating enzyme (E2), and an ubiquitin ligase (E3) *(1, 2)*. The cyclosome or Anaphase-Promoting Complex (APC) along with a WD40-repeat protein co-factor Cdc20 provides the E3 ubiquitin ligase activity necessary to execute the metaphase to anaphase transition *(1, 2)*. In budding yeast the APC is composed

Andrew D. McAinsh (ed.), *Mitosis: Methods and Protocols*, Methods in Molecular Biology 545,
DOI 10.1007/978-1-60327-993-2_17, © Humana Press, a part of Springer Science+Business Media, LLC 2009

of 13 subunits, several of which are phosphorylated in the mitotic phase of the cell cycle *(3–8)*. A budding yeast in vitro Cdc20-dependent APC (APC^{Cdc20}) assay has previously been developed, where the APC was isolated from a nonsynchronized cycling population of cells *(9–13)*. One key insight and innovation in the development of that assay was to synthesize functional Cdc20 by coupled in vitro transcription/translation (IVT/T) *(10, 11)*

In this chapter we outline in detail a modified version of this APC^{Cdc20} assay, where we make use of a highly active form of the APC isolated from mitotic cells. The major steps in this protocol are to induce a cell cycle arrest, to harvest the cells, to purify the APC, to synthesize and purify a radiolabeled substrate; to perform an E3 ubiquitin ligase enzyme assay; to run, stain, and dry a protein gel; and to acquire the results by phosphor imaging.

2. Materials

2.1. Growing, Arresting, and Harvesting Cells

1. Yeast Medium (YPD): 50 g dextrose, anhydrous, powder (VWR International, PA), 50 g Bacto™ Peptone (VWR International, PA), and 25 g Bacto™ Yeast Extract (VWR International, PA) in 2.5 L water (*see* **Note 1**).
2. Benomyl: 0.9 g methyl 1-(butylcarbamoyl)-2-benzimidazolecarbamate (Benomyl) (Sigma-Aldrich, MO) dissolved in 30 mL dimethyl sulfoxide (DMSO) (Sigma-Aldrich, MO).
3. Ampicillin: 1.5 g ampicillin sodium salt (Sigma-Aldrich, MO) dissolved in 6 mL water.
4. Streptomycin: 0.75 g streptomycin sulfate (Sigma-Aldrich, MO) dissolved in 6 mL water.
5. 4 L flasks (Fisher Scientific, PA).
6. Autoclave tape (Fisher Scientific, PA).
7. Large autoclave-safe magnetic stir-bar (Fisher Scientific, PA).
8. Magnetic stir-plates (Fisher Scientific, PA).
9. TAP-Tagged Cdc16 yeast strain (Open Biosystems, AL).
10. Rotating Rollordrum™.
11. Spectrophotometer (Beckman-Coulter, CA).
12. Sorvall centrifuge with a large SLC-4000 rotor.
13. Magnetic wand (VWR International, PA).

2.2. APC Purification

1. Lysis buffer: 200 mM HEPES at pH 8.0, 150 mM NaCl, 10% (v/v) glycerol, 5 mM ethylenediamine tetraacetic acid (EDTA) at pH 8.0, 5 mM ethylene glycol tetraacetic acid (EGTA) at pH 7.0, 1 mM Dithiothreitol (DTT), 0.1% (v/v) Nonidet-P40 (NP-40) (Sigma-Aldrich, MO), 1 mM

$NaVO_4$ (*see* **Note 2**), 10 mM NaF (Sigma-Aldrich, MO), 5 mM sodium pyrophosphate, and 10 mM glycerol 2-phosphate.

2. Wash buffer: 200 mM HEPES pH at 8.0, 150 mM NaCl, 10% (v/v) glycerol, 1 mM DTT, 0.1% (v/v) NP-40, 1 mM $NaVO_4$, 10 mM NaF, 5 mM sodium pyrophosphate, and 10 mM glycerol 2-phosphate.
3. Elution buffer: 200 mM HEPES at pH 8.0, 150 mM NaCl, 10% (v/v) glycerol, 1 mM $NaVO_4$, 10 mM NaF, 5 mM sodium pyrophosphate, and 10 mM glycerol 2-phosphate (*see* **Note 3**).
4. Phenylmethanesulfonyl fluoride (PMSF; Sigma-Aldrich, MO): 0.87 g dissolved in 50 mL 100% ethanol (*see* **Note 4**).
5. Complete™ protease inhibitor cocktail (Roche Applied Sciences, IN).
6. IgG Sepharose™ 6 Fast Flow beads (GE Healthcare Bio-Sciences, Sweden).
7. AcTEV protease and cleavage buffer (Invitrogen, CA).
8. 0.5 mm glass beads (Biospec Products, OK).
9. Bead-Beater chamber (Biospec Products, OK).
10. Gravity-flow Bio-Spin® columns (Bio-Rad Laboratories, CA).
11. Beckman Ultra-Centrifuge at 50,000 rpm (Beckman-Coulter, CA).
12. Ti70.1 rotor and tubes (Beckman-Coulter, CA).

2.3. Preparation of ^{35}S-Methionine-Labeled Pds1 Substrate

1. TNT® T7 Coupled Reticulocyte Lysate System (Promega, WI).
2. EASYTAG™ Methionine, L-[^{35}S]-, at 43.5 TBq/mmol and 377.4 MBq/mL (PerkinElmer Life and Analytical Sciences, MA; *see* **Note 5**).
3. IgG Sepharose™ 6 Fast Flow beads (GE Healthcare Bio-Sciences, Sweden).
4. IVT/T wash buffer: 200 mM HEPES at pH 8.0, 150 mM NaCl, 10% (v/v) glycerol, 1 mM DTT, and 0.1% (v/v) NP-40.
5. AcTEV protease and cleavage buffer (Invitrogen, CA).

2.4. APC^{Cdc20} Assay

1. TNT T7 Coupled Reticulocyte Lysate System (Promega, WI).
2. QAH buffer: 20 mM HEPES at pH 8.0, 100 mM NaCl, and 1 mM $MgCl_2$.
3. Ubiquitin-activating enzyme (E1) and ubiquitin-conjugating enzyme (E2) (Boston Biochem, MA; *see* **Note 6**).
4. Ubiquitin stock solution: 10 mg/mL dissolved in water. Store 15 μL aliquots at –20°C.

5. Adenosine 5'-Triphosphate (ATP) stock solution: 10 mM ATP disodium salt (Sigma-Aldrich, MO), and 10 mM $MgCl_2$. Bring the final solution to pH 7.0 with 50% (w/w) sodium hydroxide (NaOH; Fisher Scientific, PA). Store 20 μL aliquots at –20°C.
6. Ubiquitin-Aldehyde (Ub-H) (Boston Biochem, MA) stock solution: the contents of one vial, 50 μg, dissolved in 20 μL of DMSO, followed by the addition of 20 μL of water. Store in aliquots at –80°C (*see* **Note 7**).
7. 2× protein sample buffer is a solution with a final concentration of 125 mM Tris-HCl at pH 6.8, 4% (w/v) sodium dodecyl-sulfate (SDS), 20% (v/v) glycerol, and 0.002% (w/v) Bromophenol Blue (Sigma-Aldrich, MO). Add 2-mercaptoethanol (Sigma-Aldrich, MO) fresh to a final of 10% (v/v) immediately before each use.

2.5. SDS-Polyacrylamide Gel Electrophoresis (SDS-PAGE)

1. Ammonium Persulfate (APS;Bio-Rad Laboratories, CA) stock solution: 10% (w/v) in water. Only make a small volume each time and store at 4°C.
2. Water-saturated isobutanol: 50% (v/v) isobutanol in water. Shake bottle to mix the two layers well. Allow the layers to separate before use. Use the top layer.
3. Separating gel buffer: 1.5 M Tris-HCl at pH 8.8.
4. Stacking gel buffer: 1.0 M Tris-HCl at pH 6.8.
5. 30% Acrylamide is Acrylagel™ (National Diagnostics, GA; *see* **Note 8**).
6. 2% N'N'-methylene bisacrylamide is Bis-Acrylagel (Na-tional Diagnostics, GA). Store at 4°C (*see* **Note 8**).
7. N,N,N,N'-Tetramethyl-ethylenediamine (TEMED) (Sigma-Aldrich, MO).
8. Gel running buffer: 30 g Trizma® base, 10 g SDS, and 144 g glycine dissolved in 10 L water.
9. Kimwipes® (Kimberly-Clark, WI).
10. Discovery™ gel system (R. Shadel, CA).

2.6. Staining and Drying Gel

1. Gel staining solution: 0.025% (w/v) Coomassie Brilliant Blue R 250 (Sigma-Aldrich, MO), 40% (v/v) methanol, and 7% (v/v) acetic acid in water
2. Gel washing solution: 40% (v/v) methanol, and 7% (v/v) acetic acid in water
3. Filter paper (Whatman, NJ)
4. Orbital shaker (Bellco Biotechnology, NJ)
5. Slab gel dryer (Hoefer Scientific Instruments, CA)

2.7. Visualizing Gel

1. Phosphor imager, such as Storm or Typhoon system (GE Healthcare, NJ)
2. Phosphor screen (GE Healthcare, NJ)
3. ImageQuant (GE Healthcare, NJ)

3. Methods

In vivo APC^{Cdc20} is only fully active in a small portion of the cell cycle. APC^{Cdc20} initiates the metaphase to anaphase transition by ubiquitinylating Pds1, the budding yeast Securin *(1, 2)*. APC^{Cdc20} remains active throughout anaphase until mitotic exit *(1, 2)*. It has also been observed that the APC is phosphorylated in mitosis, a modification that is thought to lead to an increase in activity *(6–8)*. Thus, to study the activity and regulation of APC^{Cdc20} we chose to arrest cells in mitosis at the metaphase to anaphase transition by activating the mitotic spindle checkpoint, which targets Cdc20 for inhibition *(2)*. There are many ways to activate the mitotic spindle checkpoint. Here we outline a method that uses the microtubule poison Benomyl *(2)* (*see* **Note 9**).

When comparing APC isolated from nonsynchronized cycling cells to APC isolated from cells arrested in mitosis we have observed an increase in both Cdc20-dependency and overall activity when performing APC^{Cdc20} assays (**Fig. 17.1**). Below, we outline in detail the protocol to perform these mitotic APC^{Cdc20} assays. This protocol includes two important negative controls. The first demonstrates that the poly-ubiquitinylation of Pds1 depends upon the presence of the APC. The second negative control demonstrates that the poly-ubiquitinylation of Pds1 depends upon the presence of the Cdc20 co-factor. These controls are important because there is a slight background activity in the reticulocyte lysate in which the Cdc20 is made by IVT/T.

Using mitotic APC leads to the complete consumption of the Pds1 substrate by one hour, which is in contrast to the cycling cell APC reaction. Although we do not know the reason for this difference, we do know it is critical to obtain a strong cell cycle arrest in mitosis in order to observe this increase in activity. For example, a good cell cycle arrest in Benomyl is where 90% or more of the cells are arrested as large budded cells. In addition, because it is possible that the phosphorylation state of the APC in mitosis contributes to a high level of activity it is important to include phosphatase inhibitors in the purification lysis and wash buffers.

The major steps in this protocol are to induce a cell cycle arrest; to harvest the cells; to purify the APC by Tandem Affinity Purification (TAP) *(14, 15)*; to synthesize by IVT/T and purify a radiolabeled substrate by TAP; to perform an E3 ubiquitin ligase enzyme assay; to run, stain, and dry a protein gel; and to acquire the results by phosphor imaging.

3.1. Growing, Arresting, and Harvesting Cells

1. Make 2.5 L YPD media in each of six 4 L flasks for a total of 15 L. The powder does not need to dissolve before autoclaving. Cover each flask opening with tinfoil and a piece

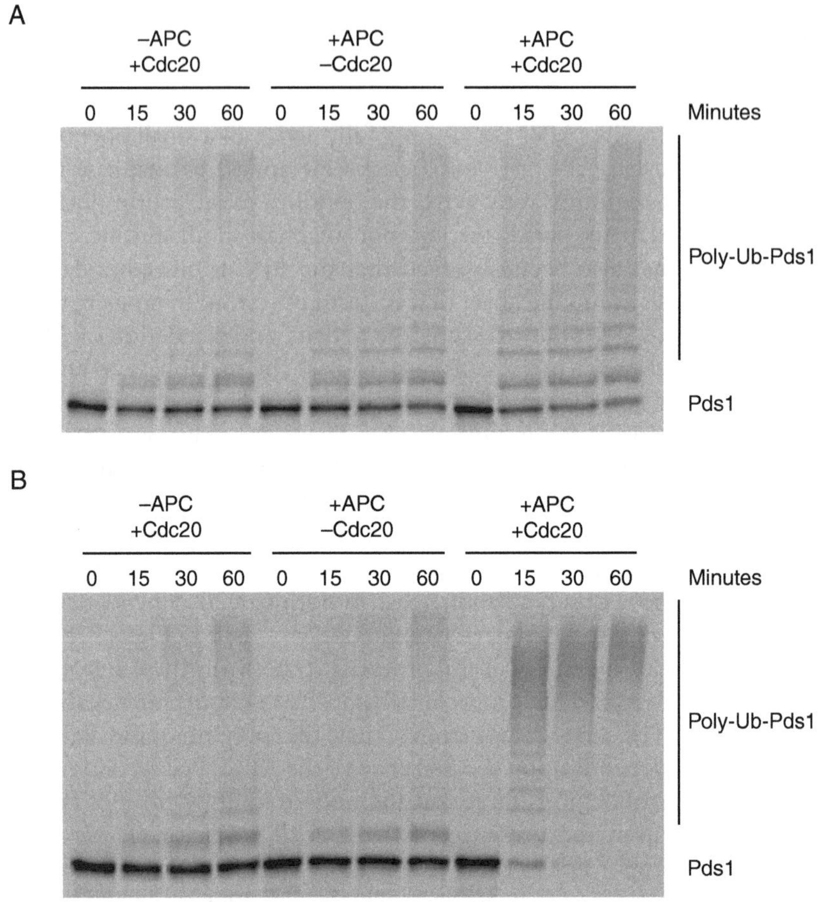

Fig. 17.1. A comparison between nonsychronized cycling cell and mitotic APC^{Cdc20} activity. APC^{Cdc20} activity is observed by the consumption of the Pds1 substrate over time coupled with an increase in ubiquitinylation of the Pds1 (Poly-Ub-Pds1) as indicated by "laddering" up the gel as more and more ubiquitin is added to the substrate. (**A**) APC assays using APC purified from cycling cells. One negative control (*left*) is to determine if the activity depends upon the addition of budding yeast APC. A second negative control (middle) is to determine if the activity depends upon the addition of the budding yeast Cdc20 co-factor. Here, a background activity can be observed. The experimental reaction (*right*) contains both APC and the Cdc20 co-factor and shows an increase in activity relative to the two negative controls. (**B**) APC assays using APC purified from Benomyl arrested cells. One negative control (*left*) is to determine if the activity depends upon the addition of budding yeast APC. A second negative control (*middle*) is to determine if the activity depends upon the addition of budding yeast Cdc20 co-factor. Here, there is a decrease in background activity in comparison to cycling cell APC. The experimental reaction (*right*) contains both APC and the Cdc20 co-factor and shows an increase in activity relative to the two negative controls, as well as in comparison to cycling cell APC.

of autoclave tape. Autoclave flasks on a liquid cycle for 15 or 30 min. Allow flasks to cool to room temperature overnight.

2. Make six additional flasks of 2.5 L YPD each as described above, except this time include a large autoclave-safe magnetic stir-bar in each flask before autoclaving. Upon completion of the autoclave cycle place the flasks on magnetic

stir-plates. Slowly increase the rate of stirring until a small vortex can be observed at the center of the media.

3. Make a fresh Benomyl solution in a 50 mL Falcon tube. Mix by vortexing on the highest setting for 30 s or until the Benomyl completely dissolves in the DMSO. Add 5 mL of the Benomyl solution to each of the six YPD flasks from step 2 on the stir plates. It is important to add the Benomyl shortly after the autoclave cycle is complete while the media are still hot. Upon addition much of the Benomyl will form a white precipitate. Continue to stir the media for 3–12 h to ensure that the Benomyl dissolves completely. Let media cool to room temperature overnight before use. It is not necessary to take the stir-bars out until immediately before use of the media (*see* **Note 10**).
4. Inoculate a TAP-tagged Cdc16 yeast strain into two 5 mL YPD cultures and grow in a rotating Rollordrum™ overnight at 30°C. Grow until the late afternoon on the next day, at which point the cultures will be saturated.
5. In the early evening, inoculate each of the six YPD flasks without Benomyl with 500 μL of the saturated overnight yeast culture. When working with these large volumes it can be difficult to maintain sterility of the media. Thus, we recommend adding 1 mL ampicillin and 1 mL streptomycin to each flask to prevent bacterial growth. Grow the cultures for 12–16 h overnight shaking at 250 rpm at 30°C.
6. At the beginning of the next day take a 1 mL sample from one flask and measure the optical density (OD) at 600 nm using a spectrophotometer. Cells are ready to collect by centrifugation when the cultures reach an OD between 0.1 and 0.5 (*see* **Note 11**).
7. In a Sorvall centrifuge with a large SLC-4000 rotor, which can hold four 1 L centrifuge bottles, spin down the cells at 4°C for 3 min at 7000 rpm, which is about 9200 × g (*see* **Note 12**). Repeat this spin four more times, each time discarding the supernatant and pouring in more of the cell culture.
8. Once all the cells have been spun down, resuspend the large cell pellet in water (*see* **Note 13**).
9. Remove the stir-bars with a magnetic wand from the YPD Benomyl flasks. Measure the total volume of the resuspended cells now in a water-cell slurry in a graduated cylinder. This volume will typically be between 500 and 600 mL. Using a second graduated cylinder, measure 1/6th of the total volume of the water-cell slurry and pour it into each one of the 6 flasks of YPD Benomyl. It is not necessary to add antibiotics to these flasks because the cells will

only be in the media for 3–4 h. Place flasks in the incubator at 250 rpm and 30°C.

10. After 3–4 h take a small sample from each flask and place it on a microscope slide and then cover with a coverslip. Look at the cells in a light microscope using a 60X objective, and count the budding morphology of 100 cells in each sample. There should be 95% or more budded cells observed, usually with 90% or more as large budded cells, which indicates a strong mitotic arrest. Collect the cells by centrifugation as described above in steps 7 and 8.
11. After resuspending the cell pellets in water spin down the water-cell slurry 50 mL at a time in 50 mL Falcon tubes in a table-top clinical centrifuge at max speed for 10 min at 4°C. Pour off the supernatant. Weigh the mass of each cell pellet in the Falcon tube using an empty tube to balance the scale. Label the mass and date on the side of the Falcon tube. Typically there are between 12 and 17 g of cell paste per 50 mL Falcon tube. From 15 L of culture one should expect to harvest between 100 and 200 g of cell paste.
12. Freeze and store the cell pellets at −80°C. It is not necessary to flash-freeze in liquid nitrogen.

3.2. APC Purification

1. Defrost about 100 g of frozen cell paste pellets for 1 h at 4°C. Push the cell paste pellets into a 200 mL Bead-Beater chamber by slicing off the bottom of the 50 mL Falcon tube containing the cells with a razor blade and then pushing the partially frozen cell paste pellet out.
2. In the Bead-Beater chamber add about 100 mL of 0.5 mm glass beads, and four Complete protease inhibitor tablets. Fill the chamber with lysis buffer and then add 2 mL of fresh PMSF. Assemble Bead-Beater as described by the manufacturer, and be sure to surround the chamber with ice water (*see* **Note 14**).
3. Perform 20 cell-lysis bead-beating cycles of 15 s "on", followed by 2 min "off", which takes a total of 45 min (*see* **Note 14**).
4. Pour the lysate, but not glass beads, into 50 mL Falcon tubes and centrifuge in a table-top clinical centrifuge at max speed for 10 min at 4°C. Pour supernatant into new 50 mL Falcon tubes and spin again for 10 min at max speed in the table-top clinical centrifuge at 4°C. Pour supernatant into new 50 mL Falcon tubes.
5. Spin this supernatant, which is typically about 150–200 mL, at 4°C for 1 h at 100,000 × g. For example, use a Ti70.1 rotor and tubes in a Beckman Ultra-Centrifuge at 50,000 rpm.

6. Pre-equilibrate the IgG Sepharose beads by washing 1 mL of the 50:50 slurry stock solution of beads twice in water, and twice in wash buffer (*see* **Note 15**).
7. At the end of the high-speed spin pour the high-speed supernatant, which is typically about 150 mL, into three 50 mL Falcon tubes. Add 333 μL of the washed IgG Sepharose bead-slurry to each tube. Cap the tubes tightly, and rotate the tubes end-over-end for 2 h at 4°C.
8. Spin down IgG beads for 15 min in a table-top clinical centrifuge at max speed at 4°C. Resuspend in 50 mL wash buffer. Wash beads in batch by spinning down twice with 50 mL of wash buffer in table-top clinical centrifuge at max speed at 4°C for 15 min. After the final wash, resuspend the IgG Sepharose in 1 mL of wash buffer.
9. Prewet two gravity-flow Bio-Spin columns with water.
10. Drip 1/2 of the beads (about 500 μL) into each gravity-flow column using wash buffer (*see* **Note 15**). Wash each column with 25 mL of wash buffer. Then wash each column with 25 mL of elution buffer, which does not contain DTT and NP-40 (*see* **Note 3**). This is the buffer in which the AcTEV protease cleavage will occur.
11. Leave about 500 μL of volume in each column by plugging the bottom of the column. Add 20 μL of AcTEV protease, mixing the solution by pipetting up-and-down in the column. Incubate the columns for 2 h at 16°C. Collect the flow-through, which is about 500 μL from each column. Freeze into 15 μL aliquots and store at –80°C (*see* **Note 16**).

3.3. Preparation of ^{35}S-Methionine-Labeled Pds1 Substrate

1. Perform a 200 μL IVT/T reaction for 1 h at room temperature using the TNT kit following the manufacturer's instructions. For example, mix 160 μL reticulocyte lysate, 10 μL L-[^{35}S]-methionine, 20 μL Pds1-TAP plasmid DNA (available upon request, or see *11, 13, 16*), and 10 μL water (*see* **Note 5**).
2. To wash and pre-equilibrate IgG Sepharose beads pipette out 60 μL of the bead-slurry from the stock vial, giving approximately a 30 μL final volume of beads. Wash the beads in 1 mL of water and then 1 mL of wash buffer. Resuspend the beads in 200 μL of wash buffer. Prewet a gravity-flow Bio-Spin column with water. Add the bead-wash buffer slurry to the column, and let the excess buffer drip out. However, in order not to let the Sepharose beads dry out, plug the bottom of the column (*see* **Note 15**).
3. Mix the IVT/T reaction with the IgG beads in the Bio-Spin column. Bind for 2 h at room temperature.

4. Wash the beads with 15 mL of the IVT/T wash buffer. Let almost all of the buffer drip out, but be sure to keep the beads moist.
5. Resuspend the beads in 185 μL of TEV cleavage buffer without the DTT. Add 5 μL of AcTEV protease to the reactions and mix well. Incubate the column for 2 h at 16°C.
6. Collect the flow-through and store at –80°C (*see* **Note 17**).

3.4. APC ^{Cdc20} Assay

1. Set up two different IVT/T reactions using nonradioactive methionine, one with Cdc20 plasmid under the control of the T7 promotor (available upon request, or *see (11, 13, 16)*), and one without DNA for the negative control.
2. For the reaction with Cdc20 set up a reaction according to the TNT kit following the manufacturer's instructions. For example, mix together 32 μL reticulocyte lysate, 4.8 μL DNA, 1.6 μL methionine, and 1.6 μL water for a total reaction volume of 40 μL. Mix well, and let reaction incubate at room temperature for 1–2 h.
3. For the reaction without Cdc20 set up a reaction according to the TNT kit following the manufacturer's instructions. For example, mix together 16 μL reticulocyte lysate, 2.4 μL water, 0.8 μL methionine, and 0.8 μL water for a total reaction volume of 20 μL. Mix well, and let reaction incubate at room temperature for 1–2 h.
4. Precharge the E1 and E2 with ubiquitin by mixing 1 μL E1, 1 μL E2, 14.4 μL ATP, 14.4 μL ubiquitin, and 17.2 μL QAH buffer. Mix well and let the reaction incubate at room temperature for 15 min.
5. Set up three APC reactions; the first is a negative control without the addition of APC, the second is a negative control without the addition of the Cdc20 co-factor, and the third is a reaction that contains both APC and Cdc20. Add in the following order the reaction contents to each of the three reaction tubes: 4.3 μL QAH buffer, 2.4 μL ubiquitin aldehyde, 20 μL reticulocyte lysate (two with the Cdc20 co-factor, and one without the Cdc20 co-factor), 4 μL APC (two with APC, and one without the APC but with 4 μL QAH buffer instead), 16 μL E1/E2/ATP/ubiquitin mix, and 13.3 μL Pds1 from purified radiolabeled stock (*see* **Note 18**). The total reaction volume should be 60 μL. Mix the reaction well and let the reaction run at room temperature. Start a timer to make note of when samples were taken.
6. Take out 15 μL at each timepoint from each reaction and mix directly with 15 μL of 2 × protein sample buffer. We typically take timepoints at 0, 15, 30, and 60 min.

3.5. SDS-Polyacrylamide Gel Electrophoresis (SDS-PAGE)

1. This protocol is for making a 10% separating gel using the Discovery gel system.
2. Clean gel plates, comb, and tubing with water and then 70% ethanol. Let all components dry before assembly. Assemble the gel plates according to the manufacturer's instructions. With the comb inserted into the gel make a mark on the outside of the front plate at about 1 cm below the comb wells. After making the mark, remove the gel comb.
3. In a 50 mL Falcon tube mix 13.4 mL 30% acrylamide, 2.6 mL 2% bisacrylamide, 10 mL 1.5 M Tris-HCL pH 8.8, and 13.8 mL water. To this mixture add 200 μL 10% APS and 20 μL TEMED. Cap and invert the tube several times to mix the solution together. Load this solution between the gel plates up until the mark that was placed 1 cm below the comb.
4. Add 1 mL of water-saturated isobutanol over the top of the gel and then fully insert the gel comb. Allow the gel to polymerize for 30 min at room temperature.
5. Remove the comb and pour off the layer of isobutanol. Be sure the wipe away any excess isobutanol by inserting a Kimwipe or paper towel between the gel plates.
6. In a new 50 mL Falcon tube mix together 3.3 mL 30% acrylamide, 1.3 mL 2% bisacrylamide, 2.5 mL 1.0 M Tris-HCL pH 6.8, and 12.9 mL water. To this mixture add 200 μL 10% APS and 20 μL TEMED. Cap and invert the tube several times to mix the solution together. Load this solution between the gel plates up to the top of the gel. Fully insert the comb. Allow the gel to polymerize for 30 min at room temperature.
7. Assemble polymerized gel into the gel running box as described by the manufacturer. Fill the top and bottom of the gel box with gel runner buffer (*see* **Note 19**).
8. Carefully load 30 μL of the APC^{Cdc20} reaction samples using thin-nosed pipette tips or a Hamilton needle. Run gel until the dye-front reaches the bottom of the gel, for example, at 200 volt for 2.5 h.

3.6. Staining and Drying Gel

1. Disassemble gel, and cut off the stacking gel portion.
2. Place the separating part of the gel in a gel-staining container, such as a Pyrex® glass container, with 100 mL Coomassie gel staining solution. Shake for 10 min at room temperature on an orbital shaker.
2. Pour off the Coomassie stain solution and add 100 mL destain solution. Shake for 10 min at room temperature.
3. Pour off the destain solution. Briefly, wash the gel once with tap water, then rehydrate the gel in 100 mL tap water, by shaking for 10 min at room temperature.

4. Pour off water and add an additional 100 mL tap water.
5. Cut 2 pieces of filter paper larger than the gel. Dry gel onto filter paper in a slab gel dryer for 1 h.

3.7. Phosphor Imaging

1. Erase phosphor screen on a lightbox for one hour as recommended by the manufacturer.
2. Tape the dried gel into phosphor cassette. Add screen, and lock tightly into place. Expose screen to the gel overnight.
3. Scan screen using phosphorimager following the manufacturer's instructions. Quantitative measurements of gel bands can be made by using manufacturer's software, such as ImageQuant.

4. Notes

1. Water that has a resistivity of 18.2 MΩ-cm is referred to as "water" in this chapter (Millipore, MA).
2. Use the following protocol to completely dissolve $NaVO_4$. Weigh out and place $NaVO_4$in a beaker with a magnetic stir-bar and place on a hot-plate/stir-plate. Add water to 90% of the final volume to be prepared, and turn on the hot-plate and the stir-plate. Add concentrated HCl to the solution until it reaches pH 10 and continue to stir and heat. Upon the addition of the HCl the solution will turn orange/yellow. Once the solution clears, add HCl again until it reaches pH 10. Repeat this HCl cycle one more time and stir until clear again. Let solution cool and add the remaining water to complete the final volume.
3. DTT and other thiol-reactive agents such as N-Ethylmaleimide (NEM) kill the APC reaction. We have also observed that the presence of NP-40 causes a slight decrease in APC activity.
4. PMSF is toxic and should be handled with care. PMSF is also unstable in water and should thus be dissolved in an anhydrous solution such as 100% ethanol right before use.
5. It should be noted that radioactive materials are strictly regulated by the government. Before ordering or handling radioactive compounds one needs to be trained and certified with regard to all safety procedures and regulations. Also, the disposal of radioactive waste is strictly regulated and all the necessary procedures should be followed.
6. Double-check the molarity of this product, because it can vary from batch to batch. The protocol outlined here assumes the E1 as a stock solution of 4.5 μM, and E2 as a stock solution of 124 μM. Instead of purchasing the

E1 and E2, one can also make the E1 and E2 recombinantly in yeast and bacteria *(13)*. In addition, one can make the E1 and E2 by TAP purification following the protocol outlined here in **Section 3.2** using TAP-tagged Uba1 and TAP-tagged Ubc4 budding yeast strains (Open Biosystems, AL) *(16)*.

7. Do not allow Ub-H to go through multiple freeze–thaw cycles. If the entire amount will not be used in one day, then aliquot it out and store it at –80°C.
8. Unpolymerized acrylamide and bisacrylamide are toxic and should be handled with care.
9. We prefer to use Benomyl instead of nocodazole, another microtubule poison. Although nocodazole is much more soluble, Benomyl is much cheaper. We have also arrested cells by overexpressing Mps1, a kinase that when overexpressed activates the mitotic spindle checkpoint, under the control of the *GAL* promoter. However, this method requires the use of raffinose and galactose, which are expensive.
10. Be sure to add Benomyl right when flasks come out of the autoclave. Be careful to set the stirring speed at an intermediate level. It is important to mix well, but sometimes if the stir-bar is spinning too fast it will "spin-out" off to the side of the flask. The final amount of Benomyl is 60 μg/mL, or 200 μM. This is a large amount of Benomyl compared to what is typically used. For example, if one wants to observe a Benomyl-sensitive phenotype on a YPD plate one would typically use 10 μg/mL. However, to get a rapid and clean arrest in liquid cultures we like to use a mixture of 30 μg/mL of Benomyl and 30 μg/mL of nocodazole. Because of the high cost of nocodazole we simply use 60 μg/mL of Benomyl.
11. If you inoculate between 6 and 8 PM, you will find that when you come in at about 9 AM the cells are at a good level for the arrest. It is very important not to let the cultures become too dense because you will not obtain a good arrest.
12. If you push the "start" button on your timer and the centrifuge at the same time this run takes about 7 min. The rotor actually never gets up to the 7000 rpm. However, the cells do form a tight pellet; we have found this 3 min setting to be the shortest we can use to get a solid cell pellet.
13. This takes a bit of work. Pour the water into one of the tubes filling the container up to the level just above the pellet. Put the lid back on and shake it vigorously for about a minute or until the cell pellet is dissolved. Then pour

this into the next tube and shake again. At the end there is usually about 500–600 mL of a water-cell slurry.

14. Ice water must be placed around the chamber. The device heats up quickly and it is important to keep it cold in order to maintain protein stability.
15. It is critical in all steps to never let the sepharose beads dry out. Great care should be taken to prevent this from happening.
16. This APC works for Cdc20-dependent reactions and Cdh1-dependent assays *(16)*.
17. One may wish to confirm the successful production and purification of the radiolabeled Pds1 by running an SDS-PAGE protein gel. To run and image this gel follow the protocol in **Sections 3.5–3.7**. However, it is important to cut off dye-front from this gel before staining and drying it.
18. The order in which the components are added is important. It is critical to add the reticulocyte lysate first to the QAH buffer/Ub-H mixture, and then add the APC and the E1/E2/ATP/ubiquitin mix. Adding the APC first to the QAH buffer/Ub-H mix kills the APC activity.
19. Be sure to remove any air bubbles at the bottom of the gel. In addition, before loading samples make sure to wash out each well with gel running buffer to remove any residual unpolymerized acrylamide.

Acknowledgments

The authors wish to thank David Morgan, Maria-Enquist Newman, Topher Carroll, and Monica Rodrigo-Brenni for training, reagents, and helpful advice. The authors also wish to thank Gregg Wildenberg for helpful comments on the manuscript. This work was supported by a U.S. National Institutes of Health (NIH) National Research Service Award Fellowship to S.C.S. and a NIH grant to A.W.M. (GM 043987).

References

1. Hershko A. (1999) Mechanisms and regulation of the degradation of cyclin B. *Philos Trans R Soc Lond B Biol Sci.* **354**:1571–5; discussion 1575–6.
2. Musacchio A., Salmon E.D. (2007) The spindle-assembly checkpoint in space and time. *Nat Rev Mol. Cell. Biol.* **8**:379–93. Epub 2007 Apr 11.
3. Zachariae W., Shin T.H., Galova M., Obermaier B., Nasmyth K. (1996) Identification of subunits of the anaphase-promoting complex of Saccharomyces cerevisiae. *Science* **274**:1201–4.
4. Zachariae W., Shevchenko A., Andrews P.D., Ciosk R., Galova M., Stark M.J., Mann M., Nasmyth K. (1998) Mass spectrometric analysis of the anaphase-promoting complex from yeast: identification of a subunit related to cullins. *Science.* Feb 20;**279**(5354):1216–9.

5. Passmore L.A., Booth C.R., Vénien-Bryan C., Ludtke S.J., Fioretto C., Johnson L.N., Chiu W., Barford D. (2005) Structural analysis of the anaphase-promoting complex reveals multiple active sites and insights into polyubiquitylation. *Mol. Cell.* Dec 22;**20** (6):855–66.
6. Kraft C., Herzog F., Gieffers C., Mechtler K., Hagting A., Pines J., Peters J.M. (2003) Mitotic regulation of the human anaphase-promoting complex by phosphorylation. *EMBO J.* Dec 15;**22** (24):6598–609.
7. Herzog F., Mechtler K., Peters J.M. (2005) Identification of cell cycle-dependent phosphorylation sites on the anaphase-promoting complex/cyclosome by mass spectrometry. *Methods Enzymol.* **398**:231–45.
8. Rudner A.D., Murray A.W. (2000) Phosphorylation by Cdc28 activates the Cdc20-dependent activity of the anaphase-promoting complex. *J. Cell Biol.* Jun 26;**149** (7):1377–90.
9. Carroll C.W., Morgan D.O. (2002) The Doc1 subunit is a processivity factor for the anaphase-promoting complex. *Nat. Cell Biol.* Nov;**4** (11):880–7.
10. Passmore L.A., McCormack E.A., Au S.W., Paul A., Willison K.R., Harper J.W., Barford D. (2003) Doc1 mediates the activity of the anaphase-promoting complex by contributing to substrate recognition. *EMBO J.* Feb 17;**22** (4):786–96.
11. Passmore L.A., Barford D., Harper J.W. (2005) Purification and assay of the budding yeast anaphase-promoting complex. *Methods Enzymol.* 398:195–219.
12. Carroll C.W., Enquist-Newman M., Morgan D.O. (2005) The APC subunit Doc1 promotes recognition of the substrate destruction box. *Curr. Biol.* Jan 11;**15** (1):11–8.
13. Carroll C.W., Morgan D.O. (2005) Enzymology of the anaphase-promoting complex. *Methods Enzymol.* **398**:219–30.
14. Rigaut G., Shevchenko A., Rutz B., Wilm M., Mann M., Séraphin B. (1999) A generic protein purification method for protein complex characterization and proteome exploration. *Nat. Biotechnol.* Oct;**17**(10):1030–2.
15. Puig O., Caspary F., Rigaut G., Rutz B., Bouveret E., Bragado-Nilsson E., Wilm M., Séraphin B. (2001) The tandem affinity purification (TAP) method: a general procedure of protein complex purification. *Methods.* **3**:218–29.
16. Rodrigo-Brenni M.C., Morgan D.O. (2007) Sequential E2s drive polyubiquitin chain assembly on APC targets. *Cell.* **130**: 127–39.

Chapter 18

In Vitro Assays for the Anaphase-Promoting Complex/Cyclosome (APC/C) in *Xenopus* Egg Extracts

Hiroyuki Yamano, Michelle Trickey, Margaret Grimaldi and Yuu Kimata

Abstract

The anaphase-promoting complex/cyclosome (APC/C), a large (20S) multisubunit E3 ligase, has an essential role to ubiquitylate numerous substrates at specific times during mitosis and G1 phase as well as in meiosis. The deregulation of the APC/C causes cell death or genomic instability, which is a hallmark of cancers. Although 13 years have passed since its discovery, the molecular mechanisms of the APC/C-dependent ubiquitylation and proteolysis are still poorly understood. The development of in vitro systems enables the identification of new substrates and investigation of the molecular mechanisms by which the APC/C recognizes its substrates. This chapter describes in vitro assays reconstituted in *Xenopus* egg extracts.

Key words: Anaphase-promoting complex, APC/C, Fizzy/Cdc20, ubiquitin, proteolysis, *Xenopus* egg, D-box, cyclin.

1. Introduction

The ubiquitin pathway is an ATP-dependent tagging system for protein degradation. It is an essential system in eukaryotic cells, where it is used for degrading damaged and misfolded proteins and also for degrading short-lived regulatory proteins during processes such as the cell cycle, transcription, signal transduction, and development. Ubiquitin tagging is highly selective and this is achieved by the sequential action of three enzymes, a ubiquitin-activating enzyme (E1), a ubiquitin-conjugating enzyme (E2), and a ubiquitin ligase (E3), and this enzymatic cascade will result in chains of four or more ubiquitin molecules on the substrates, which will then be recognized and degraded by the 26S proteasome *(1)*.

Andrew D. McAinsh (ed.), *Mitosis: Methods and Protocols*, Methods in Molecular Biology *545*,
DOI 10.1007/978-1-60327-993-2_18, © Humana Press, a part of Springer Science+Business Media, LLC 2009

The APC/C was originally discovered as an E3 ubiquitin ligase of cyclin B but it also regulates noncyclin proteins, and controls degradation of numerous proteins at specific times in the cell division cycle *(2–5)*. These include securin/Pds1/Cut2 promoting the separation of sister chromatids, proteins controlling spindle function Xkid1, Ase1, Kip1, Cin8, Aurora A, as well as geminin and Cdc6 which regulate origin licensing for DNA replication. The APC/C is also an important molecular target of the spindle assembly checkpoint (SAC) *(6)*. Aneuploidy arises from the attenuation of this checkpoint pathway *(7)*, and it is defective in certain types of cancers, such as T-cell leukemia and colorectal cancers. Moreover, in human colon cancer cells, alterations of the APC/C subunit genes have been reported. Thus, it is really important to understand how the APC/C works in cells to coordinate the cell cycle progression and maintain the genetic integrity.

The activation of the APC/C is mediated in part by the association of the Fizzy family of WD40-containing activator proteins *(3, 4, 8)*, although the precise roles of the activators remain to be elucidated. There are two major activators during the cell cycle, Fizzy/Cdc20 and Fizzy-related/Cdh1, which are required for the APC/C activity in anaphase or in G1-phase, respectively. From the perspective of the substrates, it is reasonable to assume that a number of additional APC/C substrates still exist that have not been identified yet. Therefore, to thoroughly screen APC/C substrates genomewide, both the Fizzy/Cdc20-APC/C-dependent and the Fizzy-related/Cdh1-APC/C-dependent destruction assays are essential. Here, we introduce the cell-free destruction assays using *Xenopus* egg extracts.

Substrate recognition is a key issue in understanding how the APC/C ubiquitylates its substrates. APC/C substrates identified to date contain at least one of the following destruction motifs: destruction box (D-box), KEN-box, A-box, O-box, GxEN, or MR motif *(9–14)*, however, there is still no conclusive model for substrate recognition and capture. Thus to this end, we have developed a method in *Xenopus* egg extracts to ask whether the substrates are recognized by APC/C and/or Fizzy/Cdc20.

Both the destruction and the substrate binding assays are simple and reliable as well as robust using *Xenopus* egg extracts. Destruction assays have been reported by several groups as well as us *(15–23)*, but we have recently improved both assays. For example, specific antibodies against ubiquitylation factors enable us to control levels of such factors in the *Xenopus* egg extracts. We have made mouse monoclonal Cdc27/Apc3 antibody *(23)* and rabbit Fizzy/Cdc20 antibodies *(14)*. These antibodies open up many new experiments including immunodepletion of APC/C or Fizzy/Cdc20 from destruction assays and immuno-affinity purification, allowing a more in-depth and precise approach to substrate binding and recognition.

2. Materials

2.1. Xenopus Egg Extract

1. Cysteine (dejelly) solution: 2 % cysteine in 1 × CSF salt solution. Adjust to pH 7.8.
2. 10 × XB solution: 100 mM HEPES pH 7.8, 500 mM sucrose.
3. 20 × CSF salt solution: 2 M KCl, 100 mM EGTA, 40 mM $MgCl_2$.
4. XB-CSF: 1× XB solution and 1× CSF salt solution. Final concentration is 10 mM HEPES pH 7.8, 50 mM sucrose, 100 mM KCl, 5 mM EGTA, 2 mM $MgCl_2$.
5. LPC; the protease inhibitor cocktail: 30 mg/ml leupeptin, 30 mg/ml pepstatin A, and 30 mg/ml chymostatin in DMSO. Mix equal volumes of the three to make 10 mg/ml LPC and store at –20°C. Use at final concentration of 10 μg/ml.
6. Cytochalasin B (10 mg/ml) in DMSO.
7. 50 × Energy mix: 375 mM creatine phosphate, 50 mM ATP, 10 mM EGTA, 50 mM $MgCl_2$.
8. 2 M sucrose.

2.2. In Vitro Destruction Assay

1. 8 mM $CaCl_2$ in water. Use at final concentration of 0.4 mM.
2. 10 mg/ml cycloheximide in water and stored at –20°C. Use at final concentration of 0.1 mg/ml.
3. Proteasome inhibitor: 10 mM MG132 in DMSO and stored at –20°C. Use at final concentration of 100~300 μM.
4. 1 × SDS-Sample buffer: 2% SDS, 80 mM Tris-HCl pH6.8, 10% glycerol, 0.002% w/v bromphenol blue.

2.3. A Coupled Transcription and Translation Kit

TnT T7 Quick Coupled Transcription/Translation system (Promega)

2.4. Radiochemical

Pro-mix [^{35}S], in vitro cell labeling mix (GE Healthcare), or Easy-Tag Express [^{35}S], Protein Labeling Mix (Perkin Elmer)

2.5. Coupling Antibodies to Protein A Beads

1. Dimethyl pimelimidate (DMP) solution: 1% (w/v) Dimethyl pimelimidate (Sigma) in 0.1 M sodium tetraborate. pH should be between 8.5 and 9.0. At lower pH, the coupling is not very efficient. Use immediately.
2. Coupling buffer (pH 9.0): Mix 60 ml of 0.1 M sodium tetraborate and 160 ml of 0.1 M boric acid.
3. Storage buffer (pH 8.0): Mix 7 ml of 0.1 M sodium tetraborate and 100 ml of 0.1 M boric acid.

2.6. Preparation of Fizzy-Related/Cdh1 (FZR)

1. Lysis buffer: 20 mM Tris-HCl pH 7.9, 500 mM NaCl, 4 mM $MgCl_2$, 0.4 mM EGTA, 2 mM DTT, 20 mM

β-glycerophosphate, 20% glycerol, 0.4 mM PMSF, and 10 μg/ml LPC

2. Dilution buffer: 20 mM Tris-HCl pH 7.9, 10% glycerol, 0.02% NP-40
3. Wash buffer: 20 mM Tris-HCl pH7.9, 150 mM NaCl, 2 mM $MgCl_2$, 0.2 mM EDTA, 1 mM DTT, 10 mM β-glycerophosphate, 15% glycerol, 0.01% NP-40, 0.2 mM PMSF, and 10 μg/ml LPC
4. Elution buffer: 20 mM Tris-HCl pH7.9, 150 mM NaCl, 500 mM imidazole, 2 mM $MgCl_2$, 0.2 mM EGTA, 1 mM DTT, 10 mM β-glycerophosphate, 15% glycerol, 0.01% NP-40, 0.2 mM PMSF, and 10 μg/ml LPC
5. Storage buffer: 50 mM Tris-HCl pH7.9, 100 mM NaCl, 1 mM DTT, and 10% glycerol

2.7. Depletion of APC/C or Fizzy from Xenopus Egg Extracts

XB-CSF: 10 mM Hepes pH 7.8, 50 mM Sucrose, 100 mM KCl, 5 mM EGTA, 2 mM $MgCl_2$

3. Methods

3.1. CSF extracts

Cytostatic factor-arrested extracts (referred to as CSF extracts) are prepared essentially as described by Murray (1991, *(24)*; *see* **Note 1**) .

1. Collect eggs.
2. Pour off as much buffer as possible and dejelly the eggs with cysteine solution (~5 min). After dejellying, the eggs are very fragile, so be careful when handling them.
3. Wash the eggs with XB-CSF 5 times and sort to eliminate all bad eggs. Bad eggs are activated or lysed eggs.
4. Add 1 ml versilube-S-81087C oil (or Nyosil M25) into 14 ml Falcon centrifuge tube (2059). Freeze on cardice (dry ice) and overlay with 14 μl of LPC (10 mg/ml) and cytochalasin B (10 mg/ml).
5. Carefully and quickly transfer eggs to the 14 ml Falcon tube with a cut-off Pasteur pipette.
6. To pack the eggs, centrifuge briefly (10 s) at 1000 rpm using swing rotor. All of the oil and any bad eggs will come to the top. Take out the white eggs (bad eggs) and any excess buffer to minimize dilution of the extract.
7. To crush the eggs, balance tubes and centrifuge at 10,000 rpm for 10 min in Sorvall HB-6 (or HB-4) rotor at 4°C. The crushed eggs will separate into distinct layers: a top yellow layer that contains lipids and yolk, a middle layer containing cytoplasm, and a bottom dark layer that

contains pigments, granules, and cell debris. Put the tubes on ice.

8. Take a 19 gauge needle and a 10 ml syringe, pierce the tube, and carefully collect the cytoplasmic fraction. Measure the volume and transfer to SW41 ultracentrifuge tubes placed on ice. Don't be greedy. Quality is more important than quantity.
9. Add 1/50 volume of Energy mix (50×), 1/1000 volume of LPC (10 mg/ml), and 1/1000 volume of cytochalasin B (10 mg/ml) into the extract, and mix well.
10. To clarify the extract, centrifuge again at 25,000 rpm (110,000 g) for 14 min.
11. Collect the clear cytoplasm using a needle/syringe. The extract is now ready for use.
12. To store the egg extracts in liquid nitrogen, add 1/10 volume of 2 M sucrose and mix well. Freeze the extracts in small aliquots (e.g., by dropping 24 μl directly into liquid nitrogen and collecting the drops) and store them in liquid nitrogen.

3.2. Destruction Assay

In CSF extracts, the onset of anaphase is controlled by cytostatic factor (CSF), an activity whose physiological role is to maintain unfertilized *Xenopus* eggs in metaphase of meiosis II by inhibiting the APC/C, thereby blocking cyclin proteolysis *(16, 25)*. In intact eggs, fertilization leads to increased levels of Ca^{2+} in the cytoplasm, which triggers several reactions including the inactivation of CSF, the degradation of cyclin, and the inactivation of maturation promoting factor (MPF) as well as the activation of APC/C. Similarly, in a cell-free system of CSF extracts, the addition of 0.4 mM $CaCl_2$triggers cyclin destruction and the activation of APC/C *(15 23)*. APC/C substrates are otherwise extremely stable in these extracts with some exceptions (see below). Inasmuch as endogenous Fizzy-related/Cdh1 (FZR) appears to be absent in *Xenopus* egg extracts *(20)*, this is a clean and simple system, which reflects the activity of Fizzy/Cdc20-dependent APC/C. On the other hand, FZR-dependent destruction assay can be reconstituted by adding recombinant FZR into interphase egg extracts where Fizzy/Cdc20-APC/C is inactive.

Substrates are typically prepared as labeled proteins with [^{35}S]-methionine plus cysteine (Promix; GE Healthcare) in a coupled in vitro transcription–translation system (TNT, Promega). If the translation is poor, it is difficult to observe its degradation. Thus, it is critical to obtain highly labeled substrates. A guideline is that 0.2 μl of the translated product should give a decent detectable signal. Alternatively, if you have other detection systems (e.g., immunoblotting with antibodies), nonlabeled recombinant proteins such as bacterially purified proteins could be used as substrates.

3.2.1. Fizzy/Cdc20-APC/C Dependent Destruction Assay

1. Prepare substrates. Label proteins with [^{35}S]-methionine plus cysteine in a coupled in vitro transcription–translation system (TNT, Promega).
2. Check whether the translation is OK and the signal is strong enough to perform the destruction assay by SDS-PAGE followed by fluorography. If yes, go to the next step.
3. Thaw the CSF egg extracts (if they are stored in liquid nitrogen) at 23°C and pretreat them with 0.1 mg/ml cycloheximide (CHX) for 5 min to inhibit further protein synthesis during the destruction assay.
4. 1 μl of labeled substrates are mixed with 9 μl of the CHX-treated CSF extracts (*see* **Note 2**). Subsequently, 0.4 mM $CaCl_2$ is added to initiate APC/C-dependent destruction, and samples are taken at 0, 15, 30, and 60 min and boiled in the SDS-containing sample buffer.
5. The mixtures are analyzed by SDS-PAGE and fluorography or immunoblotting. An example of the results produced is shown in **Fig. 18.1 A**.

Several controls are applicable to test whether a protein is a genuine substrate of the APC/C. First, in general, APC/C substrates are stable in CSF extracts in the absence of $CaCl_2$ (**Fig. 18.1A**, lanes 1–4; *see* **Note 3**), however, a few exceptions do exist, for example, cyclin A and Nek2A are slowly degraded without Ca^{2+} addition *(14, 26)*. Second, substrates will be stabilized by the addition of N70 (**Fig. 18.1A**, lane 9–12), the amino terminal 70 residues of the *S. pombe* cyclin B/Cdc13. The N70 contains the D-box and thus a high concentration of N70 works as a competitive substrate inhibitor of the APC/C *(23, 27)*. The N-terminal fragments of cyclin B from any organism can be used in a similar manner, as long as they contain the D-box. Third, the addition of the proteasome inhibitor MG132 also inhibits the APC/C-dependent proteolysis (**Fig. 18.1A**, lane 13–16). Ubiquitylated substrates by the APC/C are subsequently recognized and destroyed by the 26S proteasome, and thus MG132 stabilizes substrates by inhibiting the proteasome. This is not surprising, but worth including as a control to distinguish between proteasome-dependent and independent proteolysis during the reaction. Fourth, depletion of Fizzy/Cdc20 from extracts (*see* the method below; **Section 3.2**) blocks Fizzy/Cdc20-dependent activation of the APC/C, thus stabilizing substrates (**Fig. 18.1A**, lanes 17–20). This is a direct test to ask whether a protein is a Fizzy/Cdc20 substrate. Finally, if the APC/C is depleted from extracts (*see* the method below; **Section 3.2**), the genuine substrate should be stabilized (**Fig. 18.1A**, lanes 21–24). This is obvious, but is an important control.

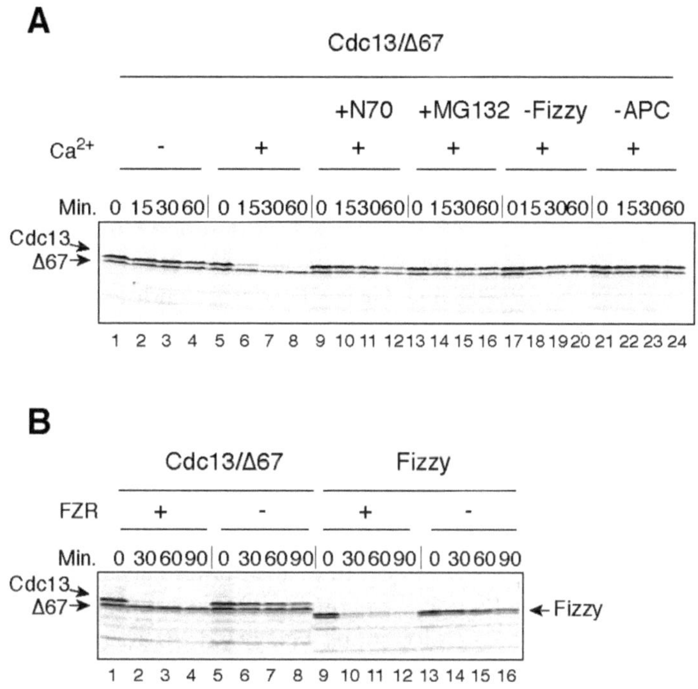

Fig. 18.1. Cell-free destruction assays reconstituted in *Xenopus* egg extracts. (**A**) Fizzy/Cdc20-APC/C dependent destruction assay. ^{35}S-labeled in vitro translated Cdc13 (cyclin B in fission yeast) together with a version of Cdc13 lacking the N-terminal 67 residues (Δ67, stable control) were used as substrates. $CaCl_2$ was added to activate Fizzy/Cdc20-APC/C. Upon addition of $CaCl_2$, Cdc13 was destroyed, whereas Δ67 was stable. Destruction of Cdc13 was blocked by either addition of the N-terminal 70 residues of Cdc13 (+N70, lanes 9–12) or a proteasome inhibitor (+MG132, lanes 13–16), or depletion of Fizzy/Cdc20 (–Fizzy, lanes 17–20) or APC/C (–APC, lanes 21–24) from the egg extracts. (**B**) Fizzy-related/Cdh1 (FZR)-APC/C dependent destruction assay. ^{35}S-labeled in vitro translated Cdc13, Δ67 and Fizzy/Cdc20 were used as substrates. Upon addition of FZR, both Cdc13 (the D-box-containing protein, lanes 1–4) and Fizzy (the KEN-box-containing protein, lanes 9–12) were destroyed. It is noteworthy that Fizzy cannot be destroyed in (**A**) Fizzy/Cdc20-APC/C dependent destruction assay.

3.2.2. Fizzy-Related/Cdh1 (FZR)-APC/C Dependent Destruction Assay

1. Prepare interphase extracts (*see* **Note 4**). Addition of 0.4 mM $CaCl_2$ to CSF extracts releases from the arrest, allowing cyclin destruction, in the presence of CHX to prevent *de novo* protein synthesis.
2. After 105 minutes of incubation at 23°C (*see* **Note 5**), purified *Xenopus* FZR (*see* Methods **Section 3.6**) is added for a further 15 min. Now, FZR-APC/C is active.
3. 1 μl of labeled substrates are mixed with 9 μl of the FZR-added interphase extracts. Samples are taken at 0, 30, 60 and 90 min (these can be extended if necessary, e.g., 0, 1, 2, and 3 h) and boiled in the SDS-containing sample buffer.

4. The mixtures are analyzed by SDS-PAGE and fluorography or immunoblotting. An example of the results produced is shown in **Fig. 18.1B**.

3.3. Anti-Cdc27/Apc3 Antibodies

To immunodeplete endogenous APC/C from *Xenopus* egg extracts, mouse monoclonal anti-Cdc27/Apc3 antibody (mAB, AF3.1) is used as described in Yamano et al. (1998) *(23, 28)*. The AF3.1 is available from several dealers such as Sigma or Santa Cruz. To detect endogenous Cdc27/Apc3 in the *Xenopus* egg extracts by immunoblotting, a second monoclonal anti-Cdc27/Apc3 antibody (BD Biosciences) is used.

3.4. Anti-Fizzy/Cdc20 Antibodies

GST and His6-tagged N-terminal 260 residues of *Xenopus* Fizzy/Cdc20 (FzyN260) were expressed in *E. coli* and purified using Ni-NTA beads as described by the supplier (Qiagen). The FzyN260 protein was further gel-purified and used to immunize rabbits. These antibodies can inhibit cyclin destruction in a cell-free *Xenopus* egg destruction assay, and are also able to deplete endogenous Fizzy/Cdc20 from *Xenopus* egg extracts *(14)*. To detect endogenous Fizzy/Cdc20 in the *Xenopus* egg extracts by immunoblotting, monoclonal anti-Fizzy/Cdc20 antibody (mAB, BA8.1) is used *(14, 28)*.

3.5. Preparation of Anti-APC/C or Anti-Fizzy Beads

To deplete APC/C or Fizzy/Cdc20 from *Xenopus* egg extracts, anti-Cdc27/Apc3 or anti-Fizzy/Cdc20 antibodies beads are prepared as follows (up to **step 3**). In contrast, to prepare antibody beads for substrate binding assay, we covalently couple antibodies to beads using the crosslinker dimethyl pimellimidate (DMP).

1. 1 ml of the suspension of protein A-conjugated magnetic beads (Dynabeads Protein A; Invitrogen) are washed three times in PBS and collected by magnetic stand.
2. Beads are then incubated with 250 μg of the anti-Cdc27/Apc3 (AF3.1) antibody or 25 μl of the anti-Fizzy/Cdc20 antibodies in 1 ml of PBS at room temperature for 1 h (*see* **Note 6**).
3. The beads are washed three times with PBS (*see* **Note 7**).
4. The antibody-bound Protein A beads are equilibrated with coupling buffer (0.1 M sodium borate/boric acid buffer, pH 9.0).
5. Dimethyl pimelimidate solution (*see* **Section 2.5, step 1**) is added into the beads, mixed and incubated at 4°C for 8–12 h.
6. The coupled beads are washed three times with the coupling buffer.
7. The beads are washed twice with 1 M Tris-HCl (pH 9.0) and equilibrated into 1 M Tris-HCl (pH 9.0) for 10 min to block free reactive sites.

8. The beads are washed three times with storage buffer, suspended in 1 ml of the storage buffer, and then stored at 4°C.

3.6. Preparation of Fizzy-Related/Cdh1 (FZR)

1. Grow 200 ml of Sf9 cells to log-phase in suspension in TC-100 medium (Invitrogen) plus 10% heat-inactivated fetal bovine serum.
2. Infect the Sf9 cells with recombinant His6-tagged *Xenopus* FZR baculovirus (BD BaculoGold) at an MOI of 10.
3. 48 h after infection, the cells are harvested and washed twice with cold PBS.
4. Suspend in 8 ml of lysis buffer and disrupt the cells with a dounce homogenizer (with a tight pestle: 10 strokes, 10 min on ice; 10 strokes, 10 min on ice; 5 strokes).
5. Spin and insoluble material is pelleted in a Sorvall SS-34 rotor (11,000 rpm for 10 min, 4°C). The supernatant is combined with 7 ml dilution buffer.
6. Add 800 μl of Ni-NTA beads (prewashed with PBS and lysis buffer) to the diluted supernatant and incubate at 4°C for 2 h with rotating.
7. The beads are washed 4 times with 12 ml wash buffer.
8. The protein is eluted with 2.4 ml elution buffer at 4°C.
9. The eluate is applied to a PD10 column, which has been equilibrated with 50 ml of storage buffer and eluted with 3.5 ml storage buffer. The yield is approximately 1 mg, at a concentration of 300 ng/μl.

3.7. Depletion of APC/C or Fizzy from Xenopus Egg Extracts

To investigate the contribution of APC/C and Fizzy/Cdc20 to protein destruction or cell cycle progression in *Xenopus* egg extracts, APC/C or Fizzy can be depleted using specific antibodies. Furthermore, by adding back recombinant proteins or purified proteins to the depleted extracts, a functional analysis of the protein/complex is feasible, proving that the *Xenopus* egg extract is a valuable tool and an excellent model system.

1. 15 μl of the suspension of the anti-Cdc27/Apc3 beads or 60 μl of that of anti-Fizzy/Cdc20 beads, which are prepared as described in Methods (**Section 3.5**), are collected by magnetic stand and washed twice with XB-CSF buffer.
2. The beads are collected by magnetic stand and mixed with 30 μl of *Xenopus* egg extracts, followed by incubation at 23°C for 20 min with gentle shaking.
3. Place tube in a magnetic stand and place supernatant into a new tube containing the same amount of the fresh antibody beads to repeat this depletion process.
4. After further 20-min incubation, the supernatant is transferred to an empty tube, and residual magnetic beads are further removed by placing tube in a magnetic stand.

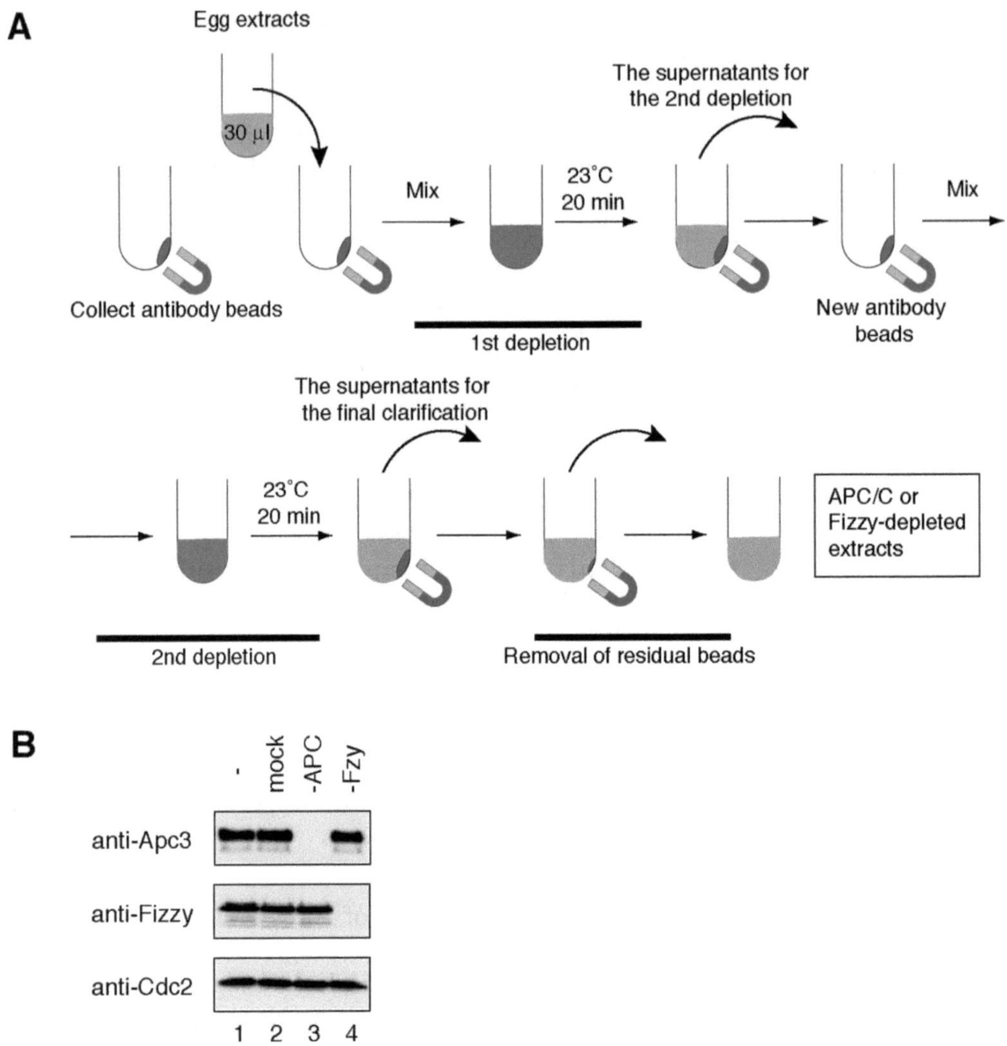

Fig. 18.2. Immunodepletion of APC/C and Fizzy/Cdc20 from *Xenopus* egg extracts. (**A**) Flowchart showing the immunodepletion of APC/C or Fizzy/Cdc20 from the egg extracts. (**B**) Untreated egg extract (lane 1), mock-, APC/C- or Fizzy/Cdc20-depleted egg extracts using specific antibodies were immunoblotted with the indicated antibodies. APC/C and Fizzy/Cdc20 were successfully depleted from egg extracts (lanes 3 and 4).

5. The supernatants are now ready to use as APC/C or Fizzy-depleted extracts (**Fig. 18.2**). With this protocol, more than 95% depletion can be achieved (*see* **Note 8**), which is analyzed by quantitative immunoblotting.

3.8. Substrate Binding Assay

The assay described here is designed for the detection of the protein–protein interactions under physiological conditions and is suitable for analyzing relatively high-affinity substrate to the APC/C or Fizzy/Cdc20 (*see* **Note 9**).

1. Prepare mock-, APC/C- and Fizzy/Cdc20-depleted egg extracts as described above.

2. Add 2 μl of labeled substrates to be tested into 20 μl of each egg extract, which is pretreated with 300 μM of a proteasome inhibitor MG132 to avoid degradation of the substrates. Incubate at 23°C for 5 min.
3. The mixtures are added to collected beads from 40 μl of suspension of the anti-Cdc27/Apc3 or anti-Fizzy/Cdc20 beads using the magnetic stand.
4. Incubate at 23°C for 30 min with gentle shaking.
5. The APC/C are Fizzy/Cdc20 are immuno-affinity purified by collecting the beads in the magnetic stand.
6. The beads are washed 3 times with 300 μl of XB-CSF buffer, then boiled in 20 μl of 2 × SDS sample buffer.

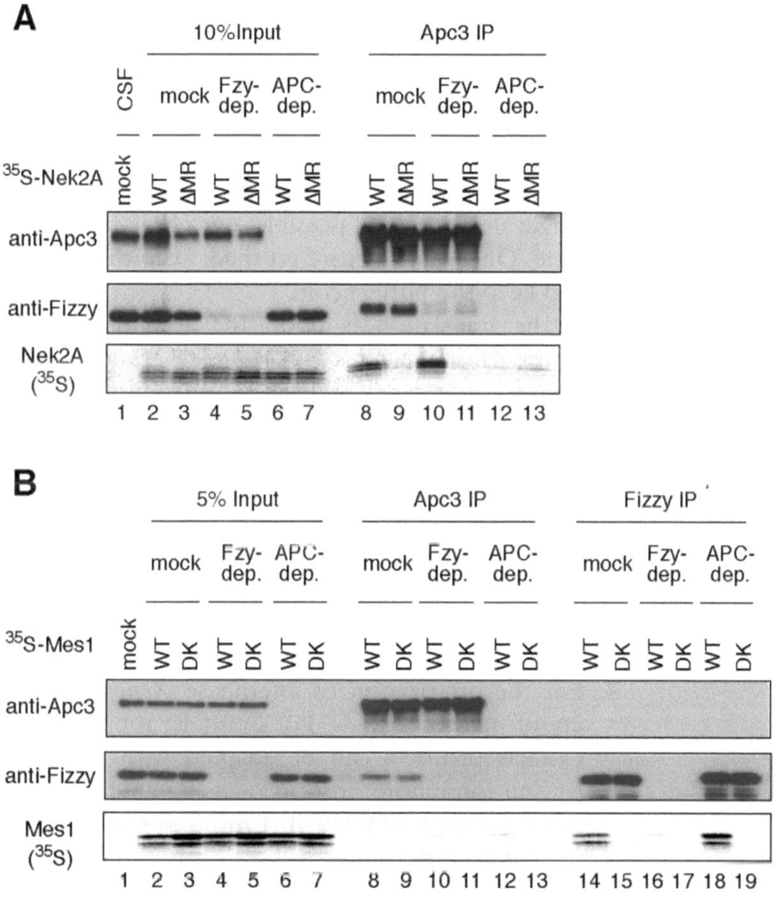

Fig. 18.3. Substrate binding assay in *Xenopus* egg extracts. (**A**) Nek2A binds to APC/C independently of Fizzy/Cdc20. ^{35}S-labeled Nek2A and ΔMR (MR motif truncation mutant) were incubated with mock-, Fizzy/Cdc20- or APC/C-depleted CSF extracts. Samples were taken before (10% input, lanes 1–7) or after Apc3 (Apc3 IP, lanes 8–13) immunoprecipitations and were analyzed by immunoblotting with anti-Apc3 or anti-Fizzy, or by fluorography (^{35}S). (Reproduced from *14.)* (**B**) Mes1 binds to Fizzy/Cdc20 independently of APC/C. ^{35}S-labeled Mes1WT and DK (D-box and KEN-box mutant) were incubated with mock-, Fizzy/Cdc20- or APC/C-depleted CSF extracts. Samples were taken before (5% input, lanes 1–7) or after Apc3 (Apc3 IP, lanes 8–13) or Fizzy/Cdc20 (Fizzy IP, lanes 14–19) immunoprecipitations and were analyzed by immunoblotting with anti-Apc3 or anti-Fizzy, or by fluorography (^{35}S). (Reproduced from *28.*)

7. The binding proteins are analyzed by SDS-PAGE and fluorography. An example of the results produced is shown in **Fig. 18.3**.

A centrosomal protein of the NIMA-related kinase, Nek2A directly binds to the APC/C via the C-terminal MR motif *(14)*. In contrast, a meiotic APC/C inhibitor, Mes1 directly binds to Fizzy/Cdc20 via the D-box and KEN-box *(28)*. Thus, it seems that the mode of substrate recognition varies depending on substrates, although further investigation will be required to address the molecular mechanisms.

4. Notes

1. First check how many frogs have laid good eggs and prepare the buffers according to the amount of eggs. Both cysteine (dejelly) solution and CSF-XB are kept at 16°C. Do every single step as fast as possible; eggs only become worse with time. Once the eggs are crushed, keep the cytoplasm on ice and keep the buffers cold.
2. If the translation is poor, the amounts of labeled substrates can be increased. It is, however, noteworthy that the destruction assay is dilution-sensitive; it apparently needs at least 70% CSF extracts in the reaction volume.
3. As a standard control, we recommend the use of mixed substrates consisting of cyclin B (Cdc13, degradable control) and a version of cyclin B lacking the N-terminal 67 residues (Δ67, stable control) to monitor both the D-box-dependent and independent destruction during the reaction. For example, if you see destruction of Δ67, D-box independent proteolysis is occurring in the extracts.
4. For the preparation of interphase *Xenopus* extracts a variety of methods exists. For example, the A. Murray cycling extracts protocol can be adapted with the addition of 2 μg/ml ionophore, instead of using an activating chamber, in the presence of CHX (0.1 mg/ml) prior to crushing spinning at 10,000 rpm.
5. We have examined how long the activity of Fizzy/Cdc20-APC/C remains active after addition of $CaCl_2$, and found that the activity is absent or very low after the 2 hr incubation. Thus, when *Xenopus* FZR is added to the interphase extracts, the FZR activates endogenous APC/C and initiates FZR-APC/C dependent ubiquitylation and destruction.
6. The amount of the anti-Fizzy/Cdc20 antibodies required is largely dependent on the antibody titer and the ability to

immunodeplete Fizzy/Cdc20. A titration will be required initially to find the right amount.

7. The beads are suspended in 1 ml of PBS and stored at 4°C. The antibodies can be crosslinked to beads; however, it sometimes reduces the depletion efficiency. Thus, we normally do not crosslink for the purpose to immunodeplete APC/C or Fizzy/Cdc20 from egg extracts.
8. Notably, the volume of extracts in a tube during the depletion process makes a difference in the efficiency of depletion. In our hands, the best depletion will be achieved when the volume is less than 30 μl in each tube.
9. To follow more dynamic or weaker association such as the D-box-dependent binding of cyclin B to APC/C, other methods can be employed *(27)*.

References

1. Hershko, A., and Ciechanover, A. (1998). The ubiquitin system. *Annu. Rev. Biochem.* *67*, 425–479.
2. Morgan, D. O. (2007). The Cell Cycle: Principles of Control. London: New Science Press.
3. Peters, J. M. (2006). The anaphase promoting complex/cyclosome: a machine designed to destroy. *Nat. Rev. Mol. Cell Biol.*, 644–656.
4. Thornton, B. R., and Toczyski, D. P. (2006). Precise destruction: an emerging picture of the APC. *Genes Dev.* ***20***, 3069–3078.
5. Acquaviva, C., and Pines, J. (2006). The anaphase-promoting complex/cyclosome: APC/C. *J. Cell Sci.* ***119***, 2401–2404.
6. Musacchio, A., and Salmon, E. D. (2007). The spindle-assembly checkpoint in space and time. *Nat. Rev. Mol. Cell Biol.* *8*, 379–393.
7. Kops, G. J., Weaver, B. A., and Cleveland, D. W. (2005). On the road to cancer: aneuploidy and the mitotic checkpoint. *Nat. Rev. Cancer* *5*, 773–785.
8. Yu, H. (2007). Cdc20: a WD40 activator for a cell cycle degradation machine. *Mol. Cell* *27*, 3–16.
9. Araki, M., Yu, H., and Asano, M. (2005). A novel motif governs APC-dependent degradation of Drosophila ORC1 in vivo. *Genes Dev.* *19*, 2458–2465.
10. Castro, A., Vigneron, S., Bernis, C., Labbe, J. C., and Lorca, T. (2003). Xkid is degraded in a D-box, KEN-box, and A-box-independent pathway. *Mol. Cell Biol.* *23*, 4126–4138.
11. Glotzer, M., Murray, A. W., and Kirschner, M. W. (1991). Cyclin is degraded by the ubiquitin pathway. *Nature* ***349***, 132–138.
12. Littlepage, L. E., and Ruderman, J. V. (2002). Identification of a new APC/C recognition domain, the A box, which is required for the Cdh1-dependent destruction of the kinase Aurora-A during mitotic exit. *Genes Dev.* ***16***, 2274–2285.
13. Pfleger, C. M., and Kirschner, M. W. (2000). The KEN box: an APC recognition signal distinct from the D box targeted by Cdh1. *Genes Dev.* ***14***, 655–665.
14. Hayes, M. J., Kimata, Y., Wattam, S. L., Lindon, C., Mao, G., Yamano, H., and Fry, A. M. (2006). Early mitotic degradation of Nek2A depends on Cdc20-independent interaction with the APC/C. *Nat. Cell Biol.* *8*, 607–614.
15. Lohka, M. J., and Maller, J. L. (1985). Induction of nuclear envelope breakdown, chromosome condensation, and spindle formation in cell-free extracts. *J. Cell Biol.* ***101***, 518–523.
16. Murray, A. W., Solomon, M. J., and Kirschner, M. W. (1989). The role of cyclin synthesis and degradation in the control of maturation promoting factor activity. *Nature* ***339***, 280–286.
17. Lorca, T., Devault, A., Colas, P., Van Loon, A., Fesquet, D., Lazaro, J. B., and Doree, M. (1992). Cyclin A-Cys41 does not undergo cell cycle-dependent degradation in Xenopus extracts. *FEBS Lett.* ***306***, 90–93.
18. van der Velden, H. M., and Lohka, M. J. (1993). Mitotic arrest caused by the amino terminus of Xenopus cyclin B2. *Mol. Cell Biol.* *13*, 1480–1488.
19. Stewart, E., Kobayashi, H., Harrison, D., and Hunt, T. (1994). Destruction of Xenopus

cyclins A and B2, but not B1, requires binding to p34cdc2. *EMBO J.* *13*, 584–594.
20. Lorca, T., Castro, A., Martinez, A. M., Vigneron, S., Morin, N., Sigrist, S., Lehner, C., Doree, M., and Labbe, J. C. (1998). Fizzy is required for activation of the APC/cyclosome in Xenopus egg extracts. *EMBO J.* *17*, 3565–3575.
21. King, R. W., Glotzer, M., and Kirschner, M. W. (1996). Mutagenic analysis of the destruction signal of mitotic cyclins and structural characterization of ubiquitinated intermediates. *Mol. Biol. Cell* *7*, 1343–1357.
22. Yamano, H., Gannon, J., Mahbubani, H., and Hunt, T. (2004). Cell cycle-regulated recognition of the destruction box of cyclin B by the APC/C in Xenopus egg extracts. *Mol. Cell* **13**, 137–147.
23. Yamano, H., Tsurumi, C., Gannon, J., and Hunt, T. (1998). The role of the destruction box and its neighbouring lysine residues in cyclin B for anaphase ubiquitin-dependent proteolysis in fission yeast: defining the D-box receptor. *EMBO J.* *17*, 5670–5678.
24. Murray, A. W. (1991). Cell cycle extracts. *Methods Cell Biol.* *36*, 581–605.
25. Masui, Y., and Markert, C. L. (1971). Cytoplasmic control of nuclear behavior during meiotic maturation of frog oocytes. *J. Exp. Zool.* *177*, 129–145.
26. Geley, S., Kramer, E., Gieffers, C., Gannon, J., Peters, J. M., and Hunt, T. (2001). Anaphase-promoting complex/cyclosome-dependent proteolysis of human cyclin A starts at the beginning of mitosis and is not subject to the spindle assembly checkpoint. *J. Cell Biol.* **153**, 137–148.
27. Yamano, H., Gannon, J., and Hunt, T. (1996). The role of proteolysis in cell cycle progression in Schizosaccharomyces pombe. *EMBO J.* **15**, 5268–5279.
28. Kimata, Y., Trickey, M., Izawa, D., Gannon, J., Yamamoto, M., and Yamano, H. (2008). A mutual inhibition between APC/C and its substrate Mes1 required for meiotic progression in fission yeast. *Dev. Cell*, **14**, 446–454.

Chapter 19

Preparation of Synchronized Human Cell Extracts to Study Ubiquitination and Degradation

Adam Williamson, Lingyan Jin and Michael Rape

Abstract

Ubiquitination and protein degradation regulate cell cycle progression in all eukaryotes. During mitosis, ubiquitination by the Anaphase-Promoting Complex/Cyclosome (APC/C) triggers sister chromatid separation and mitotic exit. The APC/C is tightly regulated by phosphorylation, ubiquitination, association of activators or inhibitors, and competitive binding of substrates. Much of our understanding of the mechanism of APC/C-dependent ubiquitination has been obtained from studies using extracts of *Xenopus laevis* eggs or synchronized human tissue culture cells. Here, we describe protocols to prepare extracts of synchronized human cells, and discuss experiments to use extracts for the biochemical analysis of APC/C-dependent ubiquitination.

Key words: Anaphase-promoting complex/cyclosome (APC/C), ubiquitin, proteasome.

1. Introduction

Protein degradation is essential for the correct progression through the human cell cycle *(1)*. During mitosis, the timely degradation of cyclins inactivates CDK1 and promotes mitotic exit *(2)*. In a similar manner, degradation of the anaphase-inhibitor securin contributes to the appropriate initiation of sister chromatid separation. A failure to degrade these cell cycle regulators arrests progression through mitosis, and their premature degradation can result in aneuploidy and, potentially, cancer *(3)*.

The specific degradation of eukaryotic cell cycle regulators depends on their modification with a chain of ubiquitin molecules *(4)*, which requires a cascade of at least three enzymes (ubiquitin-activating enzyme, E1; ubiquitin conjugating enzyme, E2; and ubiquitin ligase, E3). The specificity of ubiquitination

Andrew D. McAinsh (ed.), *Mitosis: Methods and Protocols*, Methods in Molecular Biology *545*,
DOI 10.1007/978-1-60327-993-2_19, © Humana Press, a part of Springer Science+Business Media, LLC 2009

is determined by E3s, which recruit both ubiquitin-charged E2s and substrates. Many of the ~1000 human E3s contain a catalytic RING-finger domain, which activates E2s and promotes the direct transfer of ubiquitin from the E2 to a lysine residue in the substrate.

Ubiquitination by the RING-finger E3 Anaphase-Promoting Complex/Cyclosome (APC/C) triggers the sequential degradation of cell cycle regulators, including mitotic cyclins and securin *(2)*. APC/C-substrates contain short recognition motifs referred to as D-boxes or KEN-boxes. The human APC/C further discriminates between substrates based on a kinetic proofreading mechanism *(5)*. The activity of the APC/C is controlled in a cell cycle-dependent manner by phosphorylation, ubiquitination, and regulated expression of substrate targeting factors and E2s (reviewed in *(2)*).

Much of our understanding of the catalytic mechanism and the regulation of the APC/C has been obtained from studies using concentrated extracts of *Xenopus laevis* eggs or human tissue culture cells. Synchronized human cell extracts provide an important tool to study cell cycle-dependent protein degradation under near physiological conditions *(6)*. In addition to providing insight into cell cycle regulation, degradation in extracts is a starting point for the identification of novel substrates of cell cycle-dependent ubiquitination. Note that extracts also permit the purification of E3s, such as the APC/C, to dissect the mechanism of ubiquitin chain formation during mitosis. Here, we describe procedures of preparing synchronized cell extracts and discuss experiments to study cell cycle-dependent ubiquitination and degradation.

2. Materials

2.1. Cell Culture and Synchronization

1. Dulbecco's Modified Eagle's Medium (DMEM) (Gibco/Invitrogen) supplemented with 10% fetal bovine serum (HyClone) and 1× antibiotics/antimucotics (Gibco).
2. Trypsin (0.25%)/EDTA (1 mM) (Gibco).
3. PBS (for cell culture; Gibco).
4. Thymidine (Sigma); prepare fresh 200 mM stock in sterile water; filter-sterilize it through 0.22 μm filter.
5. Nocodazole (Sigma); prepare 5 mg/ml stock in DMSO; store in small aliquots at –80°C.
6. Cell lines: HeLa S3 (for spinners and mitotic and G1-extracts); T24 or T98G (for quiescent cell extracts).
7. Spinner flasks.

2.2. Extract Preparation

1. PBS: 8 g NaCl, 0.2 g KCl, 1.44 g Na_2HPO_4, 0.24 g KH_2PO_4 pH 7.4.

2. SB-buffer: 24.3 ml ice-cold water, 500 μl 1 M HEPES pH 7.5, 37.5 μl 1 M $MgCl_2$, 25 μl 1 M DTT, 125 μl 1 M KCL 1 tablet EDTA-free protease-inhibitor (Roche; prepare fresh for each extract preparation).
3. Energy mix: 375 mM creatine phosphate, 50 mM ATP, pH8, 50 mM $MgCl_2$ (store in small aliquots at –80°C).
4. Sterile syringe needles (BD; 20G1 and 25G3/8).

2.3. Degradation and Ubiquitination Assays

1. Ubiquitin (Boston Biochem; stock 10 mg/ml in SB).
2. E2 (UbcH10 or UbcH5c); purified as recombinant protein from bacteria or commercially available from Boston Biochem.
3. E1; purified as recombinant protein from baculovirus infected *Sf9* cells or commercially available from Boston Biochem.
4. Recombinant APC/C-substrates and Emi1 purified from bacteria *(5)*.
5. Rabbit Reticulocyte Lysate coupled transcription/translation system (Promega).
6. TranS^{35}-label (MP Biochem).
7. Gel-loading buffer (2 × stock: 1.875 ml 1 M Tris-HCl, pH 6.8, 15 ml 50% glycerol, 6 ml 10% SDS, 0.003 g bromophenol blue; fill to total volume of 30 ml; add 50 μl β-mercaptoethanol to 950 μl gel-loading buffer freshly before denaturing proteins).
8. SB/U-buffer: 24.3 ml cold water, 500 μl 1 M HEPES pH 7.5, 125 μl 1 M KCl, 37.5 μl 1 M $MgCl_2$.
9. Protein-G agarose (Roche).
10. Monoclonal anti-Cdc27 antibody (Santa Cruz).
11. UBAB-buffer (10 × stock): 25 mM Tris-HCl pH 7.5, 50 mM NaCl, 10 mM $MgCl_2$.

3. Methods

We describe protocols to generate synchronized human cell extracts, which can be obtained from cells grown in suspension or from adherent cells. To produce potent mitotic or G1-extracts, it is crucial to employ synchronization protocols that limit the time cells are arrested in mitosis.

3.1. Tissue Culture and Synchronization

3.1.1. Synchronization of HeLa S3 Cells for Preparation of Mitotic and G1 Extracts

1. To inoculate a 2 L spinner flask, three almost confluent 15 cm plates of HeLa S3 cells are required. The cells are grown in DMEM-medium containing 10% fetal calf serum

and antibiotics. The best extracts are obtained if cells are split 1:2 the day before spinner flasks are inoculated.

2. 2 h before the inoculation, wash the 2 L spinner flask with deionized water, then with PBS. Autoclave the spinner 40 min on liquid cycle and let it cool down under a tissue culture hood. Add 1 L prewarmed DMEM, 100 ml fetal calf serum, and 11 ml antibiotic/antimucotic mix to the spinner.
3. Aspirate the medium off the HeLa S3 cells, and wash the cells with PBS. Aspirate the PBS, add 5 ml 0.25% Trypsin/EDTA per dish, and incubate for 5 min at 37°C. Combine the resuspended HeLa S3 cells of three plates and add them to the prewarmed medium in the 2 L spinner.
4. Incubate at 37°C, 5% CO_2 with slow stirring for three days. A constant supply of CO_2 is essential for obtaining high-quality extracts. Add 11 ml of a 200 mM thymidine solution (sterilized under the hood using a 0.22 μm filter) to the cell suspension. Incubate at 37°C, 5% CO_2 for 24 h.
5. Autoclave 500 ml centrifuge tubes for 25 min, liquid cycle. Let the tubes cool down in the tissue culture hood. Transfer the HeLa S3 cell suspension into two centrifuge tubes and spin at 1000 rpm (100 g) for 5 min. Discard the supernatant and remove any residual medium. Resuspend the cells in 30 ml prewarmed DMEM for each tube. Centrifuge and discard supernatant as before. Combine the resuspended cells of two centrifuge tubes in 20 ml prewarmed DMEM and inoculate the spinner prepared as described below.
6. During the centrifugation, wash the spinner three times with 30 ml prewarmed DMEM. Try to remove as much medium as possible. Transfer 1 L DMEM and 100 ml FBS into the spinner. At this time, do not add any antibiotics. The washed HeLa S3 cells are added to this medium.
7. Incubate the cells at 37°C, 5% CO_2 with slow stirring for 3 h. Add 22 μl of 5 mg/ml nocodazole in DMSO under sterile conditions. Incubate the cells at 37°C, 5% CO_2 for 11 h. At this time, more than 95% of cells should be in mitosis, which can be tested by staining cells with DAPI and counting cells with condensed chromosomes under the microscope. Note that the majority of cells should not have been in mitosis for longer than 2 h. If mitotic extracts are prepared, cells are harvested and processed for extraction as described below.
8. For the preparation of G1-extracts, transfer the HeLa S3 cells to autoclaved centrifuge tubes, and spin at 1000 rpm (100 g) for 5 min. Carefully aspirate any supernatant. Resuspend cells in 30 ml prewarmed DMEM, centrifuge, and carefully remove any supernatant. Resuspend cells in 20 ml prewarmed DMEM. During the centrifugation of the cells, wash spinners with prewarmed DMEM as described before.

Add 1 L prewarmed DMEM, 100 ml FCS to the spinner. Add the resuspended cells to the medium, and incubate cells for 2–4 h at 37°C, 5% CO_2 (the exact time of nocodazole release depends on the experiments to be done with the G1-extract).

3.1.2. Synchronization of HeLa S3 Cells in S Phase

1. Inoculate DMEM, 10% FCS, antibiotics with HeLa S3 cells in autoclaved spinners and incubate cells for 3 days at 37°C, 5% CO_2 as described before. Add 1 ml of a sterile 200 mM thymidine solution, and incubate cells for 24 h.
2. Wash cells and spinners as described under **Section 3.1.1**, and add resuspended cells to 1 L DMEM, 100 ml FCS. Incubate cells for 6 h followed by addition of 11 ml sterile 200 mM thymidine solution and incubate for further 18 h at 37°C, 5% CO_2. Cells can either be harvested for extraction or released into fresh medium and harvested at later time points.

3.1.3. Synchronization of Adherent Cells in Quiescence

1. Plate T24 or T98G cells in 20 15 cm dishes in DMEM/10% FCS. When cells reach confluency, remove medium. Wash twice with prewarmed DMEM, and add prewarmed DMEM, 0.1% FCS. Incubate for 72 h.
2. To harvest quiescent cells, aspirate medium. Add 10 ml PBS to each plate, and harvest cells by gentle scraping. Collect cells in centrifugation tubes and spin at 1000 rpm (100 g), 4°C, for 5 min. Discard PBS and place cells on ice.

3.2. Cell Extract Preparation

1. To harvest HeLa S3 cells for extraction, centrifuge cells in 500 ml tubes at 1000 rpm (100 g) for 5 min. Resuspend cells in 30 ml PBS per tube, and centrifuge again. Carefully remove any supernatant. Resuspend cells in 20 ml PBS, and combine cells in a 50 ml Falcon tube. Centrifuge at 1000 rpm (100 g) for 5 min. Discard the supernatant, and place cells on ice.
2. Resuspend the cell pellet in an equal volume of ice-cold PBS. Transfer the suspension to 2 ml Eppendorf tubes. Centrifuge cells at 1200 rpm, 4°C, for 5 min. Carefully discard any supernatant, even if you lose some cells.
3. To the cells in each tube, add 70% cell volume of ice-cold SB-buffer (total volume ~1.7 ml). Add 50 μl energy mix. Incubate on ice for 30 min, flip every 5 min.
4. Freeze cells in liquid nitrogen. Quickly thaw cells in 30°C water bath. Repeat freeze–thaw procedure.
5. Shear with 20 1/2G needle 7–8 times. Then shear with 25 3/8G needle twice. Centrifuge at 5000 rpm (2300 g), 4°C, for 5 min. Take supernatant, and centrifuge it again at 14,000 rpm (16,000 g), 4°C, for 30 min. Combine all supernatants in one precooled 15 ml Falcon tube on ice.

6. Aliquot 200 μl into 0.5 ml tubes. Freeze in liquid nitrogen. Store at –80°C.

3.3. Experiments Using Synchronized Human Cell Extract

3.3.1. Degradation Assays

Extracts provide a platform for studying the APC/C-dependent degradation of important cell cycle regulators. Extracts are supplemented with ^{35}S-labeled substrate prepared by in vitro transcription/translation (IVT) and the amount of substrate remaining relative to an input is analyzed. Degradation assays can be employed to measure substrate binding to APC/C, test E2 activity with APC/C, and screen candidate APC/C substrates.

3.3.1.1. Determining the Strength of APC/C Binding Motifs

Substrates bind to APC/C through conserved motifs, such as the D- or KEN-box *(2)*. Recombinant competitor proteins or synthetic peptides containing APC/C-binding sites can be titrated into degradation assays over a wide concentration range. Binding of proteins or peptides to the APC/C interferes with the degradation of a known ^{35}S-labeled APC/C-substrate, such as securin. Proteins that bind strongly to APC/C compete at low concentrations and stabilize the tracer substrate, whereas weakly bound proteins either do not compete with the tracer substrate or do so only at the highest concentrations.

3.3.1.2. Testing E2 Specificity

APC/C works with multiple E2 enzymes *(2)*. The ability of an E2 to form a functional complex with APC/C and substrate can be tested by titrating recombinant E2 into a degradation assay in cell extract and assaying degradation of a ^{35}S-labeled substrate at a single endpoint. In these assays, we use securin$^{\Delta DBOX}$ or cyclin A as substrates, both of which are degraded with slow kinetics in the absence of exogenous E2. Only E2s that work efficiently with APC/C promote substrate degradation at low concentrations.

3.3.1.3. Strong Degradation Assay to Test Candidate APC/C Substrates

Degradation assays in extract provide a means to identify novel APC/C substrates. To maximize APC/C activity, these assays are performed at 30°C with excess ubiquitin and recombinant E2. In strong degradation assays, a 2 h degradation reaction is compared with an input and a control reaction containing the APC/C inhibitor Emi1. Candidates that are APC/C substrates show diminished levels in the degradation reaction but maintain levels close to input in the presence of Emi1. Strong degradation assays can be tailored to screen libraries of candidate substrates in a 96-well format.

3.3.1.4. APC/C Specificity Controls in Degradation Assays

Demonstrating that turnover of a substrate depends on APC/C and the 26S proteasome is crucial when testing substrate degradation in extracts. APC/C-dependent degradation in extracts is impaired by addition of APC/C-inhibitors, such as Emi1 or dominant negative UbcH10^{C114S}, or by an excess of recombinant APC/C-substrates. 26S proteasome inhibitors such as MG132 (f.c. 40 μM) are used to show that degradation in extract depends on the ubiquitin–proteasome pathway.

APC/C Degradation Assay

1. Set up the following master mix:

Human Somatic Cell Extract	20 μl
Ubiquitin (10 mg/ml)	0.6 μl
Energy Mix	0.6 μl
E2 such as UbcH10 or UbcH5c	0.6 μl (optional)
^{35}S-labeled Substrate IVT	1.25 μl

2. For specificity controls set up the same master mix, but include alternatively an APC/C-substrate as competitor (1 μl of 2.5 mg/ml stock), recombinant Emi1 (1 μl of 2.5 mg/ml stock), or MG132 (40 μM final concentration).
3. Aliquot 5 μl reactions from the master mix and incubate them at 23°C. Stop reactions at desired time points (usually 0, 40, 80, 120 min). Competitor titrations and E2 specificity reactions are stopped at a single endpoint, usually 1.5 h. Reactions are stopped by adding 10 μl gel-loading buffer and boiling 5 min. Degradation assay samples are resolved by SDS-PAGE and visualized by autoradiography.

Strong Degradation Assay to Test Candidate Substrates:

1. Set up the following master mix as described below. Set up one control master mix that includes the APC/C-inhibitor Emi1.

G1-extract	50 μl
Ubiquitin (10 mg/ml)	12 μl
Energy Mix	12 μl
UbcH10 (2.5 mg/ml stock)	3 μl
Ubiquitin aldehyde (alternative)	1 μl
Alternatively: Emi1 (2.5 mg/ml)	12 μl

2. Stop one control reaction immediately after addition of substrate. Load this sample as "input".
3. Aliquot 10 μl reactions from the master mixes above, and add 1 μl ^{35}S-labeled candidate substrate IVT to each reaction. Place the degradation reaction and the control reaction containing Emi1 at 30°C for 2 h. Stop with 15 μl gel-loading buffer, and boil samples for 5 min at 95°C. Resolve samples by SDS-PAGE and visualize by autoradiography.

4. The same reactions can be set up to test candidate substrates in a 96-well format, thereby allowing screening for APC/C-substrates using human full genome unigene libraries (Genescript).

3.3.2. APC/C-Dependent Ubiquitination Assays

Recapitulation of APC/C-dependent ubiquitination in a semi-purified system allows mechanistic study of APC/C-activity. We term this system "semi-purified" because APC/C is purified by immunoprecipitation from extracts. Here we describe how to perform in vitro ubiquitination reactions using APC/C, outline an assay to test substrate ordering by APC/C, and explain how to use the ubiquitin-chain binding protein hHR23A to measure the processivity of APC/C-dependent ubiquitination.

3.3.2.1. Standard APC/C-Dependent Ubiquitination

In APC/C-dependent ubiquitination reactions, APC/C is affinity-purified from mitotic or G1-extracts and the E1–E2–E3 cascade is reconstituted in vitro. Recombinant E1 is obtained from baculovirus infected Sf9 cells, whereas E2 enzymes can often be purified from *E. coli*. Additionally, purified ubiquitin, DTT, and energy mix are required. In the presence of all components, APC/C promotes the ubiquitination of ^{35}S-labeled substrates prepared by IVT. Addition of methyl-ubiquitin allows for monoubiquitination, but eliminates formation of ubiquitin chains, and thereby reveals the number of modified lysine residues in an APC/C-substrate. Ubiquitin mutants incapable of forming certain linkages, such as ubiquitin K48R, can be employed to reveal any preference of APC/C for a specific chain topology. As in degradation assays, recombinant proteins or peptides can be added to APC/C-dependent ubiquitination reactions to compete for substrate binding.

3.3.2.2. Substrate Ordering Assay

Substrates bind APC/C through multiple binding motifs, which determine the processivity of substrate ubiquitination *(5)*. Processive substrates receive a long ubiquitin chain in a single APC/C-binding event and are degraded early in mitosis. Distributive substrates are decorated with ubiquitin chains only after multiple binding events, which delays their degradation upon mitotic exit. The substrate ordering assay recapitulates this important aspect of APC/C activity in a semi-purified system. APC/C is presented two recombinant substrates in equimolar concentrations. The more processive substrate is still modified with long ubiquitin chains, whereas the more distributive substrate receives only few ubiquitin moieties. The substrate ordering assay is performed as the APC/C-dependent ubiquitination reaction described above, but purified recombinant substrates are used in place of ^{35}S-labeled substrate from IVT. The ubiquitination profiles of each substrate are compared by Western blotting using specific antibodies.

3.3.2.3. Testing Processivity of APC/C Ubiquitination

A semi-quantitative assay for testing the processivity of APC/C-dependent substrate ubiquitination exploits the capacity of UBA-domain containing proteins, such as HHR23A, to tightly bind ubiquitin conjugates. Distributive substrates frequently dissociate from APC/C in a partially ubiquitinated state. UBA-domains present in ubiquitination assays capture dissociated, partially ubiquitinated substrates, and block their rebinding to APC/C and further chain elongation. Conversely, processive substrates receive their ubiquitin chain in a single binding event to APC/C, and UBA-domains are incapable of interfering with chain elongation. Thus, the length of ubiquitin chains of distributive substrates, but not of processive substrates, is strongly reduced if UBA-domains are included in ubiquitination assays.

APC/C-Dependent Ubiquitination Assay

1. Wash 80 μl protein G-agarose twice with 800 μl cold SB/U. Resuspend the washed beads in 800 μl SB/U. Add 50 μl monoclonal anti-Cdc27 antibody (0.2 mg/ml; Santa Cruz). Incubate for 1 h at 4°C on a roller drum.
2. Thaw 1 ml human somatic extract and centrifuge 20 min, 10,000 rpm (10,500 g), 4°C. Discard the pellet, and repeat the centrifugation step.
3. Spin antibody-coupled beads down, and remove supernatant. Add cleared extract to the beads. Incubate on a roller drum for 4 h at 4°C.
4. Spin down beads and discard supernatant. Wash beads 3 times with 800 μl SB/U containing 0.05% Triton X-100. Wash beads twice with 800 μl SB/U with no detergent. It is crucial to remove detergent completely before proceeding with the assay.
5. Resuspend beads in 25 μl SB/U and mix gently with pipette tip. Prepare the following master mix. The scale above should be enough for ~12 reactions. Work on ice and do not allow components to sit for too long.

10X UBAB buffer		1 μl
100 mM DTT		1 μl
Energy Mix		1.5 μl
Ubiquitin (10 mg/ml stock)		1.5 μl
Purified human E1 enzyme		0.5 μl
E2 enzyme (UbcH5c or UbcH10)		0.5 μl
Optional:	Recombinant UBA-domain (2.5 mg/ml)	1 μl
	Recombinant Emil (2.5 mg/ml)	1 μl
	Substrate competitor (2 mg/ml) or Synthetic peptide (3.5 mM)	1.5 μl

6. Set up reactions in the following manner. Cut a 10 μl pipette tip and stir beads carefully. Add 5 μl beads per reaction to a 1.5 ml Eppendorf tube. Add 6 μl master mix. Add 1.5 μl ^{35}S-labeled IVT for standard ubiquitination assay or 1 μl purified recombinant substrate for substrate ordering assay. Mix gently by swirling tip.
7. Incubate in a thermoshaker at 800 rpm at 25°C. Ubiquitination time depends on the substrate, and can range between 2 min for processive substrates and 30 min for very distributive substrates.
8. Stop reactions by adding 15 μl gel-loading buffer, and denature samples for 5 min at 95°C. Resolve ubiquitination reactions by SDS-PAGE and visualize by autoradiography (^{35}S-labeled substrate) or Western blot (substrate ordering assay).

3.3.3. Dissecting Spindle Checkpoint Regulation Using Mitotic Extract

The spindle checkpoint ensures proper sister chromatid separation during mitosis by preventing APC/C-activation until all chromosomes have achieved bipolar attachment to the mitotic spindle *(7)*. The spindle checkpoint proteins Mad2 and BubR1 inhibit APC/C by binding to its activator Cdc20 *(8)*. Once all kinetochores are attached to the spindle, APC/C^{Cdc20} is activated within minutes to promote the ubiquitination of securin and cyclin B1. Activation can be achieved by the nonproteolytic ubiquitination of Cdc20 and dissociation of Mad2 and BubR1 *(9, 10)*. This process can be opposed by the Cdc20-directed deubiquitinating enzyme, Usp44 *(11)*.

The spindle checkpoint regulation of APC/C can be reconstituted in human somatic extracts. Mitotic extracts derived from cells with an active spindle checkpoint display no APC/C^{Cdc20} activity, and APC/C-substrates are stable. The spindle checkpoint inhibition of APC/C^{Cdc20} in mitotic extract can be overcome by adding recombinant UbcH10 or p31^{comet} protein *(9, 10)*. Activation of APC/C^{Cdc20} can be observed by blotting extracts for cyclin B1 levels (which decrease over time) or by autoradiography of added ^{35}S-labeled substrates. The degradation assays are performed as described above.

Multiubiquitination by the APC/C, UbcH10, and p31^{comet} triggers the dissociation of Mad2 from Cdc20, which can be studied in semi-purified systems. APC/C^{Cdc20} bound to Mad2 and BubR1 is affinity purified from mitotic extract using anti-Cdc20 antibody. UbcH10, ubiquitin, and energy mix are added to the APC/C-beads followed by incubation at 25°C for 40 min. After extensive wash, the beads are boiled in gel loading buffer. The samples are analyzed by SDS-PAGE and Western blotting using specific antibodies against Cdc20, Mad2, and BubR1. The ubiquitinated Cdc20 can also be used to study the activity of Cdc20-directed deubiquitinating enzymes, such as Usp44.

4. Notes

1. It is crucial that synchronization protocols limit the time cells spend in mitosis in the presence of nocodazole. A prolonged nocodazole-dependent mitotic arrest has several nonphysiological consequences that strongly affect APC/C-dependent ubiquitination and degradation in cells and extracts. First, APC/C-substrates, such as securin, are degraded during an arrest in mitosis in the presence of nocodazole, and thus, they are depleted from extracts *(12)*. As the sequential degradation of cell cycle regulators is controlled by their competition for APC/C, the disappearance of APC/C-substrates in cells or extracts will strongly influence the degradation kinetics of any other substrate. Second, the E2 UbcH10 is depleted from cells during a prolonged arrest in nocodazole, which results in apparently low APC/C-activity (unpublished observation). Finally, extracts prepared after prolonged mitotic arrest have caspase activity and degrade several substrates independently of the APC/C. We have found that cells should not be kept in mitosis in the presence of nocodazole for longer than 2 h. Therefore, to avoid artifacts, we regard synchronization protocols that maintain a prolonged mitotic arrest *(13)* as unsuitable for the biochemical analysis of the APC/C.
2. We and others have observed dramatic effects of epitope tagging of APC/C-substrates on their degradation in extracts and their ubiquitination in purified systems. In most cases, epitope-tagging of APC/C-substrates inhibits their degradation. We observed this for several substrates, including UbcH10, which is strongly stabilized by either amino- or carboxy-terminal epitope tags *(6)*. In a few instances, epitope-tagging can lead to artificial ubiquitination within the epitope-tag, thereby masking effects of mutations on substrate ubiquitination and degradation. An important example is GFP-tagged securin, where ubiquitination within the GFP-tag masked the strong inhibitory effects of deleting the D-box of securin on its APC/C^{Cdh1}-dependent ubiquitination. GFP-tagged securin$^{\Delta DBOX}$ is, therefore, a problematic substrate for analyzing the timing of APC/C^{Cdh1}-activation in living cells *(14)*. As GFP-tagged proteins are commonly used to measure APC/C-activity in living cells, extracts provide the opportunity to rapidly test whether epitope-tagging affects their ubiquitination or degradation.
3. Extracts can be manipulated by immunodepletion or by addition of recombinant proteins. It is crucial to ensure

the full activity of added recombinant enzymes, as inactive or partially active enzymes can produce dominant-negative effects. As an example, amino-terminally tagged UbcH10 is less active than untagged or carboxy-terminally tagged UbcH10. It is not capable of targeting all its physiological targets for degradation, and its use can lead to misleading results when studying APC/C-activity or regulation.

References

1. Sullivan, M., and Morgan, D.O. (2007) Finishing mitosis, one step at a time. *Nat Rev. Mol. Cell Biol.* **8**, 894–903.
2. Peters, J. M. (2006) The anaphase promoting complex/cyclosome: a machine designed to destroy. *Nature Rev. Mol. Cell Biol.* **7**, 644–656.
3. Sotillo, R., Hernando, E., Díaz-Rodríguez, E., Teruya-Feldstein, J., Cordón-Cardo, C., Lowe, S.W., and Benezra, R. (2007) Mad2 overexpression promotes aneuploidy and tumorigenesis in mice. *Cancer Cell* **11**, 9–23.
4. Dye, B.T., and Schulman, B.A. (2007) Structural mechanisms underlying posttranslational modification by ubiquitin-like proteins. *Annu. Rev. Biophys. Biomol. Struct.* **36**, 131–50.
5. Rape, M., Reddy, S.K., and Kirschner, M.W. (2006) The processivity of multiubiquitination by the APC determines the order of substrate degradation. *Cell* **124**, 89–103.
6. Rape, M., and Kirschner, M.W. (2004) Autonomous regulation of the anaphase-promoting complex couples mitosis to S-phase entry. *Nature* **432**, 588–95.
7. Yu, H. (2002) Regulation of APC-Cdc20 by the spindle checkpoint. *Curr. Opin. Cell Biol.* **14**, 706–14.
8. Musacchio, A., and Hardwick, K. G. (2002) The spindle checkpoint: structural insights into dynamic signalling. *Nat. Rev. Mol. Cell Biol.* **3**, 731–41.
9. Reddy, S.K., Rape, M., Marganski, W.A., and Kirschner, M.W. (2007). Mutual regulation between the spindle checkpoint and the anaphase-promoting complex ensures timely progression to anaphase. *Nature* **446**, 921–5.
10. Xia, G., Luo, X., Habu, T., Rizo, J., Matsumoto, T., and Yu, H. (2004). Conformation-specific binding of p31 (comet) antagonizes the function of Mad2 in the spindle checkpoint. *EMBO J.* **23**, 3133–43.
11. Stegmeier, F., et al. (2007) Anaphase initiation is regulated by antagonistic ubiquitination and deubiquitination activities. *Nature* **446**, 876–81.
12. Brito, D.A., and Rieder, C.L. (2006). Mitotic checkpoint slippage in humans occurs via cyclin B destruction in the presence of an active checkpoint. *Curr. Biol.* **16**, 1194–200.
13. di Fiore, B., and Pines, J. (2007). Emil is needed to couple DNA replication with mitosis but does not regulate activation of the mitotic APC/C. *J. Cell Biol.* **177**, 425–37.
14. Hagting, A., Den Elzen, N., Vodermaier, H.C., Waizenegger, I.C., Peters, J.M., and Pines, J. (2002). Human securin proteolysis is controlled by the spindle checkpoint and reveals when the APC/C switches from activation by Cdc20 to Cdhl. *J. Cell Biol.* **157**, 1125–37.

Chapter 20

Biochemical Analysis of the Anaphase Promoting Complex: Activities of E2 Enzymes and Substrate Competitive (Pseudosubstrate) Inhibitors

Matthew K. Summers and Peter K. Jackson

Abstract

The Anaphase Promoting Complex (APC) ubiquitin ligase is critical for multiple processes including cell cycle, development, meiosis, and senescence. The importance of regulation of the APC by substrate competitive (pseudosubstrate) inhibitors, such as Emi1 and BubR1, has recently been demonstrated. Substrate competitive inhibitors typically bind to enzymes via the same site as substrates, but by having any combination of increased enzyme affinity and low turnover numbers, are able to "clog" the ability of the enzyme to bind and turnover substrates. For the APC, these pseudosubstrates can both position and block the APC and have been well validated as critical regulators for the APC enzymes.

We have found that the substrate competitive mechanism of inhibition is sensitive to the E2 activity driving APC catalyzed ubiquitination events. This chapter provides detailed protocols for multiple in vitro ubiquitination assays of increasing complexity and the use of pseudosubstrate inhibitors in these assays. These assays are instrumental in examining the use of E2 enzymes by the APC and the intimate relationship this has with pseudosubstrate inhibition.

Key words: Ubiquitination, mitosis, Anaphase Promoting Complex (APC), UbcH10, UbcH5, pseudosubstrate, Emi1, BubR1.

1. Introduction

The Anaphase Promoting Complex (APC) is a multisubunit E3 ubiquitin ligase complex, which catalyzes the ubiquitination and destruction of multiple cell cycle regulators, primarily the mitotic cyclins and securin. To ubiquitinate these substrates, the APC requires an E2 enzyme to direct ubiquitin into the pathway.

Andrew D. McAinsh (ed.), *Mitosis: Methods and Protocols*, Methods in Molecular Biology 545,
DOI 10.1007/978-1-60327-993-2_20, © Humana Press, a part of Springer Science+Business Media, LLC 2009

Two E2 enzyme families, Ubc4 and E2-C, have been demonstrated to function efficiently with the APC in vitro *(1)*. UbcH5 and UbcH10, respectively, represent these families in humans. UbcH5 is a promiscuous E2 in vitro and functions with a multitude of E3s including HECT, for example, E6-AP; single polypeptide RING, for example, CHFR; and multisubunit RING, for example, SCF, APC ligases *(2–4)*. Conversely, there is little evidence for in vivo activity of UbcH5 with the APC or these other ligases. Genetic evidence, however, supports the E2-C family enzymes as required for physiological APC function *(5, 6)*. In addition, as are other positive regulators of APC activity, the levels of E2-C enzymes are also cell cycle regulated in an APC-dependent manner *(7, 8)*. The activity of an E3 can change dramatically (at least in vitro) depending on the E2 it is partnered with, including, mono- versus polyubiquitination, polyubiquitin linkages, and usage of substrate lysines *(3, 9, 10)*. Clearly, to fully understand the APC (or other E3 ligase pathways) identifying and characterizing its physiological E2, enzymatic partner, is of critical importance.

APC activity is tightly regulated by additional mechanisms to ensure the timely and coordinated destruction of its substrates *(11, 12)*. Two of the most extensively studied regulators are Emi1 and the spindle checkpoint. Recently, Emi1 and the spindle checkpoint component, BubR1/Mad3, have been demonstrated to function via a common mechanism. By virtue of KEN and destruction box (D-box) APC targeting domains, these (and potentially other) APC inhibitor proteins function, primarily, as pseudosubstrates by exhibiting a higher affinity for the APC than its substrates (**Fig. 20.1**, *13–15)*. Pseudosubstrate-APC interactions are dynamic, allowing the possibility for brief interactions of substrates with the APC. This mode of inhibition is therefore sensitive to the flux of ubiquitin to the APC provided by the E2. Increasing the transfer of ubiquitin to a substrate, either by increasing E2 concentration or by the use of an E2 with an intrinsically higher ability to transfer, decreases the potency of these inhibitors *(16)*.

In this chapter we present detailed protocols for a series of ubiquitination assays that demonstrate the biochemical difference between the activity of an E3 with a promiscuous E2 and the activity of an E3 with its cognate E2 as well as the effects that these pairings have on pseudosubstrate inhibition. These assays increase in complexity: APC11 RING-domain protein, the APC2-APC11 Cullin-RING subcomplex, the APC holoenzyme in vitro, and the APC holoenzyme in a more physiological extract. In vitro, UbcH5 functions with APC11, the APC2-APC11 subcomplex, and the holoenzyme *(16–18)*. In contrast, and in testimony to its specificity for the APC, UbcH10 (E2-C family) functions only with the holoenzyme (**Fig. 20.2A–C,** *16)*.

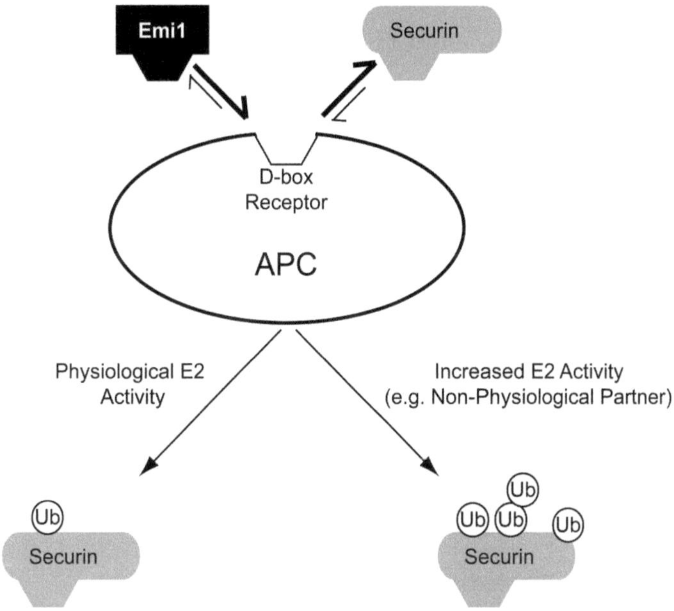

Summers and Jackson
Biochemical Analysis of the Anaphase Promoting Complex:
Activities of E2 Enzymes and Substrate Competitive ("Pseudosubstrate") Inhibitors

Fig. 20.1. Schematic representation of the relationship between E2 activity and pseudosubstrate inhibition of the APC. Pseudosubstrates, such as Emi1, exhibit higher affinity for the APC than the substrates. The pseudosubstrate will outcompete the substrate for the APC when it is active. This interaction is dynamic, however, with the pseudosubstrate engaging and disengaging the APC. Although this allows brief windows for the true substrates, such as securin, to engage the APC, the pseudosubstrates will rapidly displace these substrates. Under normal conditions E2 activity is not sufficient to ubiquitinate a substrate during this engagement. However, if the E2 activity is increased, either by concentration or by the use of a different E2 that may transfer more ubiqutin during this window, the substrate is aberrantly ubiquitinated and destroyed.

Furthermore, UbcH5 targets a greater number of substrate lysines than UbcH10 and is capable of bypassing pseudosubstrate inhibitors more readily than UbcH10 (**Fig. 20.2c**). Mitotic HeLa cell extracts recapitulate the activity of the spindle checkpoint and do not allow the ubiquitination or destruction of APC substrates, such as securin (discussed in more detail in the previous chapter). Increasing E2 activity in these extracts triggers the ubiquitination of substrates and further illustrates the specificity of UbcH10, which readily functions in the extract *(16)*.

This multitiered assay scheme in combination with E2 and, potentially, E3 mutants will facilitate the analysis of the specificity of the UbcH10 and APC interaction and allow the examination of ubiquitin transfer by this enzyme pair *(16)*. Similarly, analysis of pseudosubstrate inhibitors may also be carried out. Note that the ability to manipulate securin stability in mitotic HeLa extracts allows the sufficiency and necessity of multiple factors/activities to be tested.

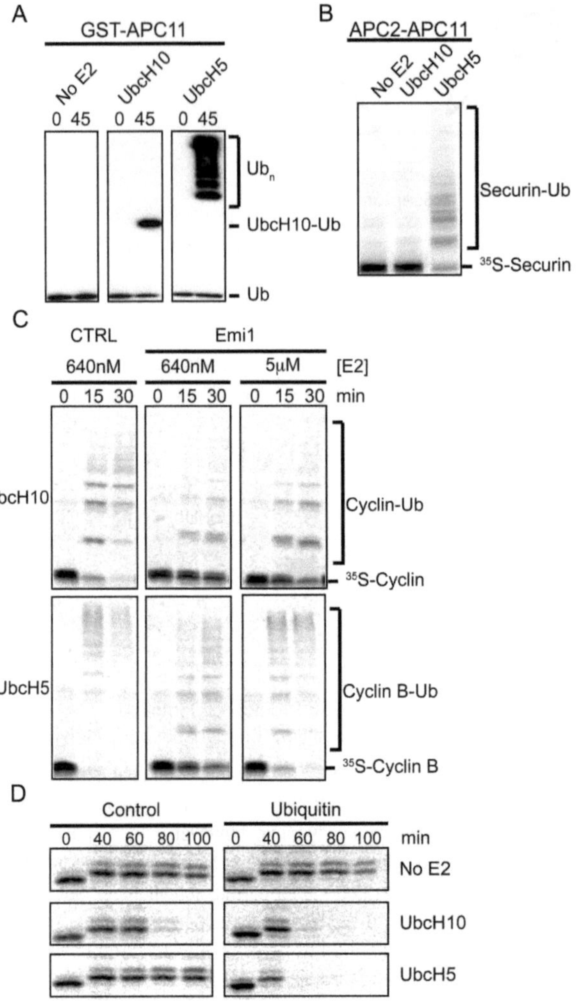

Summers and Jackson
Biochemical Analysis of the Anaphase Promoting Complex:
Activities of E2 Enzymes and Substrate Competitive ("Pseudosubstrate") Inhibitors

Fig. 20.2. E2 activity assays with various forms of the APC, performed as described. (**A**) UbcH5, but not UbcH10 is capable of forming polyubiqutin chains with GST-APC11. UbcH10 autoubiquitination is not APC11-dependent. (**B**) UbcH5, but not UbcH10, functions with APC2-APC11 in the ubiquitination of securin. (**C**) Both E2s function with the APC holoenzyme in vitro. Note that UbcH10-catalyzed ubiquitin conjugates are predominantly of lower molecular weight than UbcH5-catalyzed ubiquitin conjugates. Inclusion of Emi1 in the reactions dramatically decreases the conversion of substrate, which can be bypassed (strongly by UbcH5) by increasing the E2 concentration. (**D**) Both E2s function with the APC *in extracto.* Bypass of BubR1-mediated securin stability is observed for both E2s in the presence of excess ubiquitin. However, when ubiquitin is limiting, only UbcH10 is capable of bypass, demonstrating its specificity for the APC. Due to its ability to interact functionally with multiple E3s in the extract, UbcH5 requires a greater amount of ubiquitin for its function with the APC to be observed in this setting *(16).*

2. Materials

2.1. Preparation of Cdh1 and ^{35}S-Labeled APC/C Substrates

1. Vector containing T7 or SP6 promoter, such as pCDNA or pCS2
2. TNT Quick Coupled Transcription/Translation System (Promega)
3. RNAsin Plus (Promega)
4. Easy-Tag^{TM} L-[^{35}S]-Methionine (Perkin Elmer)

2.2. Preparation of Recombinant E1, APC11, Emi1, and E2 (UbcH10 and UbcH5) Enzymes in Bacteria

1. pET-28a vector (EMD Biosciences).
2. pGEX-6P-1 (GE Healthcare).
3. BL21(DE3)pLysS competent cells (EMD Biosciences).
4. LB-broth medium (Difco).
5. Carbanecillin and kanamycin.
6. Isopropyl-β-D-thiogalactoside (IPTG) (Roche).
7. Ni2+-NTA agarose (Qiagen).
8. Glutathione-sepharose 4B (GE Healthcare).
9. Glutathione (Sigma).
10. cOmplete Protease Inhibitor Cocktail Tablets (Roche).
11. PD-10 columns (GE Healthcare).
12. Slide-A-Lyzers 3.5 K and 10 K MWCO (Pierce).
13. Amicon Ultra Concentrators 5 K and 10 K MWCO (Millipore).
14. His-Tag lysis buffer: 20 mM HEPES (pH7.7), 500 mM NaCl, 10 mM imidazole, 10 mM β-mercaptoethanol, 0.5% Triton X-100. Immediately before use, add Complete Protease Inhibitor tablets.
15. His-Tag wash buffer: 20 mM HEPES (pH7.7), 500 mM NaCl, 30 mM imidazole, 10 mM β-mercaptoethanol, 0.5% Triton X-100.
16. His-Tag low salt buffer: 20 mM HEPES, 100 mM KCl, 10 mM β-Mercaptoethanol, 0.5% Triton X-100; adjust pH to 7.7 using NaOH.
17. His-Tag elution buffer: 20 mM HEPES (pH7.7), 100 mM KCl, 100 mM imidazole, 10 mM β-mercaptoethanol, 0.5% Triton X-100.
18. GST lysis buffer: 50 mM Tris-HCl (pH 8.0), 500 mM NaCl, 0.5% Triton X-100, 1 mM DTT. Just before use add Complete Protease Inhibitor tablet(s).
19. GST-wash buffer: 50 mM Tris-HCl (pH 8.0), 500 mM NaCl, 0.5% Triton X-100, 1 mM DTT.
20. GST-low salt buffer: 50 mM Tris-HCl (pH 8.0), 100 mM KCl, 1 mM DTT.
21. GST-elution buffer: 50 mM Tris-HCl (pH 8.0), 100 mM KCl, 1 mM DTT, 10 mM GSH.

22. Storage buffer: 20 mM HEPES (pH 7.7), 100 mM KCl, 1 mM DTT, 10% glycerol.
23. SDS-PAGE sample buffer (2×).
24. Liquid nitrogen.

2.3. Expression and Purification of APC2-APC11 Complex Using Insect Cells

1. His-Tag-APC2 and untagged APC11 in pFastBac vector (Invitrogen).
2. Bac-to-Bac Baculovirus Expression System (Invitrogen).
3. Sf9 cells (ATCC).
4. Fetal Bovine Serum (Inivtrogen).
5. Pluronic F-68 (Invitrogen).
6. Grace's Medium (Invitrogen).
7. Insect cell lysis buffer: 20 mM Hepes (pH 7.4), 500 mM NaCl, 0.5% Triton X-100, 15 mM imidazole (pH 7.5), 10 mM beta-mercaptoethanol. Add protease inhibitors just before use.
8. Insect cell low salt buffer: 20 mM Hepes (pH 7.4), 100 mM NaCl, 0.5% Triton X-100, 15 mM imidazole (pH 7.5), 10 mM beta-mercaptoethanol.
9. Insect cell elution buffer: 20 mM Hepes (pH 7.4), 100 mM NaCl, 0.5% Triton X-100, 200 mM imidazole (pH 7.5), 15 mM beta-mercaptoethanol.
10. Dialysis buffer: 10 mM Hepes (pH 7.7), 100 mM NaCl, 1 mM DTT, 1 mM $MgCl_2$, 5% glycerol.
11. Complete Protease Inhibitor Cocktail Tablets (Roche).
12. Ni^{2+}-NTA agarose (Qiagen).

2.4. In Vitro Ubiquitination Reactions

1. Substrate Independent Buffer (SID): 50 mM Tris-HCl (pH 7.5), 5 mM $MgCl_2$, 0.6 mM DTT.
2. N-terminally tagged ubiquitin and detection reagent, for example, FLAG-Ubiqutin and M2 antibody (SIGMA).
3. ^{35}S-labeled substrates.
4. Protein A-Affiprep support (BioRad).
5. Fresh dimethyl pimelimidate (DMP).
6. Polyclonal anti-Cdc27 (APC3) sera or antibody.
7. 0.2 M $Na_2B_4O_7$, pH 9.0 (SB).
8. 200 mM ethanolamine, pH 8.0–8.5. Make fresh. Adjust the pH with HCl.
9. Thimerosal (Sigma).
10. HeLa cells (ATCC).
11. Ubiquitin and ubiquitin mutant lacking lysine residues (Boston Biochem).
12. XB buffer: 20 mM HEPES (pH 7.7), 100 mM KCl.
13. XB+400 mM KCl, 0.5% NP-40.
14. 20 × Energy Regeneration Mix (NRG): 30 U/mL rabbit creatine phosphokinase type I, 7.5 mM creatine phosphate, 1 mM ATP, 1 mM $MgCl_2$, 0.1 mM EGTA.

15. Cell lysis buffer: 50 mM Tris (pH 8.0), 120 mM NaCl, and 0.5% NP40, 10 μg/mL each leupeptin, pepstatin, and chymostatin, and 1 mM DTT.
16. Liquid nitrogen.
17. SDS-PAGE sample buffer.

2.5. In Extracto Ubiquitination and Destruction Assays

1. 150 mm Tissue Culture Dishes (Corning).
2. 10-Stack CellStack Chamber (Corning).
3. HeLa cells (ATCC).
4. Dulbecco's Modified Eagle's Medium (Invitrogen).
5. 0.05% Trypsin-EDTA (Invitrogen).
6. Fetal bovine serum.
7. Nocodazole (EMD Biosciences).
8. Cell Disruption Bomb (PARR Instruments).
9. Mitotic HeLa Extract buffer: 20 mM Tris-HCl, pH 7.2, 2 mM DTT, 0.25 mM EDTA, 5 mM KCl, 5 mM $MgCl_2$.
10. 25 gauge needle.
11. Compressed nitrogen.
12. Liquid nitrogen.
13. Energy Regeneration Mix (*see* **Section 2.4**).
14. Ubiquitin (Boston Biochem).
15. ^{35}S-labeled substrates.
16. Nondestructible cyclin B fragment (Δ90).
17. Phosphorimager/Phosphorimager Screen (GE Healthcare).
18. SDS-PAGE sample buffer.

3. Methods

3.1. Preparation of Cdh1 and ^{35}S-Labeled APC/C Substrates

1. Clone Cdh1, securin, or other substrates into suitable vectors for in vitro translation (IVT). For securin make both an untagged and a version with 5 tandem Myc epitope tags at the N-terminus (*see* **Note 1**).
2. For each substrate, mix 1 μg pCS2 construct with 25 μL TNT Quick Coupled transcription/translation rabbit reticulocyte lysate, 1 μL RNAsin Plus, 2 μL TNT buffer, 1 μL minus methionine amino acids, 1 μL SP6 polymerase, and 2 μL ^{35}S-methionine. Add ddH_2O to 50 μL. For Cdh1, substitute 1 μL minus leucine amino acids and 1 μL ddH_2O for the ^{35}S-methionine.
3. Incubate at 30°C for 90 min.
4. Spin the IVT reaction at maximum speed in a microcentrifuge for 10–15 min at room temperature (*see* **Note 2**).
5. Remove the supernatant to a fresh tube.
6. Store on ice until ready to use or freeze in liquid nitrogen and store at –80°C.

3.2. Preparation of Recombinant E1, APC11, Emi1, and E2 (UbcH10 and UbcH5) Enzymes in Bacteria

1. Make bacterial expression constructs of the above genes in pGEX or pET bacterial expression vectors (*see* **Note 3**).
2. Transform expression constructs into BL21 (DE3) pLysS competent cells, plate on LB-Agar plus appropriate antibiotics, and incubate overnight at 37°C.
3. Inoculate 100 mL of prewarmed LB with appropriate antibiotics with one colony and incubate overnight at 37°C with shaking.
4. Dilute the overnight culture 1:200 into prewarmed LB plus appropriate antibiotics and continue to grow at 37°C until the culture reaches an OD_{600} of 0.6–0.8 (approximately 2 h).
5. Remove a 100 μL sample, pellet the bacteria, and resuspend in 100 μL of SDS-sample buffer.
6. Chill the cultures on wet ice to 25°C.
7. Add IPTG to 1.0 mM and incubate at 25°C with shaking for an additional 4–6 h.
8. Remove another 100 μL sample, pellet the bacteria, and resuspend in 100 μL of SDS-sample buffer. Monitor protein expression by comparison of the before and after IPTG samples by running on a 10% SDS-PAGE gel and staining with Coomassie Blue.
9. Pellet the cultures at 9000 g for 10 min at 4°C. Remove supernatant and resuspend the pellets in 30 mL of (GST- or His-Tag-) lysis buffer per liter of culture.
10. Freeze the resuspended pellets in liquid nitrogen and store at –80°C until purification.

3.2.1. Purification of GST-APC11 and GST-Emi1 C-Terminus

1. Thaw the resuspended cell pellets in room temperature water.
2. Sonicate lysate on ice using 40 s pulses, until lysate becomes clearer with a more watery consistency (approximately 4–6 pulses).
3. Centrifuge the lysate at 30,000 g for 30 min at 4°C and pass the supernatant through a 0.45 μm filter.
4. Wash GSH-sepharose three times with GST-lysis buffer.
5. Remove a sample of the cleared lysate for SDS-PAGE.
6. Add 1–2 mL of resin/L of culture to the lysates. Dilute the lysate twofold with GST-lysis buffer. Mix gently at 4°C for 1.5 h.
7. Collect a sample of the lysate flow-through for SDS-PAGE.
8. Load the resin into an empty column and wash with at least 10 column volumes of GST-wash buffer.
9. Wash with 4 column volumes of GST-low salt buffer (*see* **Note 4**).
10. Elute with GST-elution buffer. Collect 2 mL fractions. Check each fraction for protein by mixing 5 μL with 45 μL

of 1 × Bradford reagent. Collect the protein containing fractions.

11. Examine before and after IPTG samples, cleared lysate, flow-through, and 5–10 μL of the fractions for target protein by SDS-PAGE and Coomassie staining. Combine fractions containing the protein.
12. Load the protein into a 10,000 kD MWCO (Emi1) or 3,500 kD WCO (APC11) Slide-a-lyzer and dialyze into storage buffer. Dialyze for at least 2 h in two 1 L (or greater) exchanges of buffer.
13. Aliquot and freeze the protein in liquid nitrogen. Store at –80°C.

3.2.2. Purification of His-Tagged E2s

1. Thaw cell resuspended pellets in room temperature water.
2. Sonicate lysate on ice using 40 s pulses, until lysate becomes clearer with a more watery consistency (approximately 4–6 pulses).
3. Centrifuge the lysate at 30,000 g for 30 min at 4°C and pass the supernatant through a 0.45 μm filter.
4. Wash Ni^{2+}-NTA three times with His-Tag-lysis buffer.
5. Remove a sample of the cleared lysate for SDS-PAGE.
6. Add 1–2 mL of resin/L of culture to the lysates. Dilute the lysate twofold with His-Tag-lysis buffer. Mix end-over-end at 4°C for 1.5 h.
7. Collect a sample of the lysate flow-through for SDS-PAGE.
8. Load the resin into an empty column and wash with at least 10 column volumes of His-Tag-wash buffer.
9. Wash with 4 column volumes of His-Tag-low salt buffer.
10. Elute with His-Tag-elution buffer. Collect 2 mL fractions. Check each fraction for protein by mixing 5 μL with 45 μL of 1 × Bradford reagent. Collect the protein-containing fractions.
11. Exchange the proteins into storage buffer using PD10 columns.
12. Check the protein purity and expression. Examine before and after IPTG samples, cleared lysate, flow-through, 5–10 μL of the fractions by SDS-PAGE, and Coomassie staining. Combine fractions containing the protein.
13. Aliquot and freeze the protein in liquid nitrogen. Store at –80°C.

3.3. Expression and Purification of APC2-APC11 Complex Using Insect Cells

1. Prepare baculoviruses using the Bac-to-Bac Baculovirus Expression System.
2. Grow Sf9 cells to 2 × 10^6 cells/mL at 27°C with gentle shaking in Grace's medium with 10% FBS and 1% Pluronic. Remove 1 mL and place in a 6-well dish (control).
3. Coinfect cells at an MOI of 5 (*see* **Note 5**). Add virus and swirl every 10 min for 1 h at room temperature. Take 1 mL

and place in 6-well dish. Return cultures to the incubator/shaker for 48 h.

4. Observe cultures in the 6-well dish to ensure that infection has occurred (cells should appear swollen, etc.).
5. Pellet cells at 1000 g for 10 min at 4°C, wash with cold PBS, freeze pellet in liquid nitrogen and store at –80°C.
6. Thaw pellets in room temperature water.
7. Resupend the pellets in 5 volumes cold insect cell lysis buffer with protease inhibitors added just prior to use.
8. Sonicate, on ice, until lysate loses its syrupy texture and becomes watery.
9. Spin for 30 min at 30,000 g, 4°C.
10. During spin, wash 0.5 mL Ni-NTA three times with lysis buffer.
11. Take an aliquot of cleared lysate and save as "Load". Run about 10 μg/gel for Coomassie staining.
12. Incubate lysate with Ni-NTA resin. Mix end-over-end at 4°C for 1.5 h.
13. Pool resin into one 50 mL conical.
14. Wash resin 3 × 15' in lysis buffer.
15. Wash resin 2 × 5' in low salt buffer.
16. Transfer resin to column for elution.
17. Elute 5 × 300 μL elution buffer.
18. Check each elution by Bradford assay and pool those fractions with significant protein.
19. Dialyze eluates 2 × 90 min against 2 L dialysis buffer each time.
20. After dialyzing, run a 14% SDS-PAGE gel and Coomassie stain.
21. Aliquot protein and freeze in liquid nitrogen. Store at –80°C.

As illustrated in **Fig. 20.2**, the ubiquitination assays described below illustrate the specific partnership of the APC with the E2 UbcH10. Based on the experiments performed with the highly active and promiscuous UbcH5 enzyme, it would be expected that an E2 should function with its cognate RING protein in vitro. However, although the genetic data indicate UbcH10 to be the partner of the APC, it cannot function with APC11 alone and requires additional APC subunits for activity. Assays using GST-APC11 or the APC2-APC11 subcomplex may now be used to examine the effects of other factors or subunits on the activation of UbcH10, using UbcH5 as a control. Similarly, although both E2s function with the holoenzyme, they produce distinctive ubiquitin conjugates. UbcH10 produces lower molecular weight substrate–ubiquitin conjugates both by targeting fewer substrate lysines and by lower polyubiquitination activity (as visualized by using no-lysine mutant ubiquitin). The more regulated activity of UbcH10 confers a greater sensitivity to pseudosubstrate inhibitors of the APC (**Fig. 20.2C,D**), indicating that the correct E2 partner is an important aspect of proper APC regulation in vivo. By manipulating the pseudosubstrate or E2 this assay may now be used to examine critical determinants within these proteins that contribute to APC-control *(13, 16)*.

3.4. In Vitro Ubiquitination Reactions

3.4.1. Examining E2 Activity with GST-APC11

1. Make a ubiquitination cocktail on ice: 4 nM E1, 375 nM FLAG-ubiquitin, and 2 mM ATP, diluted in 1 × SID buffer (*see* **Note 6**). The final volume of the reactions will be 10 μl, however, the cocktail should be made in a final volume with 8 μl/reaction to leave room for addition of other components.
2. On ice, add 8 μL of ubiquitination cocktail to tubes.
3. Dilute E2 to 700 nM in 1 × SID buffer. Add 1 μL of E2 or buffer to each reaction.
4. Dilute GST-APC11 to 1.4 μM in 1 × SID buffer. Add 1 μL of E3 or buffer to each reaction.
5. Remove a 4 μL sample and quench with SDS-sample buffer.
6. Incubate the reactions at 30°C for 45 min. Remove another 4 μL sample, as above.
7. Resolve the reactions on by 12% SDS-PAGE gel, transfer to PVDF membrane, and probe with anti-FLAG antibody.

3.4.2. Examining E2 Activity with APC2-APC11

1. Make a ubiquitination cocktail on ice: 180 nM E1, 150 μM ubiquitin, 2 mM ATP, 1 μL ^{35}S-Myc-securin, and bring to 8 μL with 1 × SID buffer.
2. Dilute E2s to 500 μM in 1 × SID buffer.
3. Make a dilution series of APC2-APC11 protein, for example, 1:3, 1:10, 1:30, and 1:100.
4. Add 1 μL buffer or E2 and 1 μL buffer or E3 to 8 μL reactions. Remove 4 μL and quench with 4 μL sample buffer.
5. Incubate the remaining reaction at 30°C for 1 h.
6. Remove 4 μL and quench with 4 μL sample buffer.
7. Resolve the samples on 8% SDS-PAGE gel. Transfer to PVDF and expose to the phosphorimager screen for at least 6 h. Scan the screen on the phosphorimager.

3.4.3. Production of Anti-Cdc27 Beads

1. Mix equal volumes (5 mL) Anti-Cdc27 serum with SB and spin for 10 min at 20,000 g, room temperature.
2. Wash 3 mL Affiprep protein A beads (6 mL slurry) 3× with SB.
3. Mix beads with serum. Add SB to 10 bead volumes (25 mL).

4. Mix gently 1.5–2 h at room temperature.
5. Wash beads 2× with 20 vol of SB (50 mL).
6. Remove a ~10 μL aliquot of beads and add 10 μL sample buffer.
7. Add SB to 25 mL.
8. Add DMP to 20 mM (0.130 g) pH >8. Check with pH test-strip.
9. Mix 30 min at room temperature. Make ethanolamine.
10. Stop reaction by washing 2× with 25 mL 0.2 M ethanolamine.
11. Repeat step 6.
12. Resuspend the beads in 15 mL ethanolamine; mix 2 h at room temperature.
13. Wash 3 × 50 mL XB.
14. Make a 50% slurry in XB and add thimerozal to 0.1%. Store at 4°C.
15. Run pre- and postcoupling bead samples on 10% SDS-PAGE gel. Coomassie stain. You should see release of Ab heavy chain from beads before coupling, but not after.

3.4.4. In Vitro APC Reaction

1. Grow several 150 mm dishes of HeLa cells in DMEM to 70–80% confluence. Add 100 ng/mL nocodazole and incubate for 16–20 h (*see* **Note 7**).
2. Collect the mitotic cells by shake-off, freeze in liquid nitrogen, and store at –80°C until ready to make lysate.
3. Thaw the pellets at room temperature and resuspend in three pellet volumes of HeLa cell lysis buffer. Incubate at 4°C with gentle mixing for 30 min.
4. Clear the lysate by spinning at maximum speed in a microcentrifuge for 20 min at 4°C.
5. Determine the concentration of the lysate using Bradford reagent (*see* **Note 8**).
6. For each reaction add 4 μL of anti-Cdc27 beads (*see* **Note 9**) to 1.5 mg of lysate and incubate with end-over-end mixing for 1.5 h at 4°C.
7. Wash the beads 6 times with 10 volumes of XB+400 mM KCl and 0.5% NP-40 and two times with XB.
8. Add 5 μL IVT Cdh1/reaction and incubate with end-over-end mixing at 4°C for 30 min.
9. During the incubation, prepare the ubiquitination cocktail, on ice: 1 × NRG, 150 μM ubiquitin, 170 nM E1, 1 μL ^{35}S-labeled Myc-securin/reaction in XB (*see* **Note 10**). Make the calculations based on 5 μL per reaction, but only bring up in a volume using 4 μL per reaction to allow room for E2 addition.
10. Dilute E2s to 7.5 μM.
11. Set up 3 tubes for each reaction with 5 μL SDS-sample buffer.

12. Wash beads 2× with XB and immediately add 4 μL of ubiquitination cocktail to the beads, on ice.
13. Add 1 μL of E2 or XB to start the reaction, mix, and remove a 1 μL sample for the 0 min timepoint. Incubate the reactions at 30°C with end-over-end mixing. Remove additional samples at 15 and 30 min.
14. Resolve the samples on 8% SDS-PAGE gel. Transfer to PVDF and expose to the phosphorimager screen for at least 6 h. Scan the screen on the phosphorimager (*see* **Note 11**).

3.4.5. Pseudosubstrate Inhibition Using Emi C-Terminus

1. For assays of pseudosubstrate inhibition produce extract, perform IPs, and incubate with Cdh1 as described in **Section 3.4.3** (*see* **Note 12**).
2. During the incubation with Cdh1 make a serial dilution series of Emi1. For example, 0.1, 0.3, 1.0, 3.0 μM. Wash the beads 2× with XB and incubate with 5 μL of diluted Emi1 or XB for 30 min at 4°C.
3. During this incubation prepare the ubiquitination cocktail, dilute E2, and set up tubes with Sample buffer.
4. Wash 2× with XB and assay for securin ubiquitination.

3.5. In Extracto Ubiquitination and Destruction Assays

The assays described in this section build upon the assays in **Section 3.4**, but utilize a more complex system based on extracts from mitotic cells (see the previous chapter for additional details and procedures). This extract system, similar to the *Xenopus* egg extract, lends itself to biochemical manipulation, but offers the advantage of having an existing spindle checkpoint activity, and extracts can be made from siRNA-treated cells as an alternative to using antibody depletion to remove a protein from extracts. Performing the experiments in this complex and more physiological setting allows for the discovery of additional unknown characteristics and factors of the pathway. For example, consistent with the very specific functional interactions of UbcH10 and the APC observed in the assays above, UbcH5, although functioning robustly with the APC in vitro, does not do so *in extracto*, as ubiquitin is limiting for its activity with the APC (**Fig. 20.2D**, *(16))*. In addition, the presence of a functional proteasome system also allows for the examination of destruction and factors, such as isopeptidases, which may also regulate APC-mediated ubiquitination/destruction events.

3.5.1. Production of Mitotic HeLa Extracts

1. Grow four 150 mm dishes of HeLa cells in DMEM to 70–80% confluence and transfer to a cell stack and grow to 70–80% confluence (*see* **Note 13**).
2. Add 100 ng/mL nocodazole and incubate for 16–20 h.
3. Collect the mitotic cells by shake-off, freeze in liquid nitrogen, and store at –80°C until ready to make extract.

4. Thaw the pellets at room temperature and resuspend in 1.5 pellet volumes of mitotic HeLa extract buffer. Transfer to 4–5 1.5 mL microcentrofuge tubes.
5. Poke 4 holes in the lid of the tubes with a 25-gauge needle and place tubes into the prechilled nitrogen disruption chamber.
6. Add nitrogen to 1500 psi and incubate at 4°C, on ice for 40 min. Release pressure quickly to facilitate cell-breaking.
7. Spin the extracts at maximum in a microcentrofuge for 20 min.
8. Pool and aliquot the supernatants. Freeze in liquid nitrogen. Make one small aliquot and use this to determine the extract concentration after freezing. Store the extract at –80°C.

3.5.2. Ubiquitination and Destruction Assays for E2 Activity

1. For each reaction set up six 0.6 mL microcentrifuge tubes on ice.
2. Prepare the reaction cocktail: (for each reaction) 90 μg mitotic HeLa extract, 1 μL ^{35}S-labeled securin, 0.1 μg/μL cycloheximide, nondestructible cyclin B (Δ90), and 2 × NRG. Bring to a final volume of 14.4 μL with mitotic HeLa extract buffer (*see* **Notes 14–18**).
3. Dilute E2s to 30 μM and ubiquitin to 750 μM in XB.
4. To a tube add either 0.6 μL XB, 0.3 μL XB, and 0.3 μL ubiquitin, 0.3 μL XB and 0.3 μL E2, or 0.3 μL ubiquitin and 0.3 μL E2. Repeat the latter two combinations for each E2.
5. Add 14.4 μL of reaction cocktail to these tubes from step 4 and mix. Make 2 μL aliquots to the remaining tubes for each reaction. Move the tubes to a 30°C water bath.
6. Quench samples by adding 5 μL of SDS-sample buffer at 0, 40, 60, 80, and 100 min.
7. Resolve the samples on a 10% SDS-PAGE gel and analyze as in **Section 3.4.3**.

4. Notes

1. The use of the tandem Myc-epitope tags serves two purposes. First, the tag introduces several extra methionine residues, increasing the ^{35}S-labeling of each securin molecule and facilitating the detection of ubiquitin conjugates. Second, for reasons that are not fully understood, the N-terminal fusion of this epitope causes a stabilization of the securin–ubiquitin conjugates and allows monitoring of ubiquitination in extracts.

2. Spinning the IVT substrates removes the misfolded and aggregated proteins. This enhances the efficiency of the assays described below, enhancing the interpretation of the results. We do not find this to be required for the activator proteins.
3. Modification of E2 N-termini can affect the function of these enzymes. For this reason we prefer to use the smaller hexahistidine tag fused to the C-terminus. The Emi1 C-terminus (299–447) is a very efficient inhibitor of the APC, its GST fusion is more readily expressed than full-length versions, and as it does not include the Emi1 degron, it is more effective for use in mitotic extract assays. The GST fusion to APC11 is required for its independent function as an E3.
4. For the E1 protein, wash beads 2× in storage buffer, resuspend in 960 μL 40 μL of Precission protease and incubate overnight at 4°C with end-over-end mixing. Place the beads in an empty column and collect the supernatant. Wash beads with 2 column volumes storage buffer and concentrate.
5. We find that placing the His-tag on APC2, rather than APC11, results in a functional complex. The correct MOIs should be determined on a small scale first by varying the ratios of infection, for example, 1:1, 5:1, 1:5, and so on. Expression can easily be examined by Western blotting. The preps often contain contaminating proteins and may need to be checked by Western blotting as well. For this reason, titrating the activity in the initial ubiquitination reactions for the preparation is better than assigning an amount to use. Alternatively, the proteins may be expressed in bacteria. However, to simplify co-expression and complex formation we utilize the baculovirus system.
6. The use of low salt buffers, such as 1 × SID, is necessary for the function of the GST-APC11 and APC2-APC11 reactions. Increasing the salt content greatly reduces E3 activity. The ubiquitination cocktails for these two assays can be swapped to monitor ubiquitination of securin or ubiquitin transfer for either E3.
7. The APC can be purified from cells in other phases of the cell cycle and may also be used with Cdh1. By washing only with lysis buffer, which should allow the associated activators and inhibitors to remain bound, the intrinsic activity of the enzyme from that cell cycle stage can also be monitored.
8. Alternatively, the APC can be immunopurified via standard immunoprecipitation procedures without conjugated beads.

9. Unused lysate can be frozen in liquid nitrogen and stored at –80°C in single-use aliquots.
10. To assay differential lysine usage by each E2, use the ubiquitin mutant with all lysines mutated to arginine. Because no chains can be formed, the increase in molecular weight of ubiqtuitin conjugates represents an increased number of lysines modified.
11. The efficiency of the reactions can be monitored in two ways. First, the amount of unmodified substrate can be measured and the percentage converted over time can be determined for each reaction. Second, the total signal for each timepoint can be measured and the percentage of ubiquitinated substrate determined.
12. We describe here the assay of pseudosubstrate inhibition of E3 activity. The effect of E2 activity on pseudosubstrate inhibition can be assayed similarly, holding the concentration of inhibitor constant and varying the concentration of E2 used. Emi1 may also be added with the ubiquitination cocktail rather than as a separate step. Similar experiments may be carried out for BubR1 or other potential inhibitors. For BubR1, Cdc20 should be used instead of Cdh1 and care should be taken to preserve the mitotic (phosphorylated) state of the APC to enhance the interaction with Cdc20.
13. Four 150 mm dishes generally provide enough extract for multiple experiments, however, for convenience and increased reproducibility, we recommend producing the extracts on a larger scale. Excess extract can also be utilized for ubiquitination assays as well.
14. A concentration of 6 μg/μL is the minimum for extract function. Below this concentration the reaction is simply too dilute. Increasing the concentration may enhance the kinetics of the assay.
15. Securin is used as the substrate for the monitoring of destruction and Myc-securin for the monitoring of ubiquitination activity in extracts. Cyclin A can be used as a positive control for extract activity as it is readily destroyed without any further manipulation.
16. The inclusion of cyclin B Δ90 is not absolutely required for the assay to function properly. Its inclusion maintains the mitotic state of the extract, which is otherwise lost as a result of the APC-mediated destruction of cyclin B. Examining the phosphorylation state of Cdc27 will indicate the mitotic state. Maintaining the extracts in mitosis is useful when discrimination between a mitotic event and an event triggered by mitotic exit is required. Because its inclusion is not mandatory, its production is not described here, but has been described. The amount of Δ90 used is

determined by titration for each preparation. Do not spin reactions after the addition of Δ90, as the protein is not very soluble and will precipitate.

17. The addition of Δ90, by increasing the kinase activity in the extract, also increases the energy consumption and requires the use of NRG at 2× rather than 1×. Failure to maintain the ATP levels mimics a loss of mitotic kinase activity and spontaneous mitotic exit (monitored as in **Note 16**).
18. The ability of pseudosubstrates to stabilize APC substrates in this assay can be examined by titrating GST-Emi1 C-terminus into the reaction cocktail. The E2 can be similarly titrated.

References

1. Yu, H., King, R. W., Peters, J. M., and Kirschner, M. W. (1996) Identification of a cullin homology region in a subunit of the anaphase-promoting complex. *Curr Biol* **6,** 455–66.
2. Scheffner, M., Huibregtse, J. M., and Howley, P. M. (1994) Identification of a human ubiquitin-conjugating enzyme that mediates the E6-AP-dependent ubquitination of p53. *Proc Natl Acad Sci USA* **91,** 8797–801.
3. Bothos, J., Summers, M. K., Venere, M., Scolnick, D. M., and Halazonetis, T. D. (2003) The Chfr mitotic checkpoint protein functions with Ubc13-Mms2 to form Lys63-linked polyubiquitin chains. *Oncogene* **22,** 7101–7.
4. Wu, K., Chen, A., Tan, P., and Pan, Z. Q. (2002) The Nedd8-conjugated ROC1-CUL1 core ubiquitin ligase utilizes Nedd8 charged surface residues for efficient polyubiquitin chain assembly catalyzed by Cdc34. *J Biol Chem* **277,** 516–27.
5. Mathe, E., Kraft, C., Giet, R., Deak, P., Peters, J. M., and Glover, D. M. (2004) The E2-C vihar is required for the correct spatiotemporal proteolysis of cyclin B and itself undergoes cyclical degradation. *Curr Biol* **14,** 1723–33.
6. Osaka, F., Seino, H., Seno, T., and Yamao, F. (1997) A ubiquitin-conjugating enzyme in fission yeast that is essential for the onset of anaphase in mitosis. *Mol Cell Biol* **17,** 3388–97.
7. Rape, M., and Kirschner, M. W. (2004) Autonomous regulation of the anaphase-promoting complex couples mitosis to S-phase entry. *Nature* **432,** 588–95.
8. Yamanaka, A., Hatakeyama, S., Kominami, K., Kitagawa, M., Matsumoto, M., and Nakayama, K. (2000) Cell cycle-dependent expression of mammalian E2-C regulated by the anaphase-promoting complex/cyclosome. *Mol Biol Cell* **11,** 2821–31.
9. Kirkpatrick, D. S., Hathaway, N. A., Hanna, J., Elsasser, S., Rush, J., Finley, D., King, R. W., and Gygi, S. P. (2006) Quantitative analysis of in vitro ubiquitinated cyclin B1 reveals complex chain topology. *Nat Cell Biol* **8,** 700–10.
10. Christensen, D. E., Brzovic, P. S., and Klevit, R. E. (2007) E2-BRCA1 RING interactions dictate synthesis of mono- or specific polyubiquitin chain linkages. *Nat Struct Mol Biol* **14,** 941–8.
11. Peters, J. M. (2006) The anaphase promoting complex/cyclosome: a machine designed to destroy. *Nat Rev Mol Cell Biol* **7,** 644–56.
12. Diaz-Martinez, L. A., and Yu, H. (2007) Running on a treadmill: dynamic inhibition of APC/C by the spindle checkpoint. *Cell Div* **2,** 23.
13. Miller, J. J., Summers, M. K., Hansen, D. V., Nachury, M. V., Lehman, N. L., Loktev, A., and Jackson, P. K. (2006) Emi1 stably binds and inhibits the anaphase-promoting complex/cyclosome as a pseudosubstrate inhibitor. *Genes Dev* **20,** 2410–20.
14. Burton, J. L., and Solomon, M. J. (2007) Mad3p, a pseudosubstrate inhibitor of APC-Cdc20 in the spindle assembly checkpoint. *Genes Dev* **21,** 655–67.
15. King, E. M., van der Sar, S. J., and Hardwick, K. G. (2007) Mad3 KEN boxes mediate both Cdc20 and Mad3 turnover, and are critical for the spindle checkpoint. *PLoS ONE* **2,** e342.
16. Summers, M. K., Pan, B., Mukhyala, K., Jackson, P.K. (2008) The unique N-terminus of the UbcH10 E2 enzyme controls the threshold for APC activation and enhances

checkpoint regulation of the APC. *Mol. Cell,* **31(4),** 544–56.

17. Gmachl, M., Gieffers, C., Podtelejnikov, A. V., Mann, M., and Peters, J. M. (2000) The RING-H2 finger protein APC11 and the E2 enzyme UBC4 are sufficient to ubiquitinate substrates of the anaphase-promoting complex. *Proc Natl Acad Sci USA* **97,** 8973–8.
18. Tang, Z., Li, B., Bharadwaj, R., Zhu, H., Ozkan, E., Hakala, K., Deisenhofer, J., and Yu, H. (2001) APC2 Cullin protein and APC11 RING protein comprise the minimal ubiquitin ligase module of the anaphase-promoting complex. *Mol Biol Cell* **12,** 3839–51.

Index

A

B

C

D

E

L

M

N

O

P

Q

R

S

T

Z

MIX
Papier aus verantwortungsvollen Quellen
Paper from responsible sources
FSC® C105338

If you have any concerns about our products,
you can contact us on
ProductSafety@springernature.com

In case Publisher is established outside the EU,
the EU authorized representative is:
Springer Nature Customer Service Center GmbH
Europaplatz 3, 69115 Heidelberg, Germany

Printed by Libri Plureos GmbH
in Hamburg, Germany